# Polymers and Their Application in 3D Printing

# Polymers and Their Application in 3D Printing

Editors

**Hamid Reza Vanaei**
**Sofiane Khelladi**
**Abbas Tcharkhtchi**

MDPI • Basel • Beijing • Wuhan • Barcelona • Belgrade • Manchester • Tokyo • Cluj • Tianjin

*Editors*

Hamid Reza Vanaei  
ALDV-ENSAM  
France

Sofiane Khelladi  
ENSAM  
France

Abbas Tcharkhtchi  
ENSAM  
France

*Editorial Office*  
MDPI  
St. Alban-Anlage 66  
4052 Basel, Switzerland

This is a reprint of articles from the Special Issue published online in the open access journal *Polymers* (ISSN 2073-4360) (available at: https://www.mdpi.com/journal/polymers/special_issues/Polymers_Their_Application_3D_Printing).

For citation purposes, cite each article independently as indicated on the article page online and as indicated below:

LastName, A.A.; LastName, B.B.; LastName, C.C. Article Title. *Journal Name* **Year**, *Volume Number*, Page Range.

**ISBN 978-3-0365-6027-4 (Hbk)**  
**ISBN 978-3-0365-6028-1 (PDF)**

© 2023 by the authors. Articles in this book are Open Access and distributed under the Creative Commons Attribution (CC BY) license, which allows users to download, copy and build upon published articles, as long as the author and publisher are properly credited, which ensures maximum dissemination and a wider impact of our publications.

The book as a whole is distributed by MDPI under the terms and conditions of the Creative Commons license CC BY-NC-ND.

# Contents

Anouar El Magri, Salah Eddine Bencaid, Hamid Reza Vanaei and Sébastien Vaudreuil
Effects of Laser Power and Hatch Orientation on Final Properties of PA12 Parts Produced by Selective Laser Sintering
Reprinted from: *Polymers* 2022, 14, 3674, doi:10.3390/polym14173674 . . . . . . . . . . . . . . . . 1

Mohammed Dukhi Almutairi, Sultan Saleh Alnahdi and Muhammad A. Khan
Strain Release Behaviour during Crack Growth of a Polymeric Beam under Elastic Loads for Self-Healing
Reprinted from: *Polymers* 2022, 14, 3102, doi:10.3390/polym14153102 . . . . . . . . . . . . . . . . 23

Razieh Hashemi Sanatgar, Aurélie Cayla, Jinping Guan, Guoqiang Chen, Vincent Nierstrasz and Christine Campagne
Piezoresistive Properties of 3D-Printed Polylactic Acid (PLA) Nanocomposites
Reprinted from: *Polymers* 2022, 14, 2981, doi:10.3390/polym14152981 . . . . . . . . . . . . . . . . 51

Irene Buj-Corral, Héctor Sanz-Fraile, Anna Ulldemolins, Aitor Tejo-Otero, Alejandro Domínguez-Fernández, Isaac Almendros and Jorge Otero
Characterization of 3D Printed Metal-PLA Composite Scaffolds for Biomedical Applications
Reprinted from: *Polymers* 2022, 14, 2754, doi:10.3390/polym14132754 . . . . . . . . . . . . . . . . 63

Arief Suriadi Budiman, Rahul Sahay, Komal Agarwal, Rayya Fajarna, Fergyanto E. Gunawan, Avinash Baji and Nagarajan Raghavan
Modeling Impact Mechanics of 3D Helicoidally Architected Polymer Composites Enabled by Additive Manufacturing for Lightweight Silicon Photovoltaics Technology
Reprinted from: *Polymers* 2022, 14, 1228, doi:10.3390/polym14061228 . . . . . . . . . . . . . . . . 75

Woo-Shik Jeong, Young-Chul Kim, Jae-Cheong Min, Ho-Jin Park, Eunju Lee, Jin-Hyung Shim and Jong-Woo Choi
Clinical Application of 3D-Printed Patient-Specific Polycaprolactone/Beta Tricalcium Phosphate Scaffold for Complex Zygomatico-Maxillary Defects
Reprinted from: *Polymers* 2022, 14, 740, doi:10.3390/polym14040740 . . . . . . . . . . . . . . . . 93

Ahmed Abusabir, Muhammad A. Khan, Muhammad Asif and Kamran A. Khan
Effect of Architected Structural Members on the Viscoelastic Response of 3D Printed Simple Cubic Lattice Structures
Reprinted from: *Polymers* 2022, 14, 618, doi:10.3390/polym14030618 . . . . . . . . . . . . . . . . 107

Matej Pivar, Diana Gregor-Svetec and Deja Muck
Effect of Printing Process Parameters on the Shape Transformation Capability of 3D Printed Structures
Reprinted from: *Polymers* 2022, 14, 117, doi:10.3390/polym14010117 . . . . . . . . . . . . . . . . 123

Zohre Mousavi Nejad, Ali Zamanian, Maryam Saeidifar, Hamid Reza Vanaei and Mehdi Salar Amoli
3D Bioprinting of Polycaprolactone-Based Scaffolds for Pulp-Dentin Regeneration: Investigation of Physicochemical and Biological Behavior
Reprinted from: *Polymers* 2021, 13, 4442, doi:10.3390/polym13244442 . . . . . . . . . . . . . . . . 145

Baye Gueye Thiam, Anouar El Magri, Hamid Reza Vanaei and Sébastien Vaudreuil
3D Printed and Conventional Membranes—A Review
Reprinted from: *Polymers* 2022, 14, 1023, doi:10.3390/polym14051023 . . . . . . . . . . . . . . . . 159

Article

# Effects of Laser Power and Hatch Orientation on Final Properties of PA12 Parts Produced by Selective Laser Sintering

Anouar El Magri [1,*], Salah Eddine Bencaid [1], Hamid Reza Vanaei [2,3] and Sébastien Vaudreuil [1]

1. Euromed Research Center, Euromed Polytechnic School, Euromed University of Fes, Route de Meknès (Rond point Bensouda), Fes 30 000, Morocco
2. Léonard de Vinci Pôle Universitaire, Research Center, 92916 Paris La Défense, France
3. Arts et Métiers Institute of Technology, CNAM, LIFSE, HESAM University, 75013 Paris, France
* Correspondence: a.elmagri@ueuromed.org

**Abstract:** Poly(dodecano-12-lactam) (commercially known as polyamide "PA12") is one of the most resourceful materials used in the selective laser sintering (SLS) process due to its chemical and physical properties. The present work examined the influence of two SLS parameters, namely, laser power and hatch orientation, on the tensile, structural, thermal, and morphological properties of the fabricated PA12 parts. The main objective was to evaluate the suitable laser power and hatching orientation with respect to obtaining better final properties. PA12 powders and SLS-printed parts were assessed through their particle size distributions, X-ray diffraction (XRD), Fourier Transform Infrared spectroscopy (FTIR), differential scanning calorimetry (DSC), a scanning electron microscope (SEM), and their tensile properties. The results showed that the significant impact of the laser power while hatching is almost unnoticeable when using a high laser power. A more significant condition of the mechanical properties is the uniformity of the powder bed temperature. Optimum factor levels were achieved at 95% laser power and parallel/perpendicular hatching. Parts produced with the optimized SLS parameters were then subjected to an annealing treatment to induce a relaxation of the residual stress and to enhance the crystallinity. The results showed that annealing the SLS parts at 170 °C for 6 h significantly improved the thermal, structural, and tensile properties of 3D-printed PA12 parts.

**Keywords:** selective laser sintering; PA12; laser power; hatch orientation; annealing

**Citation:** El Magri, A.; Bencaid, S.E.; Vanaei, H.R.; Vaudreuil, S. Effects of Laser Power and Hatch Orientation on Final Properties of PA12 Parts Produced by Selective Laser Sintering. *Polymers* **2022**, *14*, 3674. https://doi.org/10.3390/polym14173674

Academic Editor: Lilia Sabantina

Received: 10 August 2022
Accepted: 1 September 2022
Published: 4 September 2022

**Publisher's Note:** MDPI stays neutral with regard to jurisdictional claims in published maps and institutional affiliations.

**Copyright:** © 2022 by the authors. Licensee MDPI, Basel, Switzerland. This article is an open access article distributed under the terms and conditions of the Creative Commons Attribution (CC BY) license (https://creativecommons.org/licenses/by/4.0/).

## 1. Introduction

Additive manufacturing (AM), also known as 3D printing or direct digital manufacturing, has become an alternative technology that competes with more mature technologies such as casting and forging in different industrial fields including aerospace, automotive, and biomedical fields [1–4]. It is defined as the process of joining materials to create objects from 3D model data, usually layer upon layer, as opposed to subtractive manufacturing methodologies [5]. One of the major benefits of this promising technology is the freedom of its design and its facilitation of the printing of complex geometries. It gives engineers and designers the ability to innovate and create optimized parts that are too difficult or even impossible to be processed using conventional subtractive fabrication methods [6,7]. Selective laser sintering (SLS) is a common AM technology that uses a high-power laser to sinter small particles of polymer powder into a solid structure based on a 3D model [8]. Its self-supporting ability and capacity for building relatively large parts are some of the major benefits of the SLS process [9,10]. Moreover, can produce durable prototypes and end-use parts with a high dimensional accuracy afforded by the nature of the SLS process [11,12]. SLS is also limited by the raw materials available, even if some polymers with tunable properties have been produced through mineral additives [13,14]. Another limitation of SLS is its poorer mechanical properties compared to traditional manufacturing [15,16], limiting its application for major load-bearing applications [17].

SLS polymers are selected based on the presence of a super-cooling processing window in which there is a large space between the crystallization temperature ($T_c$) and melting temperature ($T_m$). Therefore, printing at a temperature slightly below $T_m$ enables the densification of the SLS powder without reaching its melting point, thus limiting the parts' distortion. Moreover, by maintaining the temperature above $T_c$, the sintered structure remains in an amorphous phase to prevent rapid crystallization, making the powder material more suitable for the production of the final part. Therefore, the parts need to be maintained within the processing window during the build process and slowly cooled down to room temperature to avoid any deformation and crack formation [18–20].

Polyamides are the most used polymers in SLS processing, and include Polyamide-11 (PA11), [21,22] Polyamide-6 (PA6) [23,24], and especially Polyamide-12 (PA12) in either its pure or reinforced form [25,26]. The semi-crystalline Polyamide-12 (also called Nylon PA12) accounts for about 95% of the SLS materials used [27] as easy laser sintering can be achieved in comparison with other polymers [28,29]. Table 1 compares the tensile strength for the three different types of polyamides. PA12 is considered a versatile thermoplastic with excellent properties such as toughness, heat, and chemical resistance [30]. Polyamide will behave as a flexible material when thin and as a rigid one when thick. The fabricated parts are usually robust, detailed, and stable for long-term use [31].

**Table 1.** Tensile strength of some types of polyamides with their references.

| Material | PA6 | PA11 | PA12 |
| --- | --- | --- | --- |
| Tensile strength (MPa) | 3.75 | 52 | 26.25 |
| References | [23] | [21] | [29] |

SLS is a complex process that usually requires great effort and control in terms of powder and post processing after fabrication to achieve successful printing and high-quality parts. During the SLS process, the material should be kept at an elevated temperature in the build chamber to avoid any deformation of the printouts. The laser provides the necessary energy to exceed the sintering point, making it possible to form the part with the desired geometry [32]. For SLS, part quality and mechanical properties are strongly affected by a large number of printing parameters such as the laser power, laser speed, scan spacing, layer thickness, bed temperature, and build orientation [33–36]. It is therefore quite important to properly adjust those parameters in order to avoid process instabilities such as a high porosity, which is largely responsible for the poor properties of the tested materials [20]. Starr et al. [37] studied the impact of the process conditions on the mechanical properties of laser-sintered nylon. A high tensile strength was obtained through a high energy density in order to fully melt the applied powder. A higher energy density is required to reach maximum elongation performance, which is more sensitive to the build orientation. The work conducted by Caulfield et al. [12] presented a detailed study about the effects of the energy density level (which comprises the influence of the laser power, hatch spacing, and laser speed) on the mechanical properties of polyamide components. They claimed that using high energy density levels exhibits a more ductile behavior than those obtained at low energy densities. The mechanical test results reported a better elasticity modulus and tensile strength values along the primary x-axis than the secondary z-axis. It was also found that parts built with 0° orientations had a higher ultimate tensile strength and less elongation at fracture relative to the parts with a 90° build orientation.

On the other hand, powder spreading is a crucial step of the SLS process. Controlling the powder quality on the bed affects the quality of the tested parts. The powder should have a good flowability in order to enable the consistent deposition of thin dense layers of powder. Decreasing the porosity content will increase the mechanical properties. The layer thickness of the SLS process is typically between 100–150 μm. Smooth particles with a high sphericity are thus preferable to obtain parts with a desirable microstructure after sintering [38,39] and an adequate surface roughness [40]. The number of crystalline phases in the microstructure also has a significant impact on the mechanical properties of SLS-PA12 parts. Young's modulus and tensile strength increase with a higher crystallinity, while the elongation at break tends to decrease. The applied process parameters and thermal properties of the material are the major factors determining the amount of crystallinity [41]. Verbelen et al. [42] investigated four commercial polyamide grades by using a new screening methodology that encompasses the complete process chain in laser sintering. They reported that the dilatometry measurements of different PA12 powders showed a reduction in the specific volume during the crystallization phase ranging from 3.9% to 4.7%.

Hofland et al. [20] studied the impact of the process parameters on the mechanical properties by applying Response Surface Methodology (RSM) to analyze the results. They used PA12 powder with a recycled/virgin mixture ratio of 50/50 to produce parts with 0° and 90° build orientations. Dupin et al. [25] compared two types of SLS polyamide 12, Duraform PA (3D systems, Rock Hill, South Carolina, USA) and Innov PA (Exeltec, France), to improve flowability. They used 1% silica in both materials, and it was found that the specimens with the Duraform PA type yielded less porosity than Innov PA, even at a lower energy density.

The present work aims to analyze the impact of the hatch orientation and laser power on the mechanical, microstructural, and morphological properties of 3D-printed PA12 parts. The effect of the heat treatment on the mechanical properties will also be evaluated and compared to the as-built samples

## 2. Materials and Methods

### 2.1. Material and Specimen Preparation

All samples were printed on a P3200HT SLS system from TPM3D (Stratasys company) equipped with a 60W $CO_2$ laser. In this study, a Polyamide 12 (PA12) powder (Precimid1171™) from TPM3D with a density of 0.95 g/cm$^3$ was used, as it is one of the most widely used materials due to its chemical and physical properties. The chemical structure of PA12 (PA 2200) is shown in Figure 1. Small amounts of fumed silica were added to the PA12 particles to improve powder flowability.

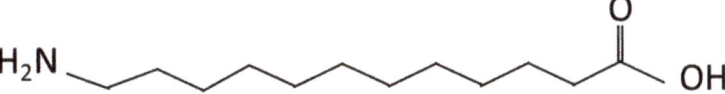

**Figure 1.** Chemical structure of Nylon (PA12), polydodecanolactam.

The software "VisCAM RP" was used to prepare the build volume and slice models into individual layers before uploading the data to the SLS machine. The main printing parameters used to produce PA12 powder samples are shown in Table 2.

As shown in Figure 2, three orientations were used during parts' placement on the XY plane in order to study the impact of hatch orientation on mechanical properties. Hatching was conducted by alternating one layer of laser scans at 0° (e.g., parallel to the X-axis) with the following layer at 90° (e.g., perpendicular to the X-axis). As this hatching strategy is applied by the software independently of the part's orientation on the XY plane, it results in the 0° and 90° orientation parts exhibiting an identical hatching strategy, albeit with a one-layer shift. Due to their hatching similarity, parts with the 0° orientation were placed in the center of the SLS build platform, while parts with a 45° and 90° orientation were positioned all around toward the sides. Three printing runs, with laser power ranging from 45 W to 57 W, were conducted with the purpose of studying the effect of the laser power on the mechanical properties of the 3D-printed parts. The laser power used in this work will be defined as percentage of the maximum laser power of the machine, which is 60 W (LP: 75% is equal to 45 W, LP: 85% is equal to 51 W, and LP: 95% is equal to 57 W).

Table 2. Main printing parameters used in selective laser sintering of PA12 powder.

| Parameters | Values | Units |
|---|---|---|
| Laser power | 75–85–95 | (%) |
| Part orientation (XY plane) | 0–45–90 | (°) |
| Layer thickness | 0.15 | (mm) |
| Platform temperature | 169 | (°C) |
| Chamber temperature | 135 | (°C) |
| Moving plate temperature | 140 | (°C) |
| Hatch spacing | 0.220 | (mm) |
| Diameter of laser beam | 0.22 | (mm) |
| Infill | 100 | (%) |
| Scanning speed | 13 | (mm.s$^{-1}$) |
| Hatch orientation (XY plane) | 0–90–0–90 | (°) |

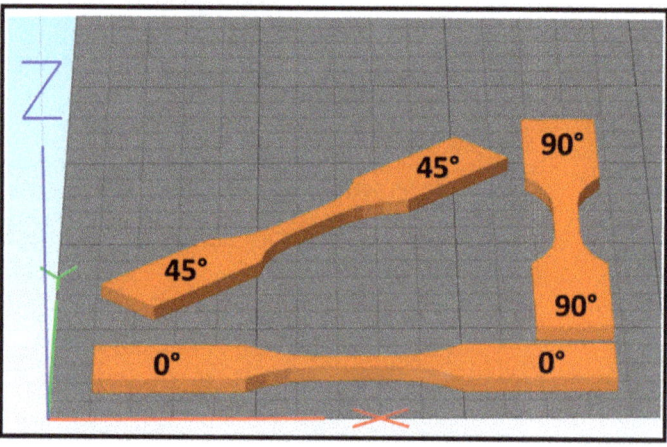

**Figure 2.** Illustration of different specimen orientations.

*2.2. Size Distribution and Particle Shape*

A dynamic image analysis measurement was performed to characterize both the size distribution and particle shape of the PA12 powder used. This analysis was performed using a Camsizer XT equipped with two digital cameras, including one optimized for the analysis of fine particles. Such a setup enables measurement of particles ranging between 2 µm and 8 mm in diameter.

Two PA12 powders were analyzed for comparison: one was the as-received powder, while the second was the un-sintered powder taken from the build volume after only one fabrication. The particle size distribution (PSD) of PA12 powder was identified as a function of percent volume. Furthermore, sphericity was chosen as a shape factor to describe the shape of particles of PA12 powder.

*2.3. Fourier-Transform Infrared Spectroscopy (FTIR)*

Fourier Transform Infrared spectroscopy (FTIR) was used in this work to analyze functional groups of SLS PA12 samples and collect infrared spectra for the structural analysis. This analysis was carried out using a NICOLET™ IS50 attenuated total reflection (ATR) spectrometer. The conditions of measurement were as follows: spectral region of 4000–400 cm$^{-1}$; spectral resolution of 4 cm$^{-1}$.

## 2.4. Differential Scanning Calorimetry (DSC)

Differential scanning calorimetry (DSC) is a common tool used for characterizing materials for laser sintering because it determines the crystallinity and quantifies the melting temperature of printed parts. This analysis was carried out on a 6.6 ± 0.1 mg powder sample using a TA Instruments DSC Q20. The measurements were carried out under a nitrogen atmosphere at a flow rate of 50 mL/min. The crystallinity, $X_c$, was calculated using the equation bellow:

$$X_c(\%) = \frac{\Delta H_m}{\Delta H_m^0} \times 100$$

where $\Delta H_m$ is the enthalpy of fusion and $\Delta H_m^0$ is the heat of fusion of 100% crystalline PA12, which is taken as 209.3 J.g$^{-1}$ [19].

The annealing process was performed as follows:

- Heating ramp of 2 °C min$^{-1}$ from room temperature to the annealing temperature Ta (130, 150, and 170 °C);
- Hold at Ta during the annealing time ta (6 h);
- Cooling ramp of 2 °C min$^{-1}$ from $T_a$ to the room temperature (25 °C);
- Heating ramp of 10 °C min$^{-1}$ to 220 °C for characterization.

After determination of the appropriate annealing temperature $T_a$, sample parts were placed directly in a natural convection oven (Dry-Line series, VWR) for annealing before mechanical testing.

## 2.5. X-ray Diffraction

XRD is a powerful tool used to analyze the atomic or molecular structure of materials. XRD was used here to identify the phase constituent of SLS PA12 powder and samples. Examination of powder and 3D-printed samples was carried out at different laser powers using a XRD X'PERT PRO MPD. Data were acquired over the range of (2θ) 0–90° with a step size of 0.0017 and a scan rate of 7°. min$^{-1}$.

## 2.6. Tensile Test

A tensile test was used to establish tensile properties of 3D-printed SLS-PA12 specimens, including tensile strength, Young's modulus, and deformation at break. The specimens used were designed according to the ASTM D638-14 "Standard Test Method for tensile properties of plastics". Three runs were conducted in series to study different laser power and different build orientation as well. Each series comprised six specimens of the D638 type-5 geometry as shown in Figure 3. Testing was carried out on a Criterion C45.105 electromechanical universal testing machine (MTS, USA) equipped with a 10 kN load cell and self-tightening jaws. A crosshead displacement speed of 5 mm min$^{-1}$ was used.

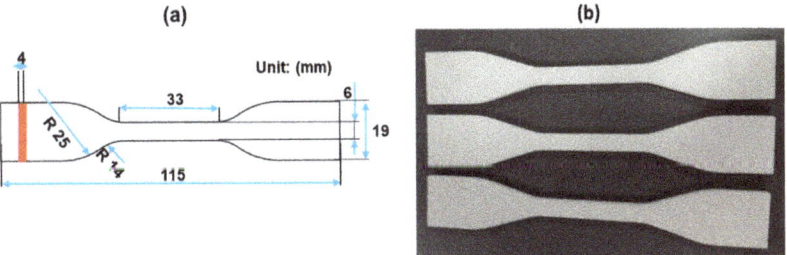

**Figure 3.** (a) tensile test bar dimensions; (b) SLS-printed specimens for tensile test.

The information obtained by the software were used to enable the calculation of tensile strength, Young's Modulus, and deformation at break, using the following equations:

$$\sigma(\text{MPa}) = \frac{F(N)}{s(\text{mm}^2)} \quad (1)$$

$$E = \frac{\sigma(\text{Mpa})}{\varepsilon} \quad (2)$$

$$\varepsilon = \frac{\Delta L}{L0}, \text{ with } \Delta L = L - L_0 \quad (3)$$

*2.7. Scanning Electron Microscopy (SEM)*

Microstructure of both powder and fracture surface was evaluated by scanning electron microscopy (SEM) using a Quanta 200 ESEM (Thermo FEI, Eindhoven, The Netherlands) configured with an EDAX (TSL) EDS/EBSD system for phase identification at high pressures. As-received PA12 powder, used-once SLS powder, and 3D-printed samples cryogenically fractured in liquid nitrogen were coated with a thin layer of electrically conducting gold (Au) to prevent surface charging. Layer arrangement and powder morphology were observed at an acceleration voltage of 5 kV in high-vacuum mode.

## 3. Results and Discussions

*3.1. Size Distribution, Particle Shape, and Morphology*

The particle size distribution (PSD) has a significant impact on the quality of SLS powder, with an ideal diameter between 20 and 80 µm. However, a large number of small diameter particles gives the powder a sticky character that limits its application in the SLS process [8]. Figure 4 illustrates the volume distribution as a function of size for the "As-received powder" and "Powder after fabrication", where a good PSD can be observed for both powders. Most of the particles for both powders fall in the 30 to 70 µm range, with an additional fraction between 20 to 40 µm. Both powders contain a low number of fine particles, with a diameter of 10 µm or less. The volume distribution for the powder after fabrication exhibits a suitable PSD despite its use in the SLS process, indicating that the powder can be reused after sieving [39]. However, it can be observed that the powder after fabrication exhibits a lesser quantity of particles larger than 60 µm compared to the as-received powder. This could be an effect of the bed-layering process, where larger particles tend to stay above and end up in the overflow bins.

This particle size decrease was confirmed through a comparison of the percentile values (D10, D50 and D90) for both powders (Table 3), where a lower diameter was observed at each percentile in the case of the powder after fabrication. These decreasing diameters imply that the powder contains a higher fraction of smaller particles than at the start. Table 2 also indicates the average particle sphericity for both the as-received powder and the powder after fabrication. It can be observed that the mean value for sphericity does not change significantly because of fabrication, ranging from 0.823 for the as-received powder to 0.818 for the powder after fabrication. While not spherical in shape, the particles for both powders are still considered of a suitable shape for the SLS process.

Figure 5 compares the morphology of the powders in the virgin state (Figure 5a) and after fabrication (Figure 5b). Both samples exhibit particles with a relatively spherical shape, although some elongated particles can be observed. All the particles exhibit a slightly wavy surface texture (similar to cauliflower), which is more pronounced for the particles exposed to the heat cycle of the SLS process. The presence of some satellites is also evident on some of these particles.

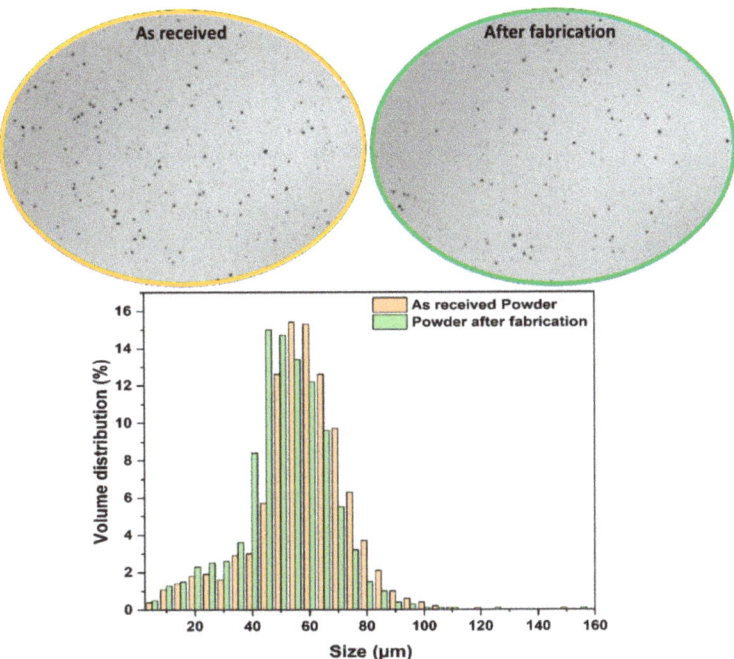

**Figure 4.** Volume distribution as function of the size for the "As-received powder" and "Powder after fabrication".

**Table 3.** Powder sample characteristics.

| Material/Characteristics | As-Received Powder | Powder after Fabrication |
|---|---|---|
| $D_{10}$ (μm) | 33.2 | 28.8 |
| $D_{50}$ (μm) | 55.7 | 49.1 |
| $D_{90}$ (μm) | 73.3 | 66.7 |
| Mean value Sphericity | 0.823 | 0.818 |

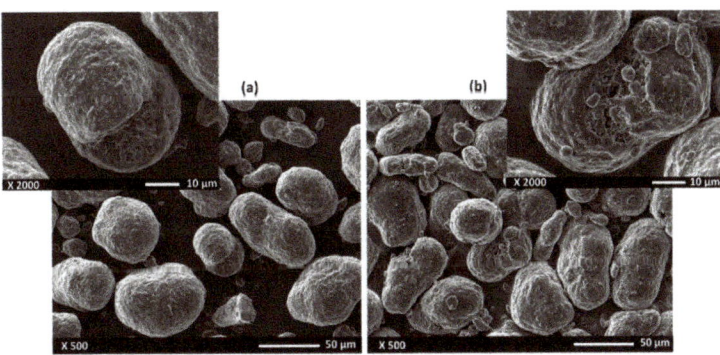

**Figure 5.** SEM micrographs of the PA12 powders: (**a**) the as-received powder; (**b**) powder after fabrication.

## 3.2. Fourier-Transformation Infrared Spectrometry (FTIR)

The ATR-FTIR spectra of the PA12 samples were recorded to provide information about the infrared bands and their roles. Figure 6 displays the spectra of both the as-received PA12 powder and the PA12 powder after fabrication. Table 4 summarizes the different vibrational bands in PA12 and their assignments [43–52]. Comparing both spectra, it can be observed that the intensities and positions of all bands are almost the same for both PA12 powders. This confirms the lack of influence of selective laser sintering on the chemical composition of PA12 powder.

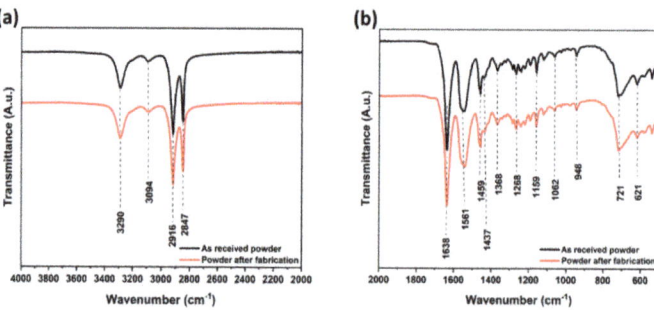

**Figure 6.** Fourier-transform infrared (FTIR) spectra of PA12 powders: As-received powder; Powder after fabrication. (**a**)-wavenumber from 2000 to 4000 cm$^{-1}$; (**b**)-wavenumber from 550 to 2000 cm$^{-1}$.

**Table 4.** Characteristic infrared bands and their assignments of SLS PA12 Powder.

| Vibrational Frequency [cm$^{-1}$]. | Assignments |
| --- | --- |
| 3290 | $\upsilon$ (N–H) stretching |
| 3094 | Fermi resonance of $\upsilon$ (N–H) stretching |
| 2916 | $\upsilon$ (CH$_2$) asymmetric stretching |
| 2847 | $\upsilon$ (CH$_2$) symmetric stretching |
| 1638 | Amide-I ($\upsilon$ (C=O) stretching and $\upsilon$ (C–N) stretching) |
| 1561 | Amide-II ($\delta$ (N–H) bending and $\upsilon$ (C–N) stretching) |
| 1459 | $\delta$ (CH$_2$) scissoring |
| 1368 | $\delta$ (CH$_2$) twisting |
| 1268 | Amide-III ($\upsilon$ (C–N) stretching and $\delta$ (C=O) in-plane bending) |
| 1159 | Skeletal motion CO–NH |
| 1062 | Skeletal motion CO–NH |
| 948 | $\delta$ (CO-NH) in-plane bending |
| 721 | $\rho$ (CH$_2$) rocking |
| 621 | Amide-IV ($\delta$ (N–H) out-of-plane bending) |

## 3.3. X-ray Diffraction

The X-ray diffraction patterns of both PA12 powders are shown in Figure 7, revealing their polymorphism and crystalline details. The XRD profile for both powder states shows that polyamide 12 exhibits two characteristic peaks at about 2θ = 20.95° and 21.50°, which are probably characteristics of the α-form of PA12 [14,53]. According to previous studies and the references cited therein, the crystal structure of polyamides has been known to be in the so-called α and γ-forms. The α-form consists of a monoclinic or triclinic lattice with chains in a fully extended planar zigzag arrangement, whereas the γ-form is a pseudo-hexagonal packing of 2$_1$ chains. Therefore, PA12 can be crystalized within structures of α and γ phases, where the major γ phase acts as a stable structure [47]. The chains in the α phase are antiparallelly oriented with an extended trans chain conformation, whilst chains in the γ form are oriented parallelly with a twisted helical conformation around the amide groups, making the γ form more stable than the α crystal structure [50].

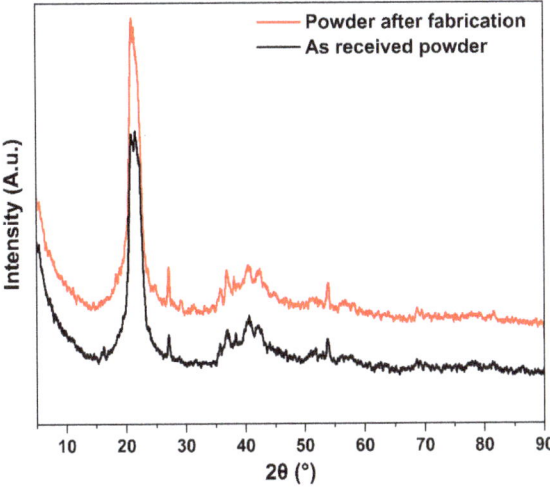

**Figure 7.** X-ray diffraction patterns of the as-received powder and powder after fabrication in range 4.5–90°.

The patterns relative to the powder after fabrication are almost similar to the patterns for the as-received powder, which confirms the possibility of re-using the powder for SLS after a suitable recycling process.

### 3.4. Differential Scanning Calorimetry

The DSC technique was applied to study the glass transition temperature, melting temperature range, and the degree of crystallinity of the PA12 material. Figure 8 shows the DSC curves for both the as-received powder and the powder after fabrication. Both curves follow almost the same shape, showing three specific thermal transitions: The first transition at around 50 °C is associated with the glass transition ($T_g$) phase where the polymer changes to a highly elastic state [54]. The second thermal transition, associated with the melting temperature ($T_m$), corresponds to the endothermic peak detected at 182 °C. During cooling, a third transition, observed at 151 °C, is attributed to the crystallization ($T_c$) of PA12, where a rearrangement of the molecular chains takes place to create crystalline lamellae inside the continuous amorphous structure. These results show that both types of PA12 powder are in a semi-crystalline state after the cooling process. As shown in Figure 8, there is a large distance between the melting and crystallization peaks, which indicates a tendency of PA12 to warp or curl during the laser-sintering process [42]. This metastable thermodynamic region of undercooled polymer is usually called the "SLS sintering window". It is important to select a sintering temperature in this temperature window range to obtain the best printing results of the material without degrading it. For both powders, the sintering temperature window of PA12 is in the range of [155 °C, 176 °C]. The crystallization temperature ($T_c$) must also be avoided as long as possible during processing.

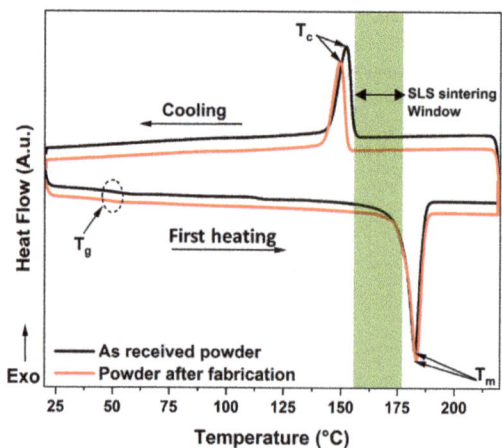

**Figure 8.** DSC thermograms of PA12 powders: as-received powder and powder after fabrication with the SLS sintering window shown in light blue.

The melting point for the as-received powder is 182.7 °C and is 183.3 °C for the powder after fabrication. The respective fusion heat values of 55.62 J·g$^{-1}$ and 48.49 J·g$^{-1}$ have been measured. After melting, the as-received powder and the powder after fabrication re-solidify with peak crystallization temperatures of 151.9 °C and 149.1 °C, respectively. The solidification (crystallization) temperature is significantly lower than the crystalline melting temperature of PA12; this phenomenon is common in crystalline polymers and is known as super cooling [18]. From Table 5, it can be noted that the as-received PA12 powder exhibit a slightly higher degree of crystallinity than PA12 after fabrication (46.62% against 44.22%). This slight decrease is possibly linked to a reduction in the polymer chain order during the printing process combined with a non-controlled cooling rate of the SLS process after fabrication. The different thermal properties and the crystallization characteristics of the DSC measurements for both powders are all collected in Table 5.

**Table 5.** DSC data corresponding to the first heating–cooling scan for the as-received powder and powder after fabrication.

| Powder State | $T_g$ (°C) | $T_m$ (°C) | $T_c$ (°C) | $\Delta H_m$ (J·g$^{-1}$) | $\Delta H_c$ (J·g$^{-1}$) | $X_c$ (%) |
|---|---|---|---|---|---|---|
| As-received powder | 51.7 | 182.7 | 151.9 | 97.59 | 55.62 | 46.62 |
| Powder after fabrication | 49.0 | 183.3 | 149.1 | 92.67 | 48.49 | 44.27 |

### 3.5. Effect of Laser Power and Hatch Orientation on Tensile Properties

The tensile properties were evaluated at different levels of laser power and hatch orientations and the results are presented in Figure 9. Figure 9a shows the tensile strength (TS) values of PA12's specimens made in various XY plane orientations when varying the laser power (LP). The measured TS values show the influence of both the XY plane orientation and laser power, with a maximum value of 25.65 MPa achieved at LP: 95% (57W) and a 0° XY plane orientation. These results clearly show that tensile properties rise with an increasing laser power for all part orientations. By increasing the laser power from LP: 75% to LP: 95% in the case of the 0° orientation, the tensile strength is increased from 19.41 MPa to 25.65 MPa. Such an increase can mostly be attributed to an improved coalescence resulting from the higher temperatures achieved by the polymer melt when a higher laser power is used [55]. These results are in good agreement with the findings of

Caulfield et al. [12], who claimed that parts built at higher energy density levels (higher laser powers) exhibited a higher tensile strength.

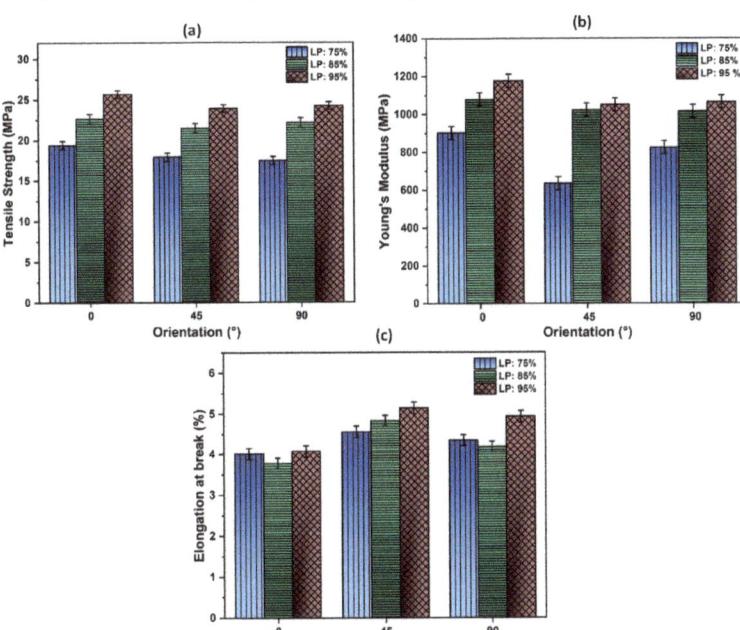

**Figure 9.** Tensile properties as a function of XY plane orientation and laser power: (**a**) Tensile Strength; (**b**) Young's Modulus; (**c**) Elongation at break.

While the effects of laser power on TS are obvious, those linked to hatch orientation are less evident. Such an isolation requires a comparison between the 0° and 90° TS values first, followed by the 45° and 90° TS values. As mentioned previously, the parts made at 0° and 90° XY plane orientation have a nearly identical hatching. This should result in similar TS values as no other SLS parameters differ, something not observed here. Parts made at 90° XY plane orientation exhibit TS values 2 to 10% lower compared to those at 0°, with the greatest difference observed at a low laser power. This decrease in TS for the 90° parts could be explained by their position on the build platen during fabrication relative to the 0° parts. It was shown that there was a need to operate with a powder layer of uniform temperature to achieve builds of multiple parts with similar mechanical properties [27,56,57]. To confirm this, thermal imaging of the preheated powder bed was performed using a Testo 890-2 IR camera. The resulting thermogram, shown in Figure 10, exhibits temperature differences of more than 10 °C between the center and sides of the bed. These gradients could mostly be attributed to a non-uniform heating by the quartz lamps used, possibly because of aging. To highlight this further, the temperature profile along the Y-axis is shown for five locations. As the 0° parts were placed in the center of the build chamber, e.g., in the center third in-between P2 and P4 (Figure 10), they enjoyed a uniform temperature with a variation less than 3 °C. For the 90° parts, their positioning at the periphery entailed more pronounced variations, with some parts partially in regions that were 10 °C lower than the set point of 159 °C. Such a difference could explain their lower TS values against the 0° parts, as the SLS process greatly relies on powder bed heating to supply most of the energy required for sintering particles. Thus, the TS is affected by the powder bed temperature, and the simultaneous production of multiple parts will require a good temperature uniformity to achieve uniform TS values.

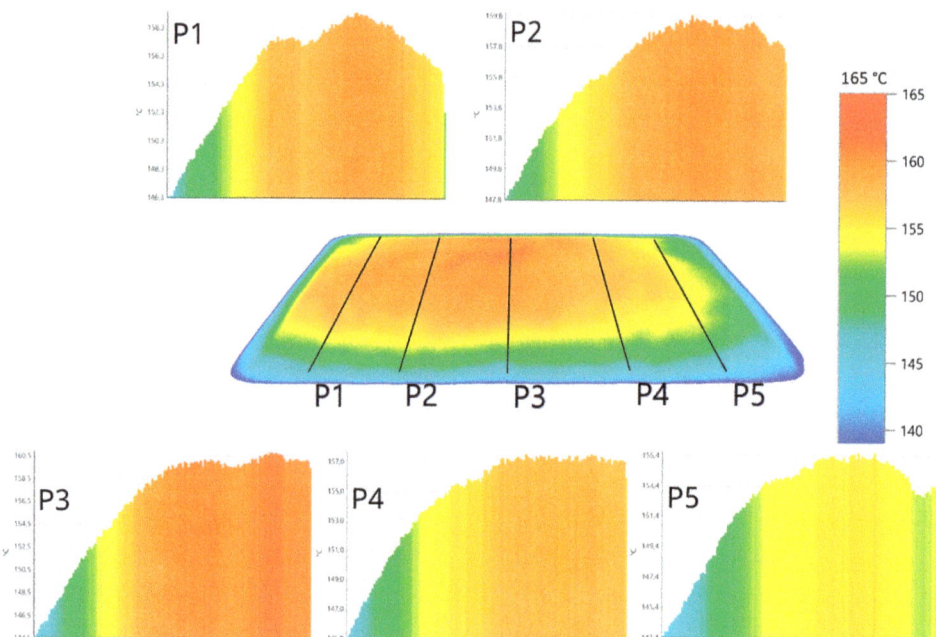

**Figure 10.** Thermal imagery of the SLS powder layer during preheating. P1 to P5 are front-to-back temperature profile taken from left to right of the platen (note that the Y-axis does not have the same numeral scale).

The parts made at a 45° XY plane orientation exhibit the overall lowest TS values compared to other orientations made in the same conditions by as much as 7.5%. As these 45° parts were positioned at the periphery, a gradient in the powder bed temperature could partially explain this. As the 45° and 90° parts were interspersed at the periphery, both orientations should exhibit similar TS values, which was not the case here. This difference could be attributed to the hatching orientation. While hatching is conducted alternatively parallel/perpendicular to the applied load in the case of the 90° parts, hatching in 45° parts is performed at an angle relative to the applied load. In a fashion similar to this well-known effect in FDM [58], more loads can be supported along the axis of hatching as it is applied to a continuous string of melted polymer and not at the joining of two strings (or hatches). This could explain the slightly lower TS values obtained for the 45° parts compared to those at 90°. These results confirm that not only laser power and hatch orientation affect the tensile strength of 3D-printed PA12 samples, but that powder bed temperature uniformity must also not be neglected.

Figure 9b displays the Young's Modulus as a function of laser power for the various XY plane orientations. It is evident that the Young's modulus values depend on laser power and XY plane orientation, with the best results (1176.6 MPa) achieved at the highest laser power and a 0° angle. This could be explained in part by the laser power and in part by the position on the build platen. Laser power can greatly affect the Young's Modulus no matter what the part's angle is, as shown by the 65% increase observed for samples produced at 45° when the laser power was increased from 75 to 95%. This increase in Young's modulus due to laser power could be attributed to the additional energy applied, which improves the particles' sintering. This will help achieve better compaction, thus increasing resulting mechanical properties as reported by Singh and al. for polyamide material [59]. Thermal non-uniformity of the powder bed can explain why the 90° parts exhibit a lower Young's Modulus than the 0° parts, with all other parameters including hatching being similar. The effects of hatching, as seen by comparing the 45° and 90° results, are limited, except when operating at a low laser power. In this instance, the lower energy applied would result in a poor joining of the hatches, which leads to lower mechanical properties.

The Elongation at break as a function of the build orientation and laser power variation was also evaluated (see Figure 9c). However, unlike the TS and Young's Modulus, the 45° orientation presents the maximum values of elongation (5.13%) for all laser power conditions. In addition, the laser power will affect the elongation at break for parts produced at various XY plane angle. Sintering at a low laser power creates weaker bonds between the powder particles, leading to decreased values of elongation at break [55]. The influence of laser power, while observed in all orientations, is more pronounced for the 45° orientation. At that angle, an increase of 12% in elongation at break was observed when raising laser power from 45W (75%) to 57 W (95%). At similar laser power, the non-uniformity of the powder bed temperature explains the difference in the elongation at break between the parts made at 0° and 90°, while hatching will explain the differences between the 45° and 90° results. In summary, tensile properties are strongly affected by laser power and to a lesser level by the temperature uniformity of the powder bed. The hatching orientation will also affect tensile properties to some extent.

In order to further understand the influence of laser power on the properties of SLS-PA12, the fracture surface of the SLS-PA12 samples was analyzed. Figure 11 shows the SEM observations of the tensile fracture surface for the SLS samples produced at LP: 75% and LP: 95%. The SEM images clearly show particles of PA12 that were melted into the dense part and the presence of some voids in between the layers, especially in the case of LP: 75% (see Figure 11a). These voids favor the delamination of sintered layers, thus explaining the decreased bending strength and inferior rigid behavior. Previous studies have reported that SLS-sintered specimens are porous due to an insufficient heat input that results in their very low mechanical properties [17,46]. The spherical particles observed in Figure 11c,d are un-melted or partially melted PA12 powders. It is clearly evident that those spherical particles are present in large amounts in the case of the sample sintered with a low laser power (LP: 75%) compared to the one sintered at a high laser power (LP: 95%). The PA12 samples with LP: 75% also exhibit a porous interior, with small bonding areas due to the insufficient laser power. This decreases the cohesion between layers while reducing the surface contact between the printed PA12's layers. The PA12 sample sintered at a 95% LP (Figure 11b) was much smoother, and the fusion effect was improved, though some un-melted powders are still visible in some areas of the cross section (Figure 11d). This adhesive enhancement in the microstructure could explain the observed increase in the tensile properties when higher laser power was used (LP: 95%).

*3.6. Effect of Laser Power on Thermal and Structural Properties of 3D-Printed Samples*

(a)   Differential Scanning Calorimetry

A DSC analysis of the printed PA12 parts at different laser powers was performed to determine the impact of this controlling parameter on the thermal characteristics of the PA12 samples manufactured by the SLS process. Figure 12 shows the first heating DSC thermograms of the printed PA12 at various laser powers. From this figure, it is clear that all the samples exhibit the same thermal transitions as powder before fabrication (see Figure 8). It is evident that all the thermograms present a similar thermal behavior and exhibit three thermal transitions. The first heat flow exchange, located at around 45 °C, is associated with the glass transition temperature ($T_g$). The second thermal transition, associated with the melting temperature ($T_m$), is the endothermic peak detected at 176 °C. During cooling, a third exothermic transition at 146 °C is attributed to the crystallization of PA12, where a re-arrangement of molecular chains takes place to create crystalline lamellae inside the continuous amorphous structure. From these results, it can be concluded that PA12 kept its semi-crystalline property after the cooling process. Table 6 summarizes all characteristic temperatures, associated enthalpies, and the degree of crystallinity of the as-printed samples at various laser powers.

**Figure 11.** SEM micrographs of the fractured surfaces of SLS-printed PA12 specimens: (**a**) LP: 75% [×100], (**b**) LP: 95% [×100], (**c**) LP: 75% [×500], and (**d**) LP: 95% [×500].

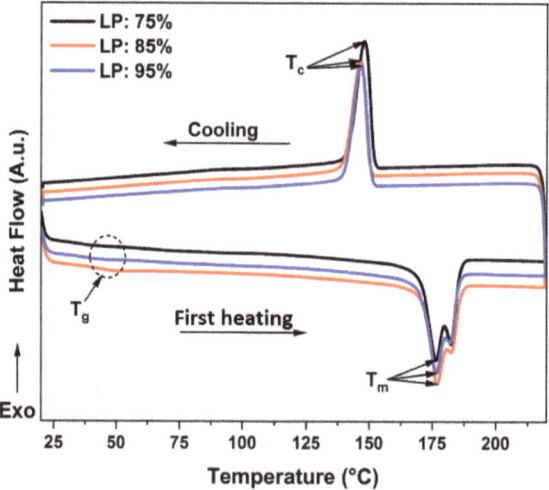

**Figure 12.** DSC thermograms of PA12 samples with the three laser powers used: LP-75%, LP-85%, and LP-95%.

Table 6. DSC data corresponding to the first heating–cooling scan for the different laser powers used.

| Specimen | $T_g$ (°C) | $T_m$ (°C) | $T_c$ (°C) | $\Delta H_m$ (J.g$^{-1}$) | $\Delta H_c$ (J.g$^{-1}$) | $X_c$ (%) |
|---|---|---|---|---|---|---|
| LP: 75% | 44.07 | 176.54 | 147.89 | 73.50 | 56.04 | 35.11 |
| LP: 85% | 45.45 | 176.93 | 146.30 | 67.50 | 50.94 | 32.25 |
| LP: 95% | 46.14 | 176.73 | 146.47 | 69.31 | 54.50 | 33.11 |

We can notice in Figure 12 a small additional peak at 180 °C in the melting transition. Zarringhalam et al. [41] found the same phenomenon and explained this additional small endotherm as resulting from the unmolten particle core after the laser-sintering process. Generally, the microstructure of SLS parts simultaneously includes fully molten particles and unmolten particle cores surrounded by spherulites. These unmolten powder particles have almost the same melting temperature as the as-received powder, leading to the additional small peak between 180 and 182 °C. The results of the DSC experiments in this study are consistent with previous studies [27,60,61]. The DSC thermograms show the absence of any significant evolution of melting temperature ($T_m$), crystallization temperature ($T_c$), and glass transition temperature ($T_g$) between the printed PA12 at different laser power. It is evident from Table 6 that the variation in laser power produces no significant changes to the degree of crystallinity of samples sintered at various laser power.

However, the degree of crystallization decreased from 46.62% (see Table 4) for the as-received powder to 35.11% for the sample sintered with 75% laser power. This difference could be attributed to a rapid cooling and less energy dissipation during the printing process. Moreover, the decrease in $T_g$ from 51.7 to 44.07 °C as a result of sintering correlates strongly with the degree of crystallization of the sample. This can be attributed to the gain in the mobility of the polymer chains as they are not partly anchored inside the crystalline domain.

(b) X-ray Diffraction

Figure 13 displays the profile of the X-ray diffraction patterns of the SLS samples produced at various laser powers (LP: 75%, LP: 85, and LP: 95%). All spectra exhibit similar diffraction peaks, with two major peaks found at 2θ of around 20° and 45°. The peak at 45° appeared after the sintering process; thus, the γ form might be more pronounced even with the presence of the α crystal form [62].

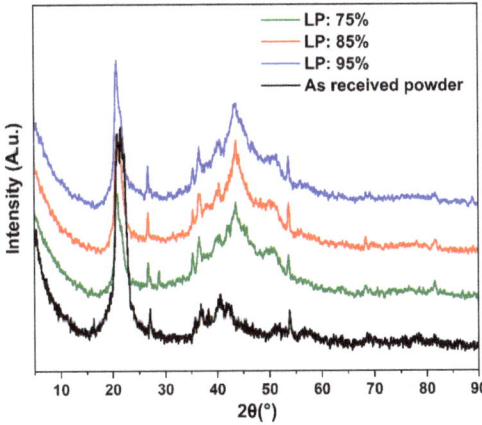

Figure 13. X-ray diffraction patterns of the SLS samples parts with different laser powers in range from 4.5–90°.

## 3.7. Annealing Impact on Thermal, Structural, and Mechanical Properties

Many studies investigating polymeric materials have considered that using heat treatment post-processing can improve the material properties and crystallinity of SLS parts made from Nylon 12 [63]. These heat treatment (annealing) studies were conducted using different settings of temperature and time. In general, better mechanical properties and crystallinity were obtained when the heat treatment was carried out close to the melting temperature [64]. During annealing, the crystallization processes highly depend on the temperature of the applied annealing procedure. A high temperature would generate an isothermal crystallization process, in which the non-crystalline polymer chains have enough energy to form more crystalline regions and an optimal arrangement [65,66]. In this work, the choice of annealing temperatures was based on this latter theory.

A DSC analysis was used to optimize the annealing temperature by determining the adequate cycle yielding the highest thermal property and degree of crystallinity. The printed parts with the selected optimized printing parameters ([0°] orientation and 95% laser power) were annealed for six hours at various temperatures (130 °C, 150 °C, and 170 °C) to allow for the relaxation of the residual stress generated during their printing process. The annealed parts were analyzed by DSC, using the first heating cycle to characterize the thermal history experienced during the annealing process. During this heating cycle, the material achieved its melting temperature ($T_m$) to characterize the melting enthalpy, and then the degree of crystallinity ($X_c$) generated during the annealing cycle. Figure 14 shows the first heating and cooling DSC thermograms of the unannealed and annealed samples under various annealing cycles. The $X_c$ and all thermal transitions were measured (see Table 6) for the samples prepared by cutting a small amount of material from the tensile test specimens (before the tests were carried out).

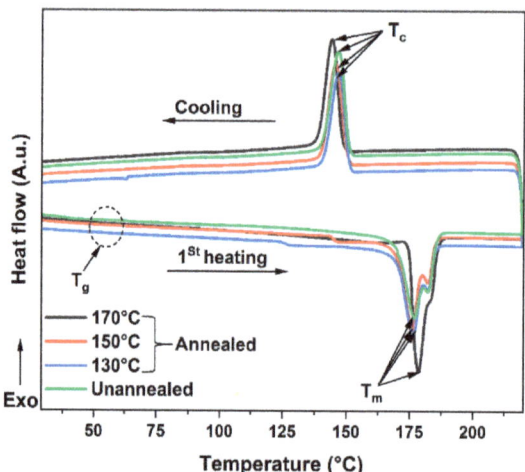

**Figure 14.** Differential scanning calorimetry curves for unannealed and annealed PA12's samples.

Table 7 summarizes the DSC results for various parameters such as $T_g$, $T_m$, and $X_c$. From this table, the DSC results show that the $T_g$ of the annealed parts shows a shift to higher temperatures with the increasing annealing temperature. When the annealing temperature was increased from 130 to 170 °C, the $T_g$ increased from 40.8 to 50.3 °C. Even though the maximum annealing temperature was achieved (170 °C), the $T_g$ value was still higher than the unannealed printed parts. It can also be noted that raising the annealing temperature from 130 to 170 °C results in an increase in the heat flow of melting from 65.51 to 76.51 J·$g^{-1}$. An increase in the relative degree of crystallinity was also observed, from 31.29% to 36.55%. This latter crystallinity enhancement could be the result of a phenomenon called secondary crystallization, which increases the lamellar form of PA12. Moreover, the increase in $T_g$ because of annealing correlates strongly with the degree of crystallinity of the printed PA12's parts. Annealing thus induces strong intermolecular interactions between polymer chains. Thus, this increase is attributed to the loss of mobility of polymer chains as they are partly anchored inside the crystalline region.

**Table 7.** DSC data corresponding to the first heating–cooling scan for the different annealing temperature used.

| Specimen State | $T_g$ (°C) | $T_m$ (°C) | $T_c$ (°C) | $\Delta H_m$ (J·g$^{-1}$) | $\Delta H_c$ (J·g$^{-1}$) | $X_c$ (%) |
| --- | --- | --- | --- | --- | --- | --- |
| Unannealed | 46.1 | 176.7 | 146.4 | 69.31 | 54.50 | 33.11 |
| Annealed at 130 °C/6 h | 40.8 | 176.3 | 146.5 | 65.51 | 49.78 | 31.29 |
| Annealed at 150 °C/6 h | 44.4 | 176.0 | 145.3 | 69.66 | 50.09 | 33.28 |
| Annealed at 170 °C/6 h | 50.3 | 178,4 | 143.9 | 76.51 | 51.79 | 36.55 |

The annealed specimens were then subjected to mechanical testing to record the Young's modulus, tensile strength, and strain at break. Figure 15 shows the typical tensile stress–strain curves of unannealed and annealed printed PA12 parts according to the various annealing conditions. All tested specimens exhibit a maximum of the stress/strain curve, followed by brittle deformation. The results indicate that tensile stress and its strain increases when the annealing temperature is increased from 130 to 170 °C. The annealed parts at 170 °C show the maximum tensile stress and strain compared to the unannealed samples. These results indicate the fact that heat treating the PLA12 parts at 170 °C for 6 h allows the material sufficient time for crystallization and the re-arrangement of the polymer chains.

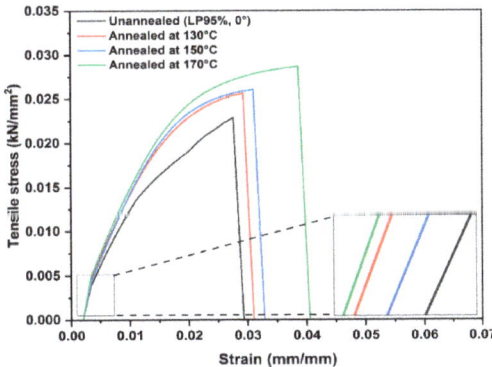

**Figure 15.** Stress–strain curves of printed PA12 in different conditions. Unannealed and annealed at: 130, 150, and 170 °C.

From Figure 16, it can be observed that annealing PA12 material favorably affects the tensile properties of the printed PA12 parts. For untreated samples, the Young's modulus and tensile strength exhibit the lowest values (1176.7 MPa and 25.5 MPa, respectively) compared to the annealed samples at 170 °C (1276.23 MPa and 29.2 MPa respectively) (see Figure 16a). This change in rigidity and strength is related to changes in the microstructures of the material, as the annealed samples exhibit a higher degree of crystallinity and glass transition temperature as discussed above. These results are in good agreement with the work of Liu et al. [67], who reported that high-temperature annealing (173 °C) can remarkably enhance the mechanical strength of printed PA12 specimens. The work by Zarringhalam and al. [64] confirmed that using heat treatment as a post-processing technique can improve the tensile strength and Young's modulus of SLS parts made from Nylon

12. However, as shown in Figure 16b, annealing at 170 °C/6 h leads to an increase in ductility where the elongation at break increases to 5.66%, while the as-printed (un-annealed) samples exhibit a value of 4.07%.

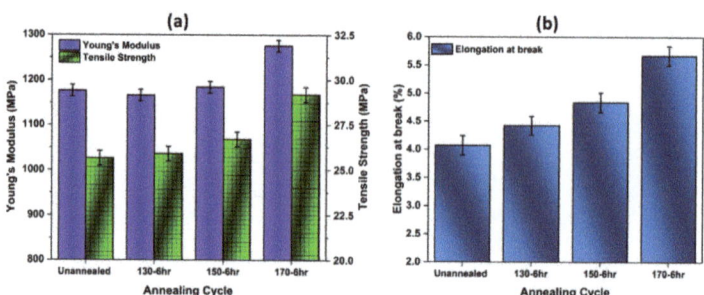

**Figure 16.** Effect of annealing on tensile properties of PA12 samples' parts: (**a**) results of Young's modulus and tensile strength; (**b**) results of elongation at break.

From the above results, it has been confirmed that high-temperature annealing (170 °C) yields the best improvement in Young's modulus (+9 MPa, or ~8.4%), tensile strength (+3.7 MPa, or ~14.5%), and elongation at break (+1.59 MPa, or ~39%) over the unannealed parts. It can be concluded from these results that annealing had a higher percent contribution to the mechanical performance over the duration of annealing. This confirms the importance of the annealing process for achieving proper chain crystallization, thus enhancing the mechanical properties of 3D parts [58].

## 4. Conclusions

This study evaluated the effect of laser power and hatch orientation on the tensile properties and morphology of the SLS PA12-produced parts. The main objective was to identify the suitable laser power and hatch orientation leading to better mechanical properties and high-quality parts. Different methods were used to study the SLS parts by considering the morphological, structural, and mechanical properties using XRD, FTIR, DSC, tensile testing, and SEM characterizations.

The results confirmed the significant impact of laser power, while the effects of hatching were almost unnoticeable when using a high laser power. A more significant condition is the uniformity of the powder bed temperature, a factor that is seldom considered. This needs to be accounted for because of its effects on the mechanical properties. However, the operator has little recourse with respect to these conditions, which are strongly dependent on the quality of the SLS system.

Operating at a high laser power minimized the presence of spherical particles normally related to un-melted powder and yielded an improved microstructure. It was also observed that reducing the laser power to LP: 75% decreases the mechanical properties, with the parts exhibiting spherical particles and a poor microstructure. Heat treating SLS-produced PA12 parts showed the positive impact of annealing, especially at 170 °C, on the tensile properties. This can be related to changes in the microstructure of the PA12 parts.

**Author Contributions:** Conceptualization, A.E.M. and S.V.; methodology, A.E.M., S.E.B., H.R.V. and S.V.; software, A.E.M., S.E.B. and S.V.; validation, A.E.M., S.E.B., H.R.V. and S.V.; formal analysis, A.E.M., S.E.B., H.R.V. and S.V.; investigation, A.E.M., S.E.B. and S.V.; resources, A.E.M., S.E.B. and S.V.; data curation, A.E.M., S.E.B. and S.V.; writing—original draft preparation, A.E.M., S.E.B., H.R.V. and S.V.; writing—review and editing, A.E.M., S.E.B., H.R.V. and S.V.; visualization, A.E.M.; supervision, A.E.M. and S.V.; project administration, A.E.M., S.E.B. and S.V.; funding acquisition, A.E.M., S.E.B. and S.V. All authors have read and agreed to the published version of the manuscript.

**Funding:** This research received no external funding.

**Institutional Review Board Statement:** Not applicable.

**Informed Consent Statement:** Not applicable.

**Acknowledgments:** The authors also gratefully acknowledge the support from the Euromed University of Fes, and the financial contribution of the Hassan II Academy of Sciences and Technology and Safran Composites (France).

**Conflicts of Interest:** The authors declare no conflict of interest.

# References

1. Abdulhameed, O.; Al-Ahmari, A.; Ameen, W.; Mian, S.H. Additive manufacturing: Challenges, trends, and applications. *Adv. Mech. Eng.* **2019**, *11*, 1–27. [CrossRef]
2. Saffarzadeh, M.; Gillispie, G.J.; Brown, P. Selective Laser Sintering (SLS) rapid protytping technology: A review of medical applications. In Proceedings of the 53rd Annual Rocky Mountain Bioengineering Symposium, RMBS 2016 and 53rd International ISA Biomedical Sciences Instrumentation Symposium, Denver, CO, USA, 8–10 April 2016; pp. 142–149.
3. Jiménez, M.; Romero, L.; Domínguez, I.A.; Espinosa, M.D.M.; Domínguez, M. Additive Manufacturing Technologies: An Overview about 3D Printing Methods and Future Prospects. *Complexity* **2019**. [CrossRef]
4. Guo, N.; Leu, M.C. Additive manufacturing: Technology, applications and research needs. *Front. Mech. Eng. Chin.* **2013**, *8*, 215–243. [CrossRef]
5. Wohlers, T. Recent Trends in Additive Manufacturing. In Proceedings of the 17th European Forum on Rapid Prototyping and Manufacturing, Paris, France, 12–14 June 2012; pp. 12–14.
6. Kozak, J.; Zakrzewski, T. Accuracy problems of additive manufacturing using SLS/SLM processes. *AIP Conf. Proc.* **2018**, *2017*, 020010. [CrossRef]
7. Goodridge, R.D.; Tuck, C.J.; Hague, R.J.M. Laser sintering of polyamides and other polymers. *Prog. Mater. Sci.* **2012**, *57*, 229–267. [CrossRef]
8. Schmid, M.; Amado, A.; Wegener, K. Materials perspective of polymers for additive manufacturing with selective laser sintering. *J. Mater. Res.* **2014**, *29*, 1824–1832. [CrossRef]
9. Cheng, J.; Lao, S.; Nguyen, K.; Ho, W.; Cummings, A.; Koo, J. SLS processing studies of nylon 11 nanocomposites. In Proceedings of the 2005 International Solid Freeform Fabrication Symposium, Austin, TX, USA, 1–3 August 2005; Volume 2005, pp. 141–149.
10. Baba, M.N. Flatwise to Upright Build Orientations under Three-Point Bending Test of Nylon 12 (PA12) Additively Manufactured by SLS. *Polymers* **2022**, *14*, 1026. [CrossRef]
11. Regassa, Y.; Lemu, H.G.; Sirabizuh, B. Trends of using polymer composite materials in additive manufacturing. In Proceedings of the 9th International Scientific Conference—Research and Development of Mechanical Elements and Systems (IRMES 2019), Kragujevac, Serbia, 5–9 September 2019; Volume 659. [CrossRef]
12. Caulfield, B.; McHugh, P.E.; Lohfeld, S. Dependence of mechanical properties of polyamide components on build parameters in the SLS process. *J. Mater. Process. Technol.* **2007**, *182*, 477–488. [CrossRef]
13. Shuai, C.; Yang, W.; Feng, P.; Peng, S.; Pan, H. Accelerated degradation of HAP/PLLA bone scaffold by PGA blending facilitates bioactivity and osteoconductivity. *Bioact. Mater.* **2021**, *6*, 490–502. [CrossRef]
14. Yan, C.Z.; Shi, Y.S.; Yang, J.S.; Liu, J.H. An organically modified montmorillonite/nylon-12 composite powder for selective laser sintering. *Rapid Prototyp. J.* **2011**, *17*, 28–36. [CrossRef]
15. Shaw, B.; Dirven, S. Investigation of porosity and mechanical properties of nylon SLS structures. In Proceedings of the 2016 23rd International Conference on Mechatronics and Machine Vision in Practice (M2VIP), Nanjing, China, 28–30 November 2016. [CrossRef]
16. Van Hooreweder, B.; Moens, D.; Boonen, R.; Kruth, J.P.; Sas, P. On the difference in material structure and fatigue properties of nylon specimens produced by injection molding and selective laser sintering. *Polym. Test.* **2013**, *32*, 972–981. [CrossRef]
17. Van Hooreweder, B.; de Coninck, F.; Moens, D.; Boonen, R.; Sas, P. Microstructural characterization of SLS-PA12 specimens under dynamic tension/compression excitation. *Polym. Test.* **2010**, *29*, 319–326. [CrossRef]
18. Vasquez, M.; Haworth, B.; Hopkinson, N. Optimum sintering region for laser sintered Nylon-12. *Proc. Inst. Mech. Eng. Part B J. Eng. Manuf.* **2011**, *225*, 2240–2248. [CrossRef]
19. Zarringhalam, H.; Majewski, C.; Hopkinson, N. Degree of particle melt in Nylon-12 selective laser-sintered parts. *Rapid Prototyp. J.* **2009**, *15*, 126–132. [CrossRef]
20. Hofland, E.C.; Baran, I.; Wismeijer, D.A. Correlation of Process Parameters with Mechanical Properties of Laser Sintered PA12 Parts. *Adv. Mater. Sci. Eng.* **2017**, *2017*, 4953173. [CrossRef]
21. Wegner, A.; Harder, R.; Witt, G.; Drummer, D. Determination of Optimal Processing Conditions for the Production of Polyamide 11 Parts using the Laser Sintering Process. *Int. J. Recent Contrib. Eng. Sci. IT* **2015**, *3*, 5. [CrossRef]
22. Leigh, D.K. A comparison of polyamide 11 mechanical properties between laser sintering and traditional molding. In Proceedings of the 2012 International Solid Freeform Fabrication Symposium, Austin, TX, USA, 6–8 August 2012; Volume 2012, pp. 574–605.
23. Zhou, W.; Wang, X.; Hu, J.; Zhu, X. Melting process and mechanics on laser sintering of single layer polyamide 6 powder. *Int. J. Adv. Manuf. Technol.* **2013**, *69*, 901–908. [CrossRef]

24. Kim, J.; Creasy, T.S. Selective laser sintering characteristics of nylon 6/clay-reinforced nanocomposite. *Polym. Test.* **2004**, *23*, 629–636. [CrossRef]
25. Dupin, S.; Lame, O.; Barrès, C.; Charmeau, J.Y. Microstructural origin of physical and mechanical properties of polyamide 12 processed by laser sintering. *Eur. Polym. J.* **2012**, *48*, 1611–1621. [CrossRef]
26. Athreya, S.R.; Kalaitzidou, K.; Das, S. Processing and characterization of a carbon black-filled electrically conductive Nylon-12 nanocomposite produced by selective laser sintering. *Mater. Sci. Eng. A* **2010**, *527*, 2637–2642. [CrossRef]
27. Bourell, D.L.; Watt, T.J.; Leigh, D.K.; Fulcher, B. Performance limitations in polymer laser sintering. *Phys. Procedia* **2014**, *56*, 147–156. [CrossRef]
28. Dizon, J.R.C.; Espera, A.H.; Chen, Q.; Advincula, R.C. Mechanical characterization of 3D-printed polymers. *Addit. Manuf.* **2018**, *20*, 44–67. [CrossRef]
29. Jain, P.K.; Pandey, P.M.; Rao, P.V.M. Experimental investigations for improving part strength in selective laser sintering. *Virtual Phys. Prototyp.* **2008**, *3*, 177–188. [CrossRef]
30. Koo, J.H.; Lao, S.; Ho, W.; Ngyuen, K.; Cheng, J.; Pilato, L.; Wissler, G.; Ervin, M. Polyamide nanocomposites for selective laser sintering. In Proceedings of the 2006 International Solid Freeform Fabrication Symposium, Austin, TX, USA, 14–16 August 2006; Volume 2006, pp. 392–409.
31. Klahn, C.; Leutenecker, B.; Meboldt, M. Design strategies for the process of additive manufacturing. *Procedia CIRP* **2015**, *36*, 230–235. [CrossRef]
32. Rajesh, R.; Sudheer, S.; Kulkarni, M. Selective Laser Sintering Process–A Review. *Int. J. Curr. Eng. Sci. Res.* **2015**, *2*, 2393–8374.
33. Zárybnická, L.; Petrů, J.; Krpec, P.; Pagáč, M. Effect of Additives and Print Orientation on the Properties of Laser Sintering-Printed Polyamide 12 Components. *Polymers* **2022**, *14*, 1172. [CrossRef]
34. Korycki, A.; Garnier, C.; Nassiet, V.; Sultan, C.T. Optimization of Mechanical Properties and Manufacturing Time through Experimental and Statistical Analysis of Process Parameters in Selective Laser Sintering. *Adv. Mater. Sci. Eng.* **2022**, 2526281. [CrossRef]
35. Drummer, D.; Rietzel, D.; Kühnlein, F. Development of a characterization approach for the sintering behavior of new thermoplastics for selective laser sintering. *Phys. Procedia* **2010**, *5*, 533–542. [CrossRef]
36. Stichel, T.; Frick, T.; Laumer, T.; Tenner, F.; Hausotte, T.; Merklein, M.; Schmidt, M. A Round Robin study for Selective Laser Sintering of polyamide 12: Microstructural origin of the mechanical properties. *Opt. Laser Technol.* **2017**, *89*, 31–40. [CrossRef]
37. Starr, T.L.; Gornet, T.J.; Usher, J.S. The effect of process conditions on mechanical properties of laser-sintered nylon. *Rapid Prototyp. J.* **2011**, *17*, 418–423. [CrossRef]
38. Amado, A.; Schmid, M.; Levy, G.; Wegener, K. Advances in SLS powder characterization. In Proceedings of the 2011 International Solid Freeform Fabrication Symposium, Austin, TX, USA, 8–10 August 2011; Volume 2011, pp. 438–452.
39. Ziegelmeier, S.; Christou, P.; Wöllecke, F.; Tuck, C.; Goodridge, R.; Hague, R.; Krampe, E.; Wintermantel, E. An experimental study into the effects of bulk and flow behaviour of laser sintering polymer powders on resulting part properties. *J. Mater. Process. Technol.* **2015**, *215*, 239–250. [CrossRef]
40. Launhardt, M.; Wörz, A.; Loderer, A.; Laumer, T.; Drummer, D.; Hausotte, T.; Schmidt, M. Detecting surface roughness on SLS parts with various measuring techniques. *Polym. Test.* **2016**, *53*, 217–226. [CrossRef]
41. Zarringhalam, H.; Hopkinson, N.; Kamperman, N.F.; de Vlieger, J.J. Effects of processing on microstructure and properties of SLS Nylon 12. *Mater. Sci. Eng. A* **2006**, *435*, 172–180. [CrossRef]
42. Verbelen, L.; Dadbakhsh, S.; van den Eynde, M.; Kruth, J.P.; Goderis, B.; van Puyvelde, P. Characterization of polyamide powders for determination of laser sintering processability. *Eur. Polym. J.* **2016**, *75*, 163–174. [CrossRef]
43. Rhee, S.; White, J.L. Crystal structure and morphology of biaxially oriented polyamide 12 films. *J. Polym. Sci. Part B Polym. Phys.* **2002**, *40*, 1189–1200. [CrossRef]
44. Czarnecki, M.A.; Wu, P.; Siesler, H.W. 2D FT-NIR and FT-IR correlation analysis of temperature-induced changes of nylon 12. *Chem. Phys. Lett.* **1998**, *283*, 326–332. [CrossRef]
45. Skrovanek, D.J.; Howe, S.E.; Painter, P.C.; Coleman, M.M. Hydrogen Bonding in Polymers: Infrared Temperature Studies of an Amorphous Polyamide. *Macromolecules* **1985**, *18*, 1676–1683. [CrossRef]
46. Yang, F.; Jiang, T.; Lalier, G.; Bartolone, J.; Chen, X. Process control of surface quality and part microstructure in selective laser sintering involving highly degraded polyamide 12 materials. *Polym. Test.* **2021**, *93*, 106920. [CrossRef]
47. Chen, P.; Tang, M.; Zhu, W.; Yang, L.; We, S.; Yan, C.; Ji, Z.; Nan, H.; Shi, Y. Systematical mechanism of Polyamide-12 aging and its micro-structural evolution during laser sintering. *Polym. Test.* **2018**, *67*, 370–379. [CrossRef]
48. Inoue, K.; Hoshino, S. Crystal structure of nylon 12. *J. Polym. Sci. Polym. Phys. Ed.* **1973**, *11*, 1077–1089. [CrossRef]
49. Li, L.; Koch, M.H.J.; de Jeu, W.H. Crystalline structure and morphology in nylon-12: A small- and wide-angle X-ray scattering study. *Macromolecules* **2003**, *36*, 1626–1632. [CrossRef]
50. Kaur, T.; Nussbaum, J.; Lee, S.; Rodriguez, K.; Crane, N.B.; Harmon, J. Characterization of PA-12 specimens fabricated by projection sintering at various sintering parameters. *Polym. Eng. Sci.* **2021**, *61*, 221–233. [CrossRef]
51. Zhang, J.; Adams, A. Understanding thermal aging of non-stabilized and stabilized polyamide 12 using 1H solid-state NMR. *Polym. Degrad. Stab.* **2016**, *134*, 169–178. [CrossRef]
52. Celina, M.; Ottesen, D.K.; Gillen, K.T.; Clough, R.L. FTIR emission spectroscopy applied to polymer degradation. *Polym. Degrad. Stab.* **1997**, *58*, 15–31. [CrossRef]

53. Li, Y.; Yan, D.; Zhou, E. In situ Fourier transform IR spectroscopy and variable-temperature wide-angle X-ray diffraction studies on the crystalline transformation of melt-crystallized nylon 12 12. *Colloid Polym. Sci.* **2002**, *280*, 124–129. [CrossRef]
54. Senatov, F.S.; Niaza, K.V.; Zadorozhnyy, M.Y.; Maksimkin, A.V.; Kaloshkin, S.D.; Estrin, Y.Z. Mechanical properties and shape memory effect of 3D-printed PLA-based porous scaffolds. *J. Mech. Behav. Biomed. Mater.* **2016**, *57*, 139–148. [CrossRef]
55. Pavan, M.; Faes, M.; Strobbe, D.; van Hooreweder, B.; Craeghs, T.; Moens, D.; Dewulf, W. On the influence of inter-layer time and energy density on selected critical-to-quality properties of PA12 parts produced via laser sintering. *Polym. Test.* **2017**, *61*, 386–395. [CrossRef]
56. Nelson, J.A.; Rennie, A.E.W.; Abram, T.N.; Bennett, G.R.; Adiele, A.C.; Tripp, M.; Wood, M.; Galloway, G. Effect of Process Conditions on Temperature Distribution in the Powder Bed During Laser Sintering of Polyamide-12. *J. Therm. Eng.* **2015**, *1*, 159–165. [CrossRef]
57. Phillips, T.; Fish, S.; Beaman, J. Development of an automated laser control system for improving temperature uniformity and controlling component strength in selective laser sintering. *Addit. Manuf.* **2018**, *24*, 316–322. [CrossRef]
58. El Magri, A.; el Mabrouk, K.; Vaudreuil, S.; Chibane, H.; Touhami, M.E. Optimization of printing parameters for improvement of mechanical and thermal performances of 3D printed poly(ether ether ketone) parts. *J. Appl. Polym. Sci.* **2020**, *137*, 49087. [CrossRef]
59. Singh, S.; Sharma, V.S.; Sachdeva, A.; Sinha, S.K. Optimization and analysis of mechanical properties for selective laser sintered polyamide parts. *Mater. Manuf. Process.* **2013**, *28*, 163–172. [CrossRef]
60. Dadbakhsh, S.; Verbelen, L.; Verkinderen, O.; Strobbe, D.; van Puyvelde, P.; Kruth, J.P. Effect of PA12 powder reuse on coalescence behaviour and microstructure of SLS parts. *Eur. Polym. J.* **2017**, *92*, 250–262. [CrossRef]
61. Majewski, C.; Zarringhalam, H.; Hopkinson, N. Effect of the degree of particle melt on mechanical properties in selective laser-sintered Nylon-12 parts. *Proc. Inst. Mech. Eng. Part B J. Eng. Manuf.* **2008**, *222*, 1055–1064. [CrossRef]
62. Wu, J.; Xu, X.; Zhao, Z.; Wang, M.; Zhang, J. Study in performance and morphology of polyamide 12 produced by selective laser sintering technology. *Rapid Prototyp. J.* **2018**, *24*, 813–820. [CrossRef]
63. Kamil, A. Post Processing for Nylon 12 Laser Sintered Components. Ph.D. Thesis, School of Mechanical and Systems Engineering Newcastle University, Newcastle, UK, November 2016.
64. Zarringhalam, H.; Hopkinson, N. Post-Processing of Duraform$^{tm}$ Parts for Rapid Manufacture. In Proceedings of the International Solid Freeform Fabrication Symposium, Austin, TX, USA, 4–6 August 2003; H. Wolfson School of Mechanical and Manufacturing Engineering, Loughborough University: Loughborough, UK, 2003. [CrossRef]
65. Yang, C.; Tian, X.; Li, D.; Cao, Y.; Zhao, F.; Changquan, S. Influence of thermal processing conditions in 3D printing on the crystallinity and mechanical properties of PEEK material. *J. Mater. Process. Technol.* **2017**, *248*, 1–7. [CrossRef]
66. Rafie, M.; Marsilla, K.K.; Hamid, Z.; Rusli, A.; Abdullah, M. Enhanced mechanical properties of plasticized polylactic acid filament for fused deposition modelling: Effect of in situ heat treatment. *Prog. Rubber Plast. Recycl. Technol.* **2019**, *36*, 131–142. [CrossRef]
67. Liu, X.; Tey, W.S.; Choo, J.Y.C.; Chen, J.; Tan, P.; Cai, C.; Ong, A.; Zhao, L.; Zhou, K. Enhancing the mechanical strength of Multi Jet Fusion–printed polyamide 12 and its glass fiber-reinforced composite via high-temperature annealing. *Addit. Manuf.* **2021**, *46*, 102205. [CrossRef]

Article

# Strain Release Behaviour during Crack Growth of a Polymeric Beam under Elastic Loads for Self-Healing

Mohammed Dukhi Almutairi [1,2,*], Sultan Saleh Alnahdi [1,3] and Muhammad A. Khan [1,2,*]

1. School of Aerospace, Transport, and Manufacturing, Cranfield University, Cranfield MK43 0AL, UK; s.alnahdi@cranfield.ac.uk
2. Centre for Life-Cycle Engineering and Management, Cranfield University, College Road, Cranfield MK43 0AL, UK
3. Sustainable Manufacturing Systems Centre, Cranfield University, College Road, Cranfield MK43 0AL, UK
* Correspondence: m.almutairi@cranfield.ac.uk (M.D.A.); muhammad.a.khan@cranfield.ac.uk (M.A.K.)

**Abstract:** The response of polymeric beams made of Acrylonitrile butadiene styrene (ABS) and thermoplastic polyurethane (TPU) in the form of 3D printed beams is investigated to test their elastic and plastic responses under different bending loads. Two types of 3D printed beams were designed to test their elastic and plastic responses under different bending loads. These responses were used to develop an origami capsule-based novel self-healing mechanism that can be triggered by crack propagation due to strain release in a structure. Origami capsules of TPU in the form of a cross with four small beams, either folded or elastically deformed, were embedded in a simple ABS beam. Crack propagation in the ABS beam released the strain, and the TPU capsule unfolded with the arms of the cross in the direction of the crack path, and this increased the crack resistance of the ABS beam. This increase in the crack resistance was validated in a delamination test of a double cantilever specimen under quasi-static load conditions. Repeated test results demonstrated the effect of self-healing on structural crack growth. The results show the potential of the proposed self-healing mechanism as a novel contribution to existing practices which are primarily based on external healing agents.

**Keywords:** 3D printing; ABS simple beam; TPU origami capsule; embedded structure; self-healing mechanism; double cantilever beam test

**Citation:** Almutairi, M.D.; Alnahdi, S.S.; Khan, M.A. Strain Release Behaviour during Crack Growth of a Polymeric Beam under Elastic Loads for Self-Healing. *Polymers* **2022**, *14*, 3102. https://doi.org/10.3390/polym14153102

Academic Editors: Hamid Reza Vanaei, Sofiane Khelladi and Abbas Tcharkhtchi

Received: 19 June 2022
Accepted: 28 July 2022
Published: 30 July 2022

**Publisher's Note:** MDPI stays neutral with regard to jurisdictional claims in published maps and institutional affiliations.

**Copyright:** © 2022 by the authors. Licensee MDPI, Basel, Switzerland. This article is an open access article distributed under the terms and conditions of the Creative Commons Attribution (CC BY) license (https://creativecommons.org/licenses/by/4.0/).

## 1 Introduction

Three-dimensional (3D) printing or additive manufacturing (AM) of smart polymers is a rapidly expanding area of technology. The variety of AM techniques available suggests it may be possible to flexibly manufacture smart but costly materials with minimum waste. On-demand or autonomous repair of forms of damage, such as cracks or scratches, can increase the operational life of products and can be facilitated using man-made polymers which are autogenous or intrinsically self-healing. A balance between healing and strong mechanical properties can be achieved by designing the architecture of the polymer to incorporate dynamic or reversible bonds [1–6]. A great deal of work still needs to be performed to successfully implement self-healing mechanisms in real applications, with most previous studies of self-curing structural damage having taken place only at a laboratory scale. The majority of reported mechanisms have been based on external disturbances such as heat-generated cracks or a chemical reaction triggering the healing mechanism within the structure. Existing mechanisms tend to depend on some form of external interference and, most of the time, work only for more significant damage. Consequently, it is virtually impossible to implement current self-healing mechanisms such as those in 3D printed products whilst they are functioning; this is a particularly important consideration in some vital applications [7–9].

An alternative approach to creating smart 3D printed products is to embed novel origami-inspired capsules into the layers of a printed component. For essential applications, in particular, such capsules could create an artificial hormone network that would make 3D printed products safer and considerably more dependable [10–12]. Standard fused deposition modelling can be utilised to embed these capsules when printing the required component, which would be a cost-effective solution for large-scale production. This is somewhat similar to the manner in which the human hormone system actuates when a virus or bacteria enters the body. The use of a strain removal-based actuation via origami-inspired capsules could radically transform the self-healing capacity within components or structures. Strain removal from an entire component could thus be initiated by any form of surface or subsurface damage. For strain removal to take place at a sub-surface level, the capsules could unfold and expand [13,14].

However, the actuation or unfolding of such capsules under strain release due to crack initiation or growth within a structure requires an understanding of its mechanical behaviour, especially in embedded conditions. To introduce and control such a process requires a workable relationship between the initial stress on the embedded capsules, the displacement of the origami folded parts in a direction to release the strain, and the magnitude of the strain released. An overall understanding of the mechanical behaviour of any selected polymer under elastic and plastic loads is necessary to assess its usefulness in the form of an origami capsule to provide the necessary strain release control.

The mechanical behaviour of ABS polymer components has been investigated for many years, and the basic features, such as stress–strain curves, are adequately known [15,16]. Such behaviour is measured elastically for very small strains and slightly larger strains when overcoming the intermolecular barriers to segmental rearrangements [17]. However, the complex properties of ABS polymer materials are temperature dependent, which has driven further investigation to determine what relationships exist between strain, stress, and temperature [18,19].

Previous research into the mechanical properties of thermoplastic polyurethane (TPU) and thermoset acrylonitrile butadiene styrene (ABS) provides help in understanding the dynamics of such beams under load. Yuan et al. investigated the behaviour of a graded origami structure under quasi-static compression. A beam was fabricated using ABS material with flat brass sheets, 0.3 mm thick, implanted between the moulds, which were then compressed. Results indicated that the proposed origami structure showed plane stiffness and higher energy absorption to external loads [20]. However, the work lacked analysis in terms of geometric optimisation and behaviour under impact load. Hernandez et al. presented a kinematic study of origami structures for both elastic and plastic polymeric beams. After assessing various design structures, it was found that the kinematic variables of the structural model could fully explain the configuration of elastic origami structures within the beam [21,22]. However, the model developed by the researcher is far more complete and needs fewer variables for efficient FEA. Li and You researched open section origami beams to demonstrate energy absorption. Their research focused on designing a beam which included origami geometries and which retained its cross-sectional height better than conventional beams when subject to large externally imposed bending deformations. Despite numerical simulation, the model did not develop origami geometries able to cope with symmetrical vertical loads; also, the energy absorption model needed to be validated [17]. Nevertheless, origami-based encapsulation has shown promising results [23], but tests of mechanical strength and healing properties tend to have been carried out on soft and weak materials [24].

The encapsulation of folded material, such as TPU in rigid and static structures, can induce self-healing properties in a structure, assisting it in overcoming extreme fatigue conditions, material degradation, and failure due to micro-cracks [25,26]. Moreover, by activating the self-healing process, the material becomes safer and more durable, saves the time and cost of replacing particular items, and reduces inefficiencies incurred due to damage [27,28].

The four-point flexure response of the ABS beam has been researched by Dhaliwal and Dundar and showed high impact resistance and toughness. Their work examined the strain rate using the Generalised Incremental Stress-State Model. Though the compressive elastic modulus of ABS is found to be much higher than its tensile elastic modulus, the Von-Mises is yielded at a much lower force [29]. This means that at higher deformations, the ABS beam may not produce the predicted theoretical results. Therefore, it is necessary to continue to research self-healing techniques of polymers using origami structures.

Lee [30,31] conducted an experiment using a large elastic bending machine to investigate the elastic energy behaviour of curved–creased origami to assess material bending behaviour. As the first step, an origami design model was developed to use different folds to produce the patterns necessary to make the 3D form required to meet a prescribed buckling criterion. The model was then used to simulate the shape of the origami capsule required, after which the results could be experimentally validated. The study by Lee [30] showed that skewed curved–creased laminated surfaces could help in assembling compliant and energy-absorbing structures, but the study itself did not provide any direct evidence for using this mechanism for self-healing. However, once the results were validated, it allowed a healing process using origami capsules to be simulated.

In this work, the response of polymeric beams of ABS and TPU materials under elastic and plastic loads is investigated. The experimentation process included the use of strain gauges of different thicknesses to determine the deflection of the cantilever beam under test [32,33]. The tests included observation of the effects of the material and binder on two types of 3D-printed beams and were designed to test their elastic and plastic responses under different bending loads. These responses were used to develop an origami capsule-based novel self-healing mechanism triggered by crack propagation due to strain release in a structure.

The origami capsules were cross-shaped and made of four small beams that could be folded or elastic deformed and embedded in the main beam structure. Under the strain released due to crack propagation in the main beam, the small beams of the origami capsule unfolded in the direction of the path of the crack and hence increased the structure's resistance to crack propagation. This increase in the crack resistance was validated in a delamination test of a double cantilever specimen under quasi-static load conditions. Repeated results demonstrated the effect of self-healing on structural strength against crack growth. The results show the potential of a proposed self-healing mechanism as a novel contribution to existing practices, which are primarily based on external healing agents

The paper is structured as follows: Section 2 describes the methodology, including the selection of materials, experimental setup and procedure of the simple beam, origami beam and origami embedded structure. Section 3 provides the results and discussion. The conclusions are presented in Section 4.

## 2. Methodology

In this section, various techniques used to prepare and characterise the samples are described. Specifically, bending loads were placed on the end of rectangular beams of the polymeric materials to gain a better understanding of their elastic and plastic behaviour. The four steps in this research are shown in Figure 1. The first step was selection of the polymeric material and included the preparation of the specimen and experimental methods. The second step was design of the polymeric structure, including the origami capsule. The third step was the design of the experiment to investigate the properties of the samples, including bending moment and delamination tests. In the final step, the tensile test machine was used to obtain strain–stress curves, bending points and delamination effects using a single bending moment, see Figure 1.

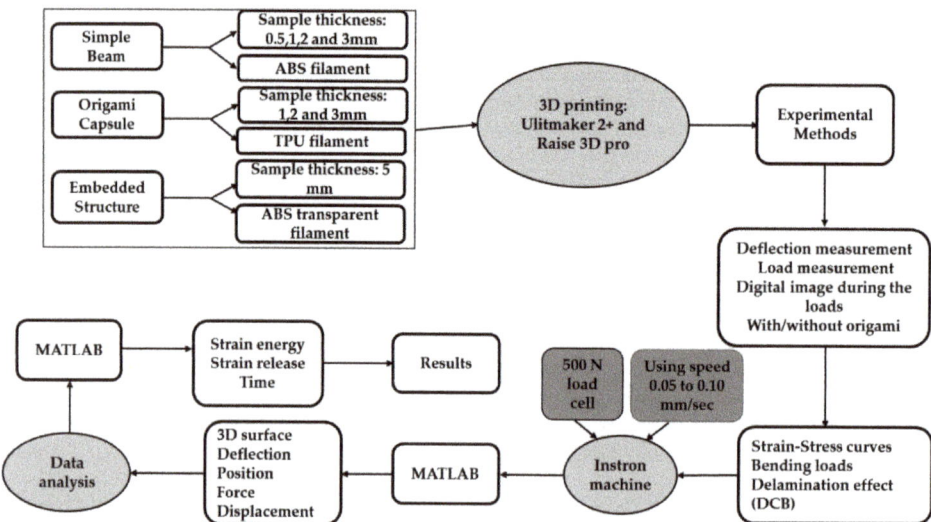

**Figure 1.** Methodology diagram.

## 2.1. Material Selection

The first polymeric material selected was ABS, one of the most common raw materials used for printing beams via fused deposition modelling. ABS has good impact resistance, high rigidity, strain resistance, etc., even at low temperatures [34], properties that make it a suitable material for the intended application. Sample parts were fabricated at variable parameters and tested for bending strength. The second material chosen was TPU. This is of interest because of its versatility in terms of a wide range of mechanical properties, good abrasion resistance and low density. TPU is more elastic than ABS and very suitable to be folded as capsules. TPU has additional benefits compared to other polymers, such as being extremely flexible, durable and smooth to the touch.

A Raise3D Pro printer was used to print the beam-based origami capsule and embedded structure beam. The 3D printed samples and capsule were produced with two printing parameters: orientation and layer thickness. The platform was heated to 80 °C with a screw speed of 50 mm/s. At least 1 kg of filaments with a diameter of 1.75 ± 0.05 mm served as the extender. During the printing process, the slicer programme used this diameter to calculate the required feed rate [35–37]. The mechanical printer parameters are presented in Table 1 and depicted in Figures 2 and 3.

**Table 1.** Printing parameters.

| Parameters | Value |
|---|---|
| Nozzle size (mm) | 0.4 |
| Layer thickness (mm) | 0.1, 0.2 |
| Build orientation | 0°, ±45°, 90° |
| Infill density (%) | 100 |

## 2.2. Specimen Preparation

The specimen was designed as a simple beam with embedded structure. The design of the embedded capsule is shown in Figure 4, which also shows its dimensions. The cantilever beam was designed using the inventor software, as shown in Figure 4a, sample thickness is 3.0 mm, length 145 mm, and width is 10 mm. Figure 4b the origami capsule thickness 3.0 mm, length 19 mm, and width 5 mm. Figure 4c shows the length of specimen,

193.0 mm, width 30 mm, and thickness 5 mm. These dimensions were maintained in all tests.

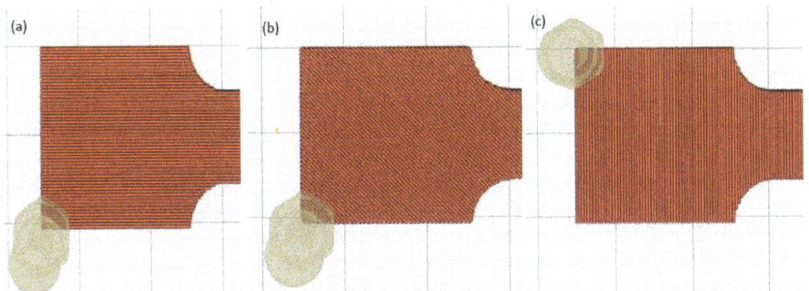

**Figure 2.** Printing Directions of ABS: (**a**) 0 orientation; (**b**) ±45 orientation; (**c**) and 90 orientation.

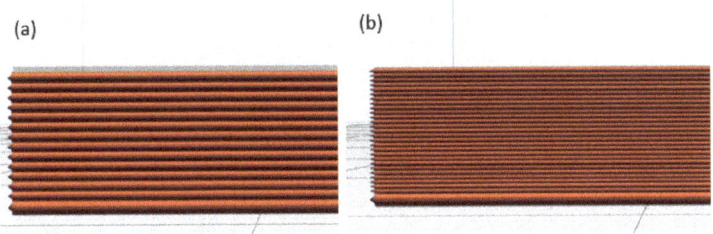

**Figure 3.** Layer thickness: (**a**) 0.10 mm (**b**) 0.20 mm.

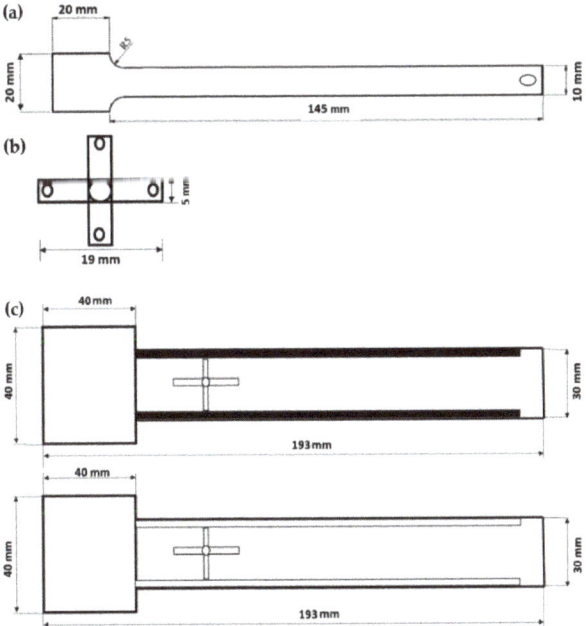

**Figure 4.** Geometry of specimens: (**a**) ABS simple beam; (**b**) origami capsule TPU; (**c**) Double cantilever beam (DCB (with hole and pillars).

G-code files for printing the above specimens on a 3D printer were created using Idea Maker software (Raise3D pro2). The 3D printer process from drawing to fused deposition is shown in Figure 5.

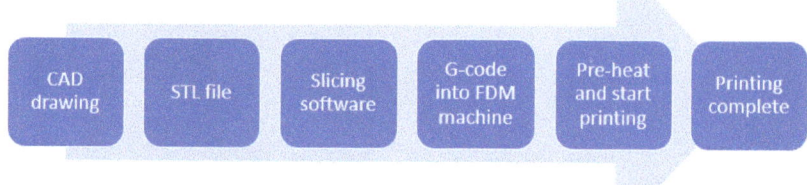

**Figure 5.** The 3D printer process from drawing to fused deposition.

At least three samples were printed of each simple beam, capsule, and embedded structure. The infill density was 100% in all cases. First, TPU and then blends containing 5, 10, and 20 wt% TPU were printed. Nozzle temperature was set to 60 °C for all capsules. Printing speed was 40 mm/s, and print bed temperature was 60oC. Similarly, ABS simple beams were also printed with infill densities of 40, 60, and 80 wt%, respectively. Here the printing speed was constant 60 mm/s with nozzle diameter 0.4 mm. The print bed temperature was 80 °C and 100 °C. For each configuration, two samples were printed [8].

*2.3. Design of Experiment*

In this experimental study bending load and delamination tests were carried out. Both sets of experiments began with the printing of samples, the simple beams, the origami capsules and the origami capsules embedded in the beams. The samples were then subjected to bending load and delamination tests using an Instron 5944 Universal Testing Machine (UTM). Specifically, the bending load was applied to better understand the elastic and plastic behaviour of ABS and TPU. The stresses were calculated according to the force provided by the UTM. In addition, video images taken during the loading determined the deflection of the beam at 15 different points along its length. This provided the overall response of the beam under bending load. Furthermore, each quasi-static double cantilever beam (DCB) test was conducted three times using the UTM. The loading value was measured with a load cell attached to the tensile test machine. The opening displacement and crack length were measured with a camera.

2.3.1. Simple Beam and Origami Beam

In this work, the parameters were set as shown in Tables 2 and 3 using design of experiment methodology. The ABS simple beam and TPU origami capsule were manufactured. Sensor calibration was performed (3 times for each beam thickness: 0.5 mm, 1.0 mm, 2.0 mm, and 3.0 mm. Deformation load and deflection data were recorded, and the data (strain, applied load, and deflection) plotted using Excel.

**Table 2.** Origami capsule designs.

| Origami Capsule Shape | Thickness (mm) | Dimensions (mm) | Loads (g) |
|---|---|---|---|
| Cross | 1.0, 2.0, 3.0 | 19L/5W | 1, 2, 4, 6, 11, 16, 26, 36, 56, 86, 106 |

**Table 3.** Experimental setup for simple beam.

| Sample Number | Beam Thickness mm | Dimensions, mm (Length/Width) | Loads (g) Attached to the Beam, See Figure 6 |
|---|---|---|---|
| 1 | 0.5 | 145/10 | 1, 2, 3, 4, 5, 6, 7, 8, 9, 10 |
| 2 | 1.0 | 145/10 | 1, 3, 5, 10, 15, 25, 35, 45, 55, 75, 100 |
| 3 | 2.0 | 145/10 | 5, 10, 20, 30, 50, 70, 90, 110, 160, 210, 310, 410 |
| 4 | 3.0 | 145/10 | 5, 10, 20, 30, 50, 70, 90, 110, 160, 210, 310, 410 |

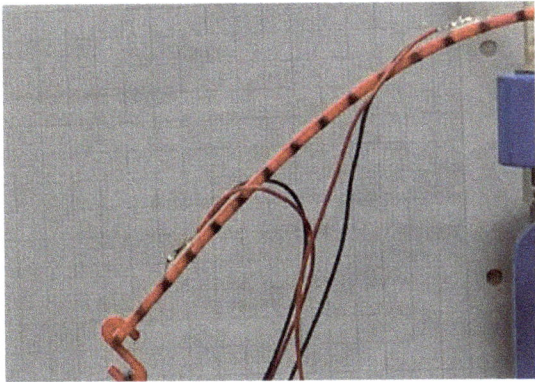

**Figure 6.** Deflection points.

### 2.3.2. Origami Beam Embedded Inside Structure of the Beam

A simple experiment was conducted to record load and displacement using a delamination test on the specimen beams printed using ABS polymer, as shown in Table 4. The stress/strain relationship with and without the origami capsule was then evaluated.

**Table 4.** Experiment Scheme in embedded structure.

| Specimen Type | Crack Length | Thickness of Capsule | Mechanical Testing |
|---|---|---|---|
| DCB origami | 40 mm | 1 mm, 2 mm, 3 mm | Delamination test |
| DCB without origami | - | - | Delamination test |

The initial hypothesis was that specimens containing the origami structures would be more resilient and exhibit higher tensile strength when loaded axially. Conversely, specimens that did not have origami structures embedded within them should exhibit lower resilience or lower tensile strength. In order to test this hypothesis, two hollow 3D beam samples were printed that could be joined later by mechanical means. One of the samples contained 3D printed origami structures embedded inside using an adhesive. The other, the control, was the same 3D printed beam but without the origami structure embedded within it. The specimens were loaded and pulled (tensile loading) axially. Force vs. displacement (F/D) curves were obtained, which corresponded to the stress/strain relationship. In order to convert F/D curves into a stress/strain relationship, force values were divided by the cross-sectional area of the beam, while D values were divided by the initial gauge length. A schematic of the experimental procedure is presented in Table 5.

**Table 5.** Specimen setup.

| Specimen | Beam Thickness mm | Dimension, mm (Length/Width) | Type of Structure |
|---|---|---|---|
| with origami capsule | 5 mm | 193L/30W | With holes and pillars |
| without origami Capsule | 5 mm | 193L/30W | With holes and pillars |

## 2.4. Experiment Setup and Procedure

In this experiment, a micrometre was used to apply a deflection to the end of a beam. Before starting the experiment, the dimensions of our simple beam and origami beam were measured using inventor software. The dimensions of the beams are given in Tables 2 and 3 above.

For each specimen, the following set of procedures was carried out.

1. First, prepare the surface of the test piece by applying conditioner and neutralisers. A step-by-step procedure was developed. The process of bonding the strain gauge should be carried out precisely without errors. Notably, the surface area of the strain gauge should be stuck together by first cleaning the surface with sandpaper and then using conditioners to neutralise the free-end and the fixed support. Finally, to complete the surface preparation of the beam, a generous volume of the neutraliser is applied and wiped out with the cotton ball.
2. To further explain the process for educational purposes, the bonding area must be cleared with alcohol/acetone. After clearing the surface, the necessary marks are placed on the bonding site, preferably with a fine graphite pencil, such that no residual deposition affect the measurement.
3. Clamping of the beam: the flat portion of the ABS beam was clamped in the test machine.
4. Place the strain gauges on the sample. One in free end and other one close to the fixed support, as shown in Figure 7.

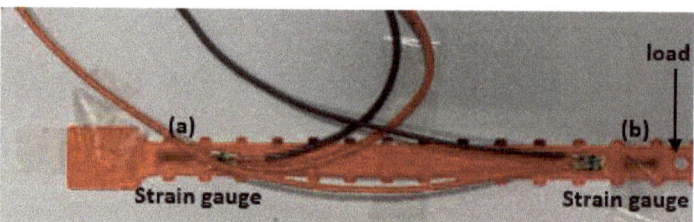

**Figure 7.** Plane view of strain gauge mounting points on test beam (**a**) fixed support and (**b**) free end.

1. Connect the strain gauges to the DAQ-card and the Signal-Express software
2. Calibrate the strain gauges with no load on the sample and set readings to zero.
3. Measure the distance between free end of each sample and the nearest strain gauge.
4. Apply loads progressively from 0 N to 4.02 N and measure the corresponding strain on each of the two strain gauges. Remove the masses in the reverse order in which they were added to produce a hysteresis plot.
5. Placing the protractor parallel to the edge of the clamping machine (i.e., starting point of the beam). The fixing should be firm, so there is no unwanted movement of the protractor.
6. Camera orientation: The camera was placed parallel to the longitudinal side of the beam such that the protractor could be easily seen. The distance between the camera and the beam was 30 cm.

### 2.4.1. ABS Simple Beam Behaviour in Normal Condition

A simple ABS beam of 145 mm length, see Figure 4, was fixed at one end as cantilever beam. HD camera was chosen for strain measurement rather than crosshead displacement because of the compliance of the loading mechanism and load cells, which is typical in such tests. HD Camera deflection measurement on both sides of the sample compensates for any lateral bending during loading. The procedure for testing the sample begins by setting the selected beam thickness. Next, the loads were applied at the free end and readings of the deflections taken. For every deformation, a picture of each point was taken, as shown in Figure 6. Specimen preparation only required a light-ordered pattern of black paint on beam, see Figure 6 on the white plastic background. Images of the samples were captured via camera, and deflection data obtained. After a sample was placed in the testing machine and a preload applied, a pair of reference images (one image per camera) were taken of each side of the sample. The applied loads ranged from 1 g to 410 g, depending on beam thickness, see Table 3. These were placed at the end of the beam. The wires used to connect the strain gauge to the DAQ (NI 9235) [38]. The D-card meter was connected to the computer via chassis (NI cDAQ-9174) for post-processing and data analysis.

This methodology proved to be efficient, and testing of a single specimen could be performed in matter of minutes, including mounting the specimen, taking initial undeflected images, and loading the specimen through to failure. The complete setup, including the camera and universal testing machine used for bending load.

### 2.4.2. Polymeric Origami Beam Behaviour

The simple ABS beam was replaced by one with an origami insert; first, a "cross", see Table 2. The loads were applied to the origami capsule, and measurements taken via the computer using the signal conditioning unit and data logger. The experiment was carried out with three tubes of thicknesses of 1.0, 2.0, and 3.0 mm. Each time the camera was set to a required value, and the corresponding strain values were recorded. After repeating the experiment three times, the average value of the results was obtained and noted. The origami capsules were designed using inventor software with different shapes to test the workability of different capsules, Figure 8. The designs of the capsules were such that their geometrical features were confined under the initial pre-stressed conditions.

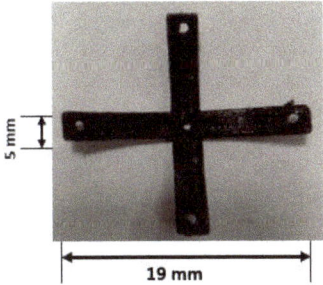

**Figure 8.** Origami capsule "cross".

### 2.4.3. Beam Behaviour with the Origami Capsule Contained within It

The setup and experimental design for the DCB test are shown in Figure 9. The beams were, as shown in Figure 4c, 30 mm wide and 193.0 mm long, with a 40 mm longitudinal pre-crack extending from the front of the specimen, see Figure 9. End tabs of 30 mm width were glued on the external faces of the specimens on either side of the pre-crack and pinned to an electromechanical uniaxial testing machine with a 500 N load cell. The DCB tests were performed on an Instron testing machine with displacement rates that could be varied between 0.05 to 0.10 mm/s

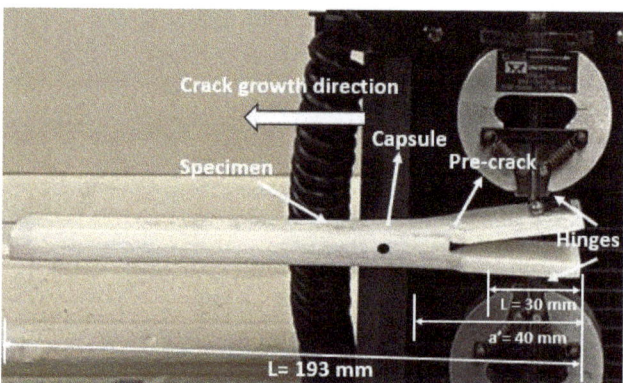

**Figure 9.** Double cantilever beam (DCB) test setup and showing pre-crack.

Because the ABS samples were transparent, crack length was directly recorded from the top using a camera. Each sample was tested 3 to 5 times. The corresponding energy release rates and critical energy release rates were calculated using a simple beam. The programmed loading history was not monotonic: indeed, eight loading and unloading cycles at the same displacement rate were programmed into the machine to verify the absence of permanent deformations, which would indicate parasite sources of energy dissipation. For each cycle, the maximum displacement at loading was defined, as well as a minimum force at unloading, set at 5 N to avoid compression of the test specimen.

Three different types of specimens were printed, one for each set of parameters, 1 mm, 2 mm, and 3 mm thickness. Three specimens were manufactured and tested under the same conditions for each thickness to confirm the experimental repeatability of the results obtained.

Mode I interlaminar toughness tests were performed on DCB beams, see Figure 4c containing an origami capsule orientated normal to the direction of crack growth [39–43]. The DCB specimens had a 40 mm long pre-crack at the front of the specimen, as shown in Figure 9. Two hinges were glued onto the top and bottom surfaces of the sample so they could hold the ends of the arms of the DCB specimen. The delamination crack growth in the direction of the origami capsule was as shown in Figure 9.

## 3. Results and Discussion

### 3.1. Results for ABS Simple Beam Behaviour

The displacement responses of the ABS materials with bending loads applied at the tip are shown in Figure 10. It is evident that for the 3.0 mm thick beam within the elastic limit, the maximum stress yielded a deflection of 56 mm. However, for the 0.5 mm thick beam, observed a deflection of 79 mm. Within the elastic limit, for small deflections, the value of the stress is directly proportional to the force and inversely proportional to the thickness:

$$Deflection \propto \frac{f(force)}{f(thickness)} \quad (1)$$

The values of force and maximum deflection were used to calculate the strain energy of the beam. It was assumed that the amount of stress applied is wholly converted into strain energy, which is represented as:

$$Maximum\ Strain\ energy\ of\ beam\ (U) = \frac{\sigma^2}{2E} Ba \quad (2)$$

where $\sigma$ represents the stress applied, $E$ is the elastic modulus of the material, $B$ is the beam's thickness, and $a$ indicates the length of the beam.

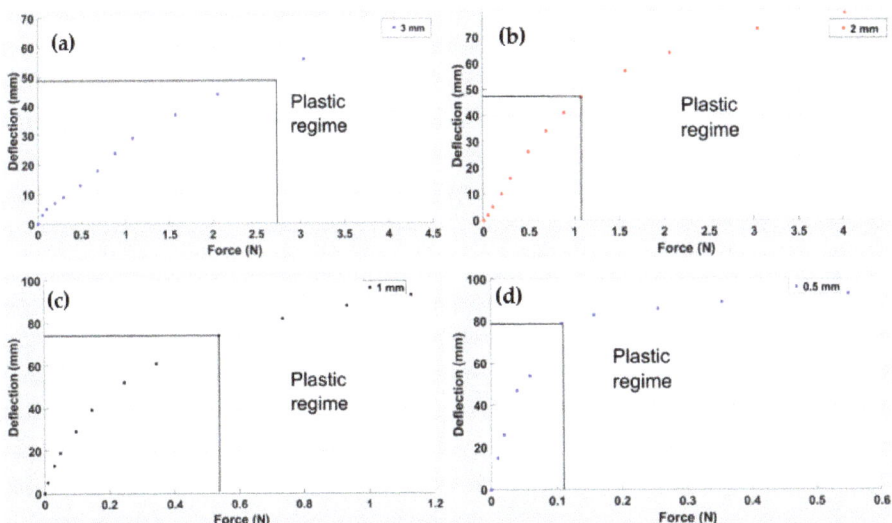

**Figure 10.** Deflection of end loaded simple ABS beam 145 mm long, 10 mm wide and thicknesses: (a) 3 mm, (b) 2 mm, (c) 1 mm and (d) 0.5 mm.

From the force–deflection curves shown in Figure 10, it is seen that the maximum deflection for the beam of 3.0 mm thickness was 64 mm at the maximum applied force of 4.022 N. The maximum deflection for the beam of 2.0 mm thickness was 78 mm, and the maximum applied force was again 4.022 N. The maximum deflection for the beams of 1.0 mm and 0.5 mm thicknesses was 93 mm at the maximum applied forces of 1.73 N and 0.55 N, respectively. Hence it follows that:

Maximum strain energy beam at 0.5 mm beam thickness

$$= \frac{191.48^2}{2 \times (1681)} \times 5 \times 145 = 7907 \text{ N.mm}$$

Maximum strain energy at 1 mm beam thickness

$$= \frac{98.19^2}{2 \times (1681)} \times 10 \times 145 = 4158 \text{ N.mm}$$

Maximum strain energy at 2 mm beam thickness

$$= \frac{87.48^2}{2 \times (1681)} \times 20 \times 145 = 6601 \text{ N.mm}$$

Maximum strain energy at 3 mm beam thickness

$$= \frac{38.88^2}{2 \times (1681)} \times 30 \times 145 = 1956 \text{ N.mm}$$

From the calculated values, it is noted that maximum strain energy is observed for the 0.5 mm thick beam, 7907 N.mm and the least value of strain energy is observed for the thickest beam, 3.0 mm, which is 1956 N.mm. This validates the findings that maximum force and beam thickness yield minimum strain energy, and the lower the magnitude of force and beam thickness, the higher the strain energy.

In Figure 10, the vertical black lines indicate the initiation of the plastic regime of each beam; it is clear that the greater the thickness of the beam, the greater the force required.

This is the reason why the plastic region for the 0.5 mm beam commenced at 0.15 N, whereas for the 3.0 mm beam, the force required was 3.04 N.

With the observed changes for the maximum deflection with respect to different loads, there is a need to analyse whether or not the values change with position. This was performed by plotting a 3D surface graph in the next section (see Figure 11) to show the parametric relationship between force, deflection, and position of the beam.

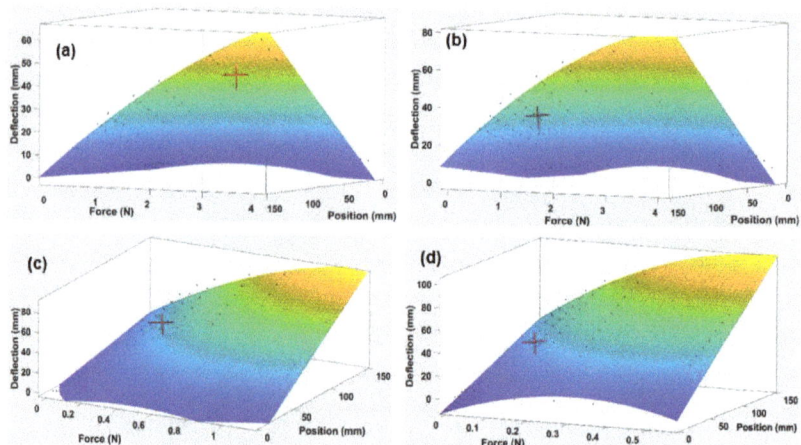

**Figure 11.** Three-dimensional gradient graph of ABS beam the showing relationship between applied Force (N), Position (mm), and Deflection (mm) for four beam thicknesses: (**a**) 3 mm, (**b**) 2 mm, (**c**) 1 mm, (**d**) 0.5 mm.

It is evident that the gradient increases in value as the thickness of the beam increases in four stages, from 0.5 mm to 2.0 mm.

With respect to different positions of the beam, we have assumed that deflection at different forces varies accordingly.

The curves shown in Figure 12 demonstrate strain energy as a function of applied force for simple ABS beams of (a) 0.5 mm, (b) 1.0 mm, (c) 2.0 mm, and (d) 3.0 mm thickness. The trend indicates that the strain energy attained its maximum value for 0.5 mm thickness (7907 N.mm), followed by 6600 N.mm for 2.0 mm, 4158 N.mm for 1 mm, and 1955 N.mm for 3 mm. It was also noted that the strain energy for all thicknesses except 2.0 mm had reached zero before 0.5 N, while the strain energy for the 2.0 mm thick beam reached zero value only as the force approached 1.0 N. This shows that the greater the value of the beam thickness, the greater the strain energy, and the more gradual will be the process of strain energy decay over time.

Mathematically,

$$deflection = f(force, position)$$

Each of the 3D surfaces shown in Figure 11 is approximated using a polynomial equation as given in Equation (3):

$$Deflection\ of\ the\ beam\ (x = Force, y = position\ at\ any\ point) = p00 + p10x + p01y + p20x^2 + p11xy \quad (3)$$

where $p00, p10, p01, p20,$ and $p11$ are the coefficients of the polynomial.

The results of the coefficients at various beam thicknesses are indicated in Table 6.

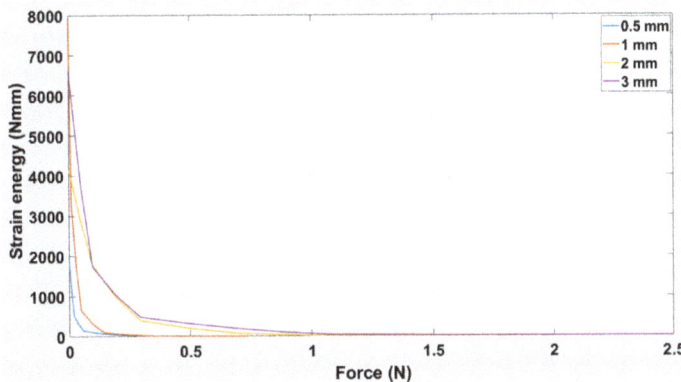

Figure 12. Strain energy vs. applied force for simple ABS beam of thickness: 0.5 mm, 1 mm, 2 mm, and 3 mm.

Table 6. Different values of coefficients at simple ABS beams and TPU Origami capsule of different thicknesses.

| Coefficients | Simple ABS Beam | | | | TPU Origami Capsule | | |
|---|---|---|---|---|---|---|---|
| | 0.5 mm | 1 mm | 2 mm | 3 mm | 1 mm | 2 mm | 3 mm |
| P00 | −10.25 | −7.013 | −6.002 | −2.196 | 0.5864 | 1.176 | −0.05332 |
| P10 | 134.8 | 37.79 | 10.3 | 4.519 | 49.29 | −18.06 | −24.81 |
| P01 | 0.3297 | 0.1968 | 0.1247 | 0.0363 | 0.5673 | 2.483 | 3.927 |
| P20 | −209.2 | −38.41 | −2.804 | −1.402 | 14.57 | 37.92 | 21.46 |
| P11 | 0.6598 | 0.4525 | 0.1259 | 0.1112 | −3.208 | −2.87 | −2.373 |
| w00 | 2.02 | −10.70 | 18.99 | −17.32 | −0.9095 | 3.318 | −1.822 |
| w01 | −0.05 | 0.32 | −0.65 | 0.58 | −0.2359 | 2.623 | −1.82 |
| w10 | −40.07 | 251.30 | 500.80 | 327.40 | −30.3 | −158.3 | 177.2 |
| w11 | 0.04 | −0.08 | −0.37 | 0.86 | 0.0795 | 0.0995 | −3.387 |
| w20 | 74.75 | −465.60 | 909.20 | −556.70 | −19.91 | 83.07 | −48.59 |

By substituting the coefficients in Equation (3), the polynomial equation for each beam thickness can easily be found.

From the above Table 6, it is clear that as the thickness increases, the absolute value of the coefficients decreases, which ultimately reduces the R-squared values. This is the reason why the 3.0 mm thick beam has the lowest values of coefficients and the highest R-squared value. The results can be further simplified in terms of reducing the variables and coefficients. This is performed by plotting the curves of coefficients against beam thickness, which allows the corresponding slopes of the curves (the coefficients w1, w2, and w3) to be determined (see Figure 13).

Each plot in Figure 13 is fitted for a third-degree polynomial, so the generalised equation becomes:

$$f(x\ thickness) = w1x^3 + w2x^2 + w3x + w4 \qquad (4)$$

where $w1$, $w2$, $w3$, and $w4$ indicate the coefficients of the polynomial equation, and $x$ indicates the thickness of the beam.

The generalised equation, Equation (4), is simpler to analyse than Equation (3), and is effective in determining the coefficient whatever the thickness of the beam.

The above equations are formulated by substituting the values in the generalised equation. The R-squared value for each coefficient is 1.00, which shows a perfect fit of the curve, as indicated in Table 6. The equations are set for a third-degree polynomial in each case, so there is virtually no discrepancy in the value of any coefficient. This equation can be used to analyse the response of a simple ABS beam of any thickness in the range of 0.5 to 3.0 mm under different loads up to the elastic limit.

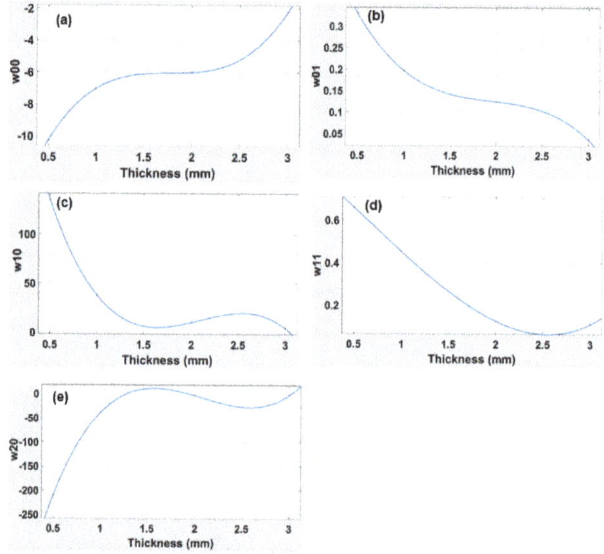

**Figure 13.** Graphs indicating the relationship of simple ABS beam thickness vs. correlation coefficient for (**a**) w00 (**b**) w01 (**c**) w10 (**d**) w11 and (**e**) w20.

## 3.2. Results with the TPU Origami Capsule

In the laboratory, it is possible to design simple experiments in order to examine the deflection of a "cross" TPU capsule held at one tip and with a load applied at the free end; this is effectively a cantilever beam of length 19 mm, width 5 mm and a thickness of 1.0 mm, 2 mm and 3.0 mm. The deflection vs. force curves obtained with the origami capsule for these three thicknesses are shown in Figure 14. The maximum deflection observed for the 3.0 mm thick capsule was 19 mm with an applied force of 1.5 N; the maximum deflection for the 2.0 mm thick capsule was 17 mm for an applied force of 1.3, and the maximum deflection for the 1.0 mm thick capsule was 15 mm for an applied force of 1.0 N.

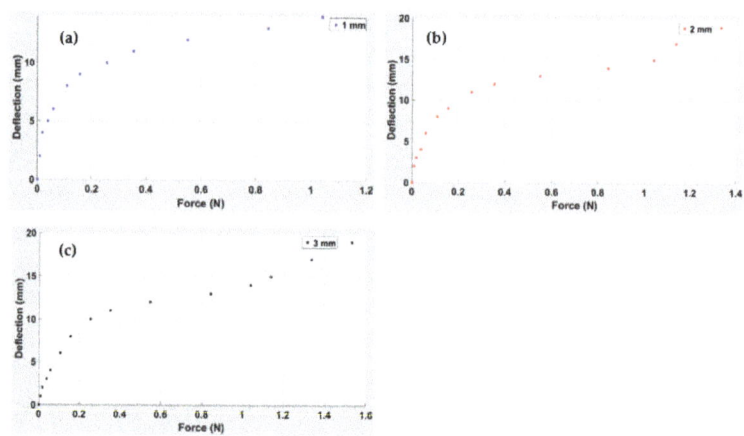

**Figure 14.** Deflection of an origami TPU "cross" capsule acting as a cantilever beam as a function of force for "capsule" thicknesses: (**a**) 3 mm, (**b**) 2 mm and (**c**) 1 mm.

Using a similar technique to that used with the simple ABS beam, 3D surface graphs for the TPU "cross" were plotted using MATLAB (see Figure 15).

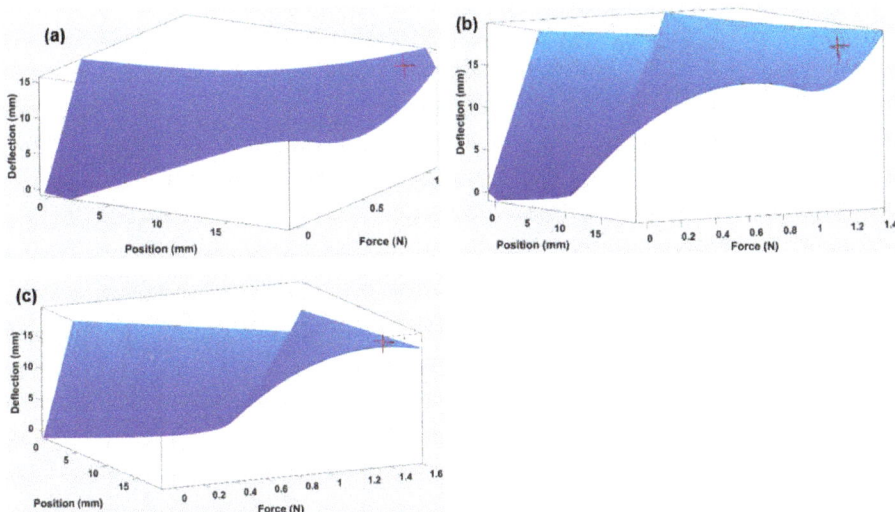

**Figure 15.** Three-dimensional gradient graph of TPU capsule "cross" showing relationship between Force (N), Position (mm), and Deflection (mm) for three beam thicknesses (**a**) 1.0 mm, (**b**) 2.0 mm, (**c**) 3.0 mm.

Figure 16 illustrates the response of strain energy with force applied to the TPU "cross" beam. It is clear from the trends that, for all beam thicknesses, the beam's responsiveness to strain energy is exponential and decreases with the increase in applied force. It is evident that a force of 0.25 N is the maximum force at which all three beam thicknesses showed the response of strain energy.

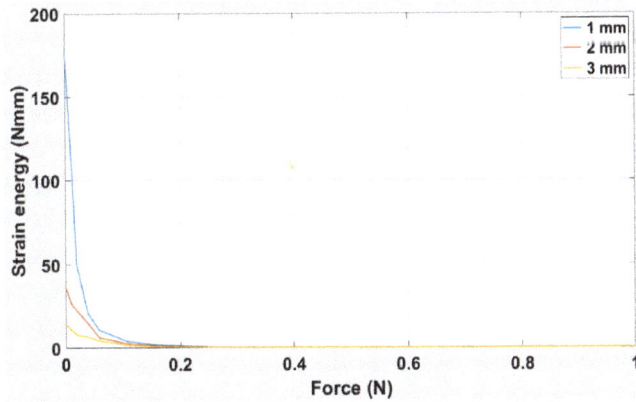

**Figure 16.** Strain energy versus applied force for TPU "cross" capsule for thicknesses: 1.0 mm, 2.0 mm, and 3.0 mm.

Deflection as a function of force and position is observed for TPU. From the 3D surface graphs, it is evident that a plastic region was achieved at the maximum values of applied load for each thickness of the TPU "cross" beam.

For the TPU "cross", the generalised equation for any arbitrary point can be presented by Equation (3).

By substituting the coefficient values in Equation (3), the polynomial equation for each thickness can easily be found.

The R-squared value for each thickness (see Table 6 above) was found to be 0.9966, 0.9818, and 0.9647, respectively, for the 1, 2, and 3.0 mm thicknesses of the "cross" capsule. This indicates that as the thickness of the TPU beam increases, the accuracy of the model equation declines. As with the ABS, the polynomial equation for the TPU "cross" can be simplified using only its thickness. By plotting the corresponding slopes on the curves, the coefficients (w1, w2, and w3) are determined as indicated in Figure 17.

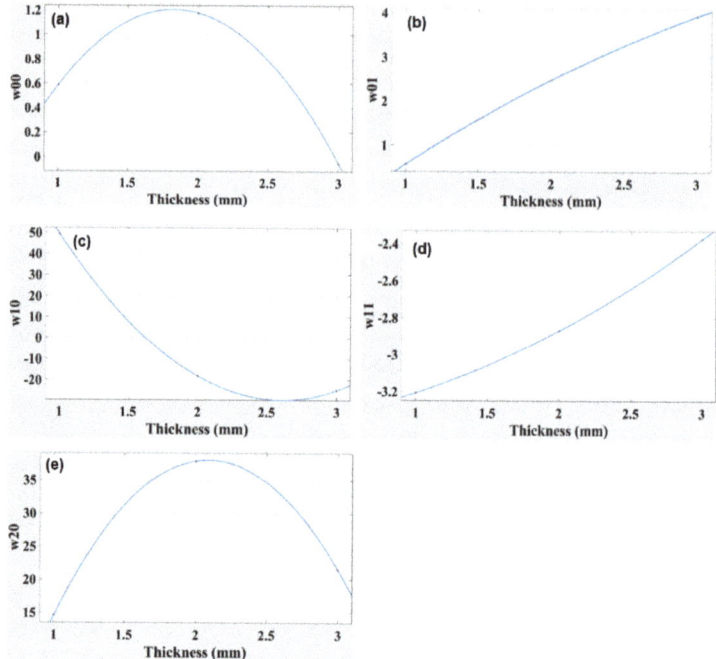

**Figure 17.** Graphs indicating the relationship of TPU "cross" beam thickness versus the coefficient for (**a**) w00 (**b**) w01 (**c**) w10 (**d**) w11 (**e**) w20.

Each graph in Figure 17 was fitted with a second-degree polynomial, so the generalised equation can be presented by Equation (4).

By substituting the values of the coefficient in Equation (4), the generalised equations can be found.

The R-squared values for all four coefficients, as indicated in Table 6 above, were 1.0 for a second-degree polynomial, suggesting a perfect fit for TPU "cross" origami capsule, whereas a perfect fit for the ABS required a third-degree polynomial.

3.2.1. Discussions of Simple ABS Beam and TPU Origami Capsule

The experimental results of the beam were collected and analysed using graphical, simulation, and statistical techniques. The deflection, position, and force vary with thickness, and a 3D gradient graph was plotted in MATLAB to show this (see Figure 11). It is evident from this figure that because the thickness of the beam increased from 0.5 mm to 2.0 mm, the gradient shifted towards a smaller deflection. The 0.5 mm thick beam reached a maximum deflection of 93 mm (see Figure 11a). The 1.0 mm beam achieved a maximum deflection of 90 mm (Figure 11b); the 2.0 mm beam reached a maximum deflection of

78 mm (Figure 11c), and the 3.0 mm thick beam showed a deflection of 64 mm (Figure 11d), were 0.9681, 0.9632, 0.9662, and 0.9819, respectively. This confirms that the thickness of the beam is a critical parameter that moderates the deflection at different values of position and applied force. These values are also significant because they demonstrate that the values of the deflection of the beam can be correlated through a regression model (Table 6).

Of all the results, the beam with the 0.5 mm thickness reported the maximum deflection value, and 3 mm reported the minimum value. This indicates that the thickness of the beam is a critical parameter that modulates the deflection at different values of position and force.

The analysis of the beam's elastic modulus helped compute the ABS' resistance to elastic deformation. From the results shown in Figure 10, it can be seen that as the thickness of the beam increased from 0.5 mm to 3.0 mm, the elastic modulus decreased from $3.6 \times 10^9$ Pa to $1.8 \times 10^8$ Pa. This shows that increasing the thickness of the beam reduces the elastic modulus of the beam. The initiation of the plastic region is indicated by red crosses, as shown in Figure 11. For a 0.5 mm thick beam, the elastic region lasted until the load was 0.108 N, and the plastic region was maintained until 0.549 N. For a beam of 1.0 mm thickness, the elastic region was maintained till 0.343 N, and the plastic region existed up to the maximum load of 1.128 N. For 2.0 mm thickness, the elastic region extended to a load of 1.08 N, significantly more than for the 0.5 mm and 1.0 mm beam thicknesses. Lastly, for a 3.0 mm thick beam, the plastic region was sustained till 4.02 N, showing that for the 3.0 mm thickness, the plastic region commenced at an end load of mass of 310 gm.

The minimum strain calculated to activate the plastic region for the four beams was 0.5 mm—$1.79 \times 10^{-6}$; for 1.0 mm—$5.70 \times 10^{-6}$; for 2.0 mm—$1.74 \times 10^{-5}$; and for 3.0 mm—$4.56 \times 10^{-5}$. This shows that as the thickness increased, the plastic region was activated at a greater magnitude of force and a greater overall strain rate. In Figure 12, it is clear that the strain energy is obtained at a maximum of 1 N for all four beams. However, the beam thickness significantly affected the decay of the overall strain energy. This shows that ABS beam may be ideal for low-stress release applications, but for higher stress, the material may not be sufficiently resilient.

In order to simplify the calculation and apply variable thicknesses, the values of all five coefficients from the model equation $w_{00}$, $w_{10}$, $w_{01}$, $w_{20}$, and $w_{11}$ were plotted against thickness in Figure 13, and the R-squared values for all four coefficients were 1.0. The R-squared was evaluated for a third-degree polynomial equation but increasing the degree placed the value out of range.

From the experimental results, it is evident that the modulus of ABS increases with strain rate. From the material point of view, the ABS beam depends on both compression and shear rates, which are different for different thicknesses. The elastic limit is reached more rapidly for thin beams and gradually increases as the thickness increases. The reasons for this are the moment of inertia and elastic modulus of the beam, which depend on the properties and cross-sectional dimensions of the material. ABS as a polymer sustained the load to 4.02 N for a 3.0 mm thick beam, indicating that the load sustainability of the designed polymeric beam is suitable for further research work.

The TPU "cross" beam was characterised in order to extend the research work and scope of the study. The analytical process was similar to that for the simple ABS beam, involving beam deflection for different beam thicknesses, in this case from 1.0 mm to 3.0 mm. From an analysis of the 3D curves, as shown in Figure 15, it is evident that the plastic regime had been achieved at the maximum values of applied load for each thickness of the TPU "cross" beam. The R-squared value was 0.9966, 0.9818, and 0.9674, respectively, for the 1.0, 2.0, and 3.0 mm capsule thicknesses. This indicates that as the TPU "cross" origami capsule thickness increases, the accuracy of the model equation decreases, as indicated in Table 6. It can be inferred that the equation is less well-adapted to thicker beams. The maximum stress values recorded for the 1.0, 2.0, and 3.0 mm capsules, respectively, were $6.24 \times 10$ Pa, $7.61 \times 10$ Pa, and $3.88 \times 10$ Pa.

Upon examination of Figure 14, it is evident that the capsule's elastic limit for the 1.0 mm thick capsule was about 1.04 N, for 2 mm, 1.33 N, and for 3 mm, 1.33 N. Compared to the ABS, the highest elastic modulus was for the 3.0 mm thick beam: $3.88 \times 10$ Pa, corresponding to the maximum load of 4.02 N. This demonstrates that for the same beam design, the ABS will yield a higher elastic modulus than the TPU "cross".

In Figure 17, w00, w10, w01, w20, and w11 are plotted against thickness using the model equation; research corresponds to the R-squared values for all coefficients being 1.0. In conclusion, from the experimental results for the TPU "cross" capsule, it is clear that the elastic modulus had a lower value than was achieved with the simple ABS beam. However, one prominent effect that was highlighted for the TPU "cross" capsule was that the beam's plastic region was activated at relatively low values of applied force. This also suggests high flexibility in the TPU "cross", which can be used to advantage in those designs where the beam needs to be folded and activated even when there is little change in the applied force. The TPU "cross" is more likely to remain elastic under deformation; this is why we chose to use the TPU "cross" as the material for the capsule and chose ABS to be the beam. It was clear from Figure 16 that the strain energy is obtained at a maximum of 0.25 N for all three capsules. However, the capsule thickness significantly affected the overall decay of strain energy. This suggests that the TPU "cross" capsule may be ideal for higher stresses because the material may be more resilient.

Thus, we inserted a TPU "cross" capsule inside the DCB and calculated the strain energy released by crack propagation in the DCB to assess whether it would activate the TPU "cross". Since the values of strain had been calculated above, it was easy to pinpoint the amount of strain energy released during crack propagation. The question is whether the amount of strain energy released would activate the TPU "cross" module. The behaviour of the "cross" is observed to be duplicated by the roller under a bending load.

3.2.2. Discussion on Error in Predictions

Once the beam modelling is completed, the validation is performed by adjusting the 3d surface graphs in model approximation and prediction. This enables us to choose the optimum model equation for different materials. As mentioned earlier, the initial process is to assess the model values with the points of the experimental design. The criteria used to test the model fit between different observations and predictions on the deflection, force, and displacement are used. Notably, during the MATLAB plot, the role of determination involved both R2 and adjusted R2, followed by Root Mean Square Error (RMSE).

For the research study, it is questionable what the probable difference between the points obtained from the predicted model versus the experimental design is. Since the number of simulations is not restricted, evaluation of Absolute Error and Root Mean Square Error (RMSE) can be considered for validation. The MSE values obtained for ABS and TPU are indicated in Figures 18 and 19, respectively.

The above-indicated Figures 18 and 19 show the thickness versus error ranges for MSE and RMSE for TPU and ABS, respectively. The highest error range is obtained for ABS, and the lowest is noted for TPU. This is due to the presence of the lowest degree polynomials in ABS model equations and higher in TPU. The MATLAB simulation converts the polynomial equation into an algebraic equation and then carries out the calculations. Therefore, neglecting a higher degree in a calculation in any algebraic equation reduces the model's accuracy. The only method to reduce error difference is to conduct an experimental research study with precision, as it will reduce analytical and experimental differences.

Theoretically, the proposed numerical model converts the continuous function into a piece-wise function by dividing the domain of the graph into discrete elements. Within this phenomenon, when we try to approximate the continuous function to discrete function, this leads to the generation of error, which generally accounts for a numerical error, in the case of surface graphs, which involve the modelling of the continuous system through discrete elements. With this inherited error, MATLAB does simulate the solution known as a numerically converged solution. However, there also exists a solution that is numerically uncoverged since MATLAB has ignored the inherited error, so the difference between numerically converged and unconverged produces a differential error [44]. Specific to the model equations we proposed, it is presumed the error leading to MSE is the differential error.

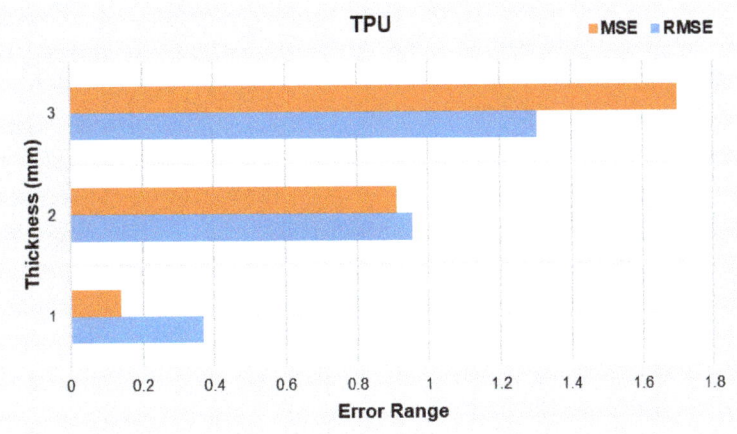

**Figure 18.** Thickness versus error ranges for MSE and RMSE for TPU beam.

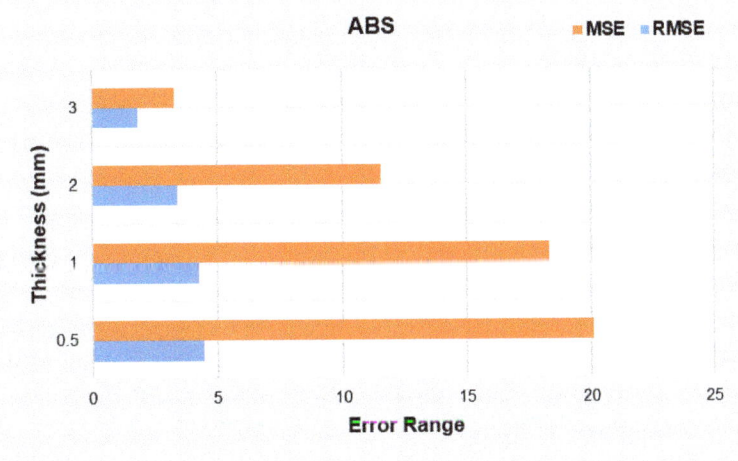

**Figure 19.** Thickness versus error ranges for MSE and RMSE for ABS beam.

### 3.3. Results for Origami "Cross" Module Embedded Structure

For the DCB test, instead of having interlaminar crack growth in the DCB, one of the arms was broken (see Figure 20). The responses of the DCB model can help us to analyse the behaviour of the origami capsule and whether it activates a self-healing mechanism. The analytical process seeks to estimate how much stress is released when the beam is deflected due to the application of a force. With the DCB, it is assumed that the force is dependent on the strain release phenomenon. The response of force vs. displacement for the DCB is presented in Figure 21. Here, the proposed standard was modified by adding a video recorder and camera to the test setup: a picture, which coincided with the force and displacement measurements, was taken every 10 s and was used to visually evaluate and measure the position of the crack tip during the tests with the TPU origami capsule in place.

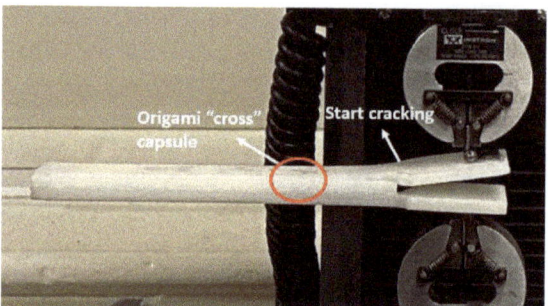

**Figure 20.** Measuring DCB crack length and displacement during the test.

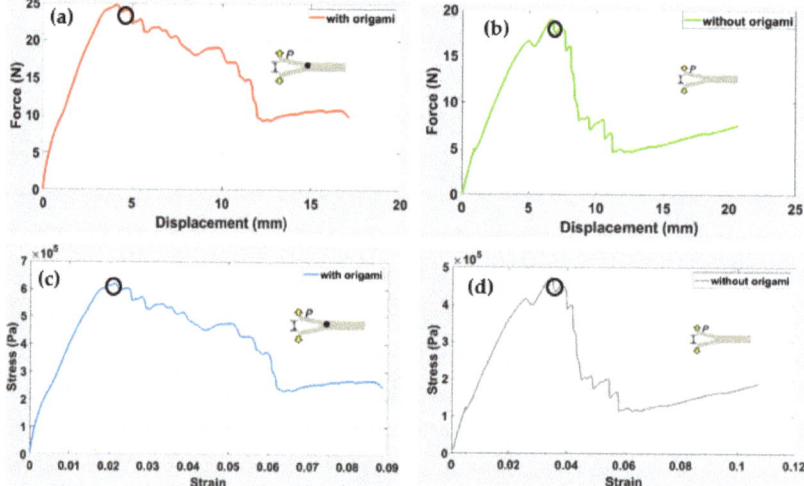

**Figure 21.** Crack length in DCB under the quasi-static conditions. The load–displacement graphs obtained for the beams with origami capsules and without origami capsules are shown in (**a**–**d**).

The load–displacement graphs obtained for the beams with origami capsules and without origami capsules are shown in Figure 21a–d. Nonlinearities in the load–displacement relation were observed for the specimen with the origami capsule. In Figure 21a, the maximum resistant force is 25 N. For the specimen without origami, Figure 21b, it is clear that the maximum force that can be resisted is 19 N at a total displacement of 6.4 mm.

Noticeable is the sudden and substantial drop in force that occurs in both cases, with and without the origami capsule. When the capsule is present, the drop is from about 15 N to 5.4 N, starting at a displacement of about 11 mm. When the capsule is absent, the drop is from about 15 N to 8 N, starting at a displacement of about 8 mm. This sudden failure precedes the full collapse of the DCB. However, the maximum displacement of the beam without the capsule reaches 20 mm, which is significantly higher than that for the beam with the capsule.

We can estimate the percentage error in the experimental deviation using the time and strain released.

$$\text{Strain release due to crack } (U) = \frac{\sigma^2}{2E} B\pi a^2 \qquad (5)$$

where $\sigma$ represents the stress applied, $E$ is the elastic modulus of the materials, $B$ is the area length, $\pi$ is the area from middle DCB until the open area, and $a$ indicates the length of the beam.

Table 7 indicates the experimental and theoretical values for the strain released (see Equation (5)) during the beam test with and without origami. The results indicate that the difference of strain in the beam with origami is 0.0871, and without origami is 0.0267, which comes to 8.71% and 2.67%, respectively. This shows that the strain released without origami is greater than that with origami. The response of the TPU, which have observed separately, is the same as that of the DCB. In order to use the model equation in calculations relating to self-healing behaviour, we have to include the calculated deviations.

**Table 7.** Experimental and theoretical model of the beam (a) with origami, (b) without origami.

|  | with Origami (5 mm) | without Origami (5 mm) |
| --- | --- | --- |
| Experimental value of strain release | $\epsilon = 8.87 \times 10^{-2}$ | $\epsilon = 2.62 \times 10^{-2}$ |
| Theoretical value of strain release | $\epsilon = 1.53 \times 10^{-3}$ | $\epsilon = 4.61 \times 10^{-6}$ |
| Percent Deviation in strain release | $\Delta\epsilon = 0.0871$ (8.71%) | $\Delta\epsilon = 0.0267$ (2.67%) |

Discussion:

This research assessed the behaviour of origami capsules embedded in the DCB structure. The DCB tests were carried out on an Instron test machine with displacement rates varying from 0.05 to 0.10 mm/s. The energy release rate and values of critical energy were calculated on the basis of the configurations. The material of the specimen was ABS, which was used to evaluate the stress–strain relationship for the DCBs with and without the presence of origami capsules. The results show that the presence of an origami capsule results in a more robust and resilient beam that can withstand greater fluctuations than a beam without an origami capsule. A mathematical analysis of the force vs. displacement and stress vs. strain curves was performed to help assess whether the hypothesis and research arrangement were valid.

From the experimental results, two pairs of graphs were obtained for the beams with origami and without origami, as shown in Figure 21a–d. Figure 21a, shows that with the origami capsule present, the graph proceeds as an almost straight line from 0 N to the maximum force of 24 N, at which the total displacement was 4.8 mm. This denotes the elastic limit of the beam and the resistance at the maximum load. After this point, the beam continued to extend, and the displacement increased, reaching a maximum displacement of 17.5 mm. This was the plastic region of the beam.

For the specimen without origami, as indicated in Figure 21b, it is clear that the maximum force that can be resisted is 19 N at a total displacement of 6.4 mm. The sudden and substantial drop in force from 15N to 5.4 N after this failure was obvious; it was enough to damage the overall beam before it fully collapsed. The maximum displacement of the beam reached 20 mm, which was significantly higher than for the beam with origami.

This is evident from a comparison with the results of previous research work by Simon et al. [40], who carried out load vs. displacement tests on specimens with and without a laminate lay-up. There was a clear difference in delamination lengths for the two specimens, with a rapid drop in load for the non-laminated lay-up compared to a gradual decline in force for the laminated lay-up. This result is similar to our beam results, as shown in Figure 21.

These results indicate that a beam with the origami capsule resists failure better than a beam without the capsule. The beam dimension may also play a significant role in defining the strength of the material. A study by Brunner et al. [42] on the applicability of delamination resistance of different materials indicates that multi-directional lay-ups pose issues due to crack branching and deviation from the plane. The delamination resistance seen in DCB tests depended on the fibre orientation. Alternating the orientations of the cross-ply composites in the beam from 0° to 90° yielded a 50% deviation from the mid-plane.

Figure 21c,d presents graphs of stress and strain, with and without the origami capsule. With the capsule present, the maximum stress was 6 × 106 Pa, and the strain was 0.09, whereas for the specimen without the capsule, the maximum stress was 4.5 × 106 Pa and the strain was 0.11. The stress–strain relation has also been studied by Chen et al. [45] using high-density stitched beams. When the load vs. displacement curve were compared, it was evident that the load increased linearly with the displacement, but when the stitches broke, crack initiation caused a sudden drop in load. Results closer to those in our study are reported by Kato et al. [46]; they found reported an example of crack propagation in DCB made from a satin weave E-glass fabric. The results are comparable in the sense that the delamination was 1.0 mm in width. Additionally, the load fluctuation was also steady, such as that of origami, which indicates the high tensile strength of the material sufficient to bear the increasing load, even during crack formation.

In conclusion, the proposed TPU "cross" origami capsule tends to absorb a sudden fluctuation in load and retards the displacement that may lead to failure. This contrasts with the results of the beam without an origami capsule: these showed a rapid decrease in load with an excessive displacement that led to the failure of the beam. Therefore, the hypothesis that specimens which integrate origami structures are more resilient and exhibit higher tensile strength when loaded axially is supported. Similarly, DCB beams that do not have origami structures embedded within them exhibit lower resilience or lower tensile strength.

### 3.4. Results for a Comparative Study for Strain Energy Activation in an Embedded Structure Versus a Simple TPU Capsule

The results presented in Figure 22 indicate a correlation between time and strain energy for simple TPU capsules 1, 2, and 3.0 mm in thickness. It is seen that the magnitude of the strain energy increases with the thickness of the beam since 3.0 mm showed the highest and 1.0 mm showed the least responsiveness over time.

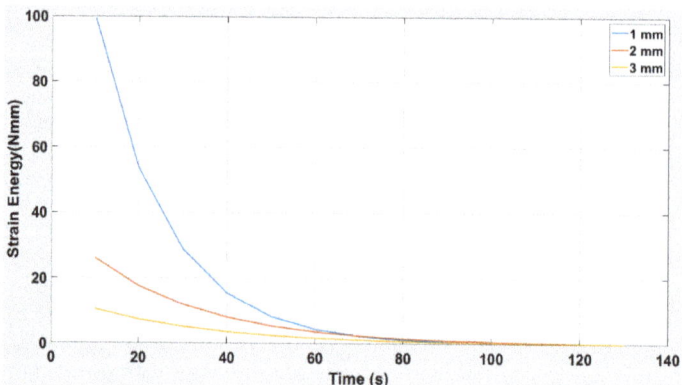

**Figure 22.** Strain Energy versus time for TPU "cross" beam structure of thicknesses: 1.0 mm, 2.0 mm, and 3.0 mm.

Figure 23 indicates the strain release over time for the beam with an embedded capsule. It is clear from the trends that the embedded beam of greatest thickness has the highest strain release and vice versa. The shape of the curves was exponential for all thicknesses. Compared to the normal TPU beam structure in which the strain energy lasted for 75 s in the TPU inside an embedded structure, the strain release covered a period of 140 s. The average strain release for 1, 2, and 3 mm beam thickness were 0.00038, 0.01153, 0.0473. Similarly, the standard deviations were 0.00075, 0.00507, and 0.05626 for beams of 1, 2, and 3 mm thickness, respectively.

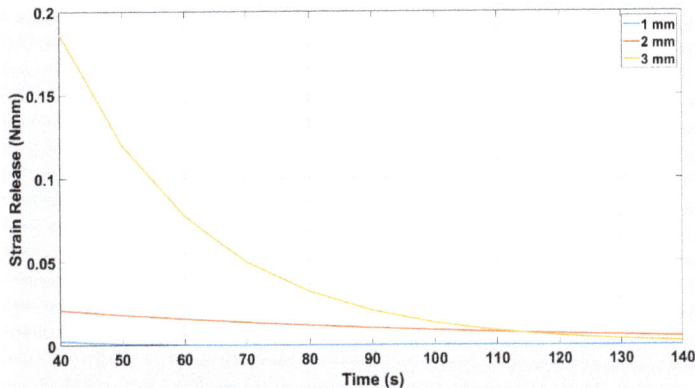

**Figure 23.** Strain release versus time for TPU inside an embedded DCB beam for thicknesses: 1.0 mm, 2.0 mm, and 3.0 mm.

The generalised equation for any thickness of the beam is represented as:

$$y = Strain\ energy = f(x = time) = ae^{(bx)} \qquad (6)$$

where $y$ indicates the strain energy (N.mm), a and b are coefficients, and $x$ represents time.

By substituting the values of a and b into Equation (6), the corresponding equation for thickness can be represented.

Figure 24 indicates the responsiveness of the strain release. It is evident that strain release for the DCB with and without the origami capsule has different response times. We see that maximum strain release was attained at 40 s for the beam containing the capsule and 3.5 s for the beam with no capsule. This shows that strain release in the beam with a capsule is higher than for the beam without a capsule.

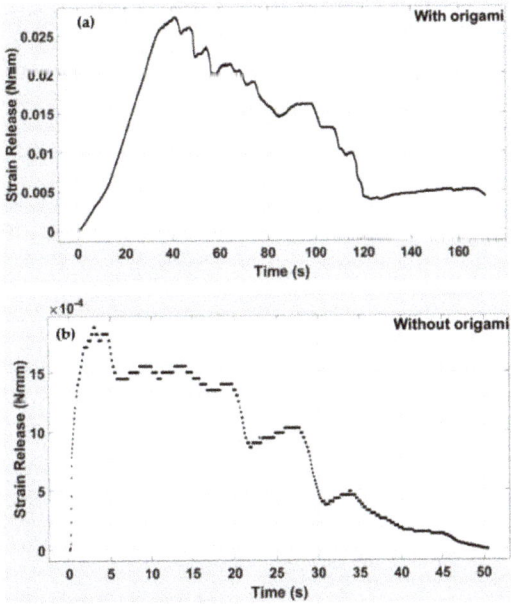

**Figure 24.** Strain release vs. time for DCB (**a**) with origami capsule and (**b**) without capsule.

From the graphs presented in Figures 22 and 23, a third-degree polynomial was used to fit the data points, so the generalised equation becomes:

$$f(x = time) = s1x^3 + s2x^2 + s3x + s4 \tag{7}$$

where $s1$, $s2$, $s3$, and $s4$ indicate the coefficients of the polynomial equation and $x$ indicates the time of the beam (see Table 8).

**Table 8.** Values of the linear equation coefficients and parametric coefficients for origami and non-origami capsules.

| Coefficient | S1 | S2 | S3 | S4 | R-Squared |
|---|---|---|---|---|---|
| With origami capsule | $4.971 \times 10^{-8}$ | $-1.479 \times 10^{-5}$ | 0.001144 | $-0.003213$ | 0.9200 |
| Non origami capsule | $5.415 \times 10^{-8}$ | $-4.429 \times 10^{-6}$ | $6.075 \times 10^{-5}$ | 0.001356 | 0.9333 |

The R-squared values for with-origami and non-origami capsules are found to be 0.9200 and 0.9333, respectively. This indicates high accuracy for the model equation.

Discussion:

In a structural analysis of the beam's response to strain energy, it is notable that the beam resists external actions by developing internal stresses induced in the material of the beam by the external forces and their subsequent displacements. The response of these internal stresses also changes with secondary parameters, such as those of geometry and dimensions. In the present research, it was evident that the strain energy for both the embedded structure DCB and the simple TPU beam produces an exponential decay curve. The generalised equation of the curve is indicated in Equation (6), and as shown in Table 9, For both the embedded structure and the simple TPU origami capsule, the value of the constant, a, increases and the value of the exponent, b, decreases with increasing thickness.

**Table 9.** Coefficients for DCB with an embedded TPU capsule and simple TPU beam, for three beam thicknesses.

| Thickness | Coefficients | Embedded Structure DCB (Average) | TPU Beam Thickness (Average) |
|---|---|---|---|
| 1 mm | a | 0.1199 | 184.3 |
|  | b | $-0.09576$ | $-0.06174$ |
| 2 mm | a | 0.03747 | 38.61 |
|  | b | $-0.01419$ | $-0.03903$ |
| 3 mm | a | 1.0784 | 14.96 |
|  | b | $-0.04389$ | $-0.03509$ |

Given the similarity in trends for the simple TPU origami capsule and embedded structure, it is evident from Figures 22 and 23 that the magnitude of strain energy increases with the thickness of the beam. In Figure 23, it is observed that the embedded beam with the greatest thickness has the highest strain release, and the converse also holds. Compared to the simple TPU beam structure in which the strain energy lasted for 75 s, the strain release in the TPU-embedded structure extended over a period of 140 s. This shows that strain release was more slowly dissipated in the beam embedded with a TPU capsule than in the simple TPU beam., The response of the 3.0 mm beam is also significant; it shows that the embedded structure can sustain higher strain energy values than the same structure without an embedded capsule. Therefore, when including a healing mechanism in a beam where high strain energy is required, it is necessary to select the thickness of the highest value.

To ascertain the healing rate, we can calculate from Figures 22 and 23 the difference between a TPU in an embedded structure and a simple TPU origami capsule from the extended rate of strain energy. It was found that the embedded structure had an extended dissipation time of 55 secs for all three beams before the trend reached zero. In Figure 24, the time response to the strain release was plotted for beams with and without origami capsule inserts. It is evident that the strain release had a different response time depending on whether an insert was present. For instance, the beams attained maximum strain release at.40 s and 3.5 s, with the larger value corresponding to the beam with the insert, meaning that a crack will not propagate so fast when it is being healed. Because the strain release is dependent on how far the crack has propagated, the strain released due to the presence of the origami capsule acts to resist the crack's tendency to propagate.

## 4. Conclusions

The research has sought to determine the behaviour of self-healing beams under the elastic and plastic loads for ABS and TPU materials.

- The study calculated the strain via strain energy and strain release for beams with and without origami capsules.
- Origami capsules were made in the shape of a cross with the four small beams comprising the arms, folded or elastically deformed and embedded in the main beam structure.
- Regarding crack propagation in the main beam, once the strain is released due to the crack, the small beams comprising the origami capsules open in the direction of the crack path and increase the crack resistance of the structure.
- When ABS transparent was used as the beam and TPU as the embedded capsule, it was found that the capsule worked as a self-healing mechanism, healing the crack before it occurred at a force of 24 N.
- From the results, it is evident that the properties of the TPU allowed deformation to remain flexible, which is why it was considered a suitable material for a novel self-healing mechanism that can be triggered by crack propagation due to strain release in a structure.
- Structural analysis has shown that the greater the beam thickness, the greater force required to attain the plastic region.
- The delamination tests showed the presence of a capsule suppressed the high compressive stresses induced by the bending moment in the vicinity of the crack tip and prevented the specimen from breaking and allowing crack propagation.
- The results show the potential of origami capsules as a novel self-healing mechanism to extend existing practice, which is primarily based on external healing agents.

**Author Contributions:** Conceptualization, M.D.A. and M.A.K.; methodology, M.D.A.; software, M.A.K.; validation, M.D.A., S.S.A. and M.A.K.; formal analysis, M.D.A.; investigation, M.D.A.; resources, M.A.K.; data curation, M.D.A.; writing—original draft preparation, M.D.A.; writing—review and editing, M.D.A., S.S.A. and M.A.K.; visualization, M.D.A. and M.A.K.; supervision, M.A.K.; project administration, M.A.K.; funding acquisition, M.A.K. All authors have read and agreed to the published version of the manuscript.

**Funding:** This research received no external funding.

**Institutional Review Board Statement:** Not applicable.

**Informed Consent Statement:** Not applicable.

**Data Availability Statement:** The data presented in this study are available on request from the corresponding author.

**Conflicts of Interest:** The authors declare no conflict of interest.

## References

1. Khoo, Z.X.; Teoh, J.E.M.; Liu, Y.; Chua, C.K.; Yang, S.; An, J.; Leong, K.F.; Yeong, W.Y. 3D Printing of Smart Materials: A Review on Recent Progresses in 4D Printing. *Virtual Phys. Prototyp.* **2015**, *10*, 103–122. [CrossRef]
2. Shafranek, R.T.; Millik, S.C.; Smith, P.T.; Lee, C.U.; Boydston, A.J.; Nelson, A. Stimuli-Responsive Materials in Additive Manufacturing. *Prog. Polym. Sci.* **2019**, *93*, 36–67. [CrossRef]
3. Zhang, Z.; Demir, K.G.; Gu, G.X. Developments in 4D-Printing: A Review on Current Smart Materials, Technologies, and Applications. *Int. J. Smart Nano Mater.* **2019**, *10*, 205–224. [CrossRef]
4. Garcia, S.J. Effect of Polymer Architecture on the Intrinsic Self-Healing Character of Polymers. *Eur. Polym. J.* **2014**, *53*, 118–125. [CrossRef]
5. He, F.; Thakur, V.K.; Khan, M. Evolution and New Horizons in Modeling Crack Mechanics of 3D Printing Polymeric Structures. *Mater. Today Chem.* **2021**, *20*. [CrossRef]
6. He, F.; Khan, M. Effects of Printing Parameters on the Fatigue Behaviour of 3d-Printed Abs under Dynamic Thermo-Mechanical Loads. *Polymers* **2021**, *13*, 2362. [CrossRef]
7. MacDonald, E.; Wicker, R. Multiprocess 3D Printing for Increasing Component Functionality. *Science* **2016**, *353*, aaf2093. [CrossRef]
8. Invernizzi, M.; Turri, S.; Levi, M.; Suriano, R. 4D Printed Thermally Activated Self-Healing and Shape Memory Polycaprolactone-Based Polymers. *Eur. Polym. J.* **2018**, *101*, 169–176. [CrossRef]
9. Belowich, M.E.; Stoddart, J.F. Dynamic Imine Chemistry. *Chem. Soc. Rev.* **2012**, *41*, 2003–2024. [CrossRef]
10. Willocq, B.; Odent, J.; Dubois, P.; Raquez, J.M. Advances in Intrinsic Self-Healing Polyurethanes and Related Composites. *RSC Adv.* **2020**, *10*, 13766–13782. [CrossRef] [PubMed]
11. Fleet, T.; Kamei, K.; He, F.; Khan, M.A.; Khan, K.A.; Starr, A. A Machine Learning Approach to Model Interdependencies between Dynamic Response and Crack Propagation. *Sensors* **2020**, *20*, 6847. [CrossRef]
12. Zhang, Z.P.; Rong, M.Z.; Zhang, M.Q. Mechanically Robust, Self-Healable, and Highly Stretchable "Living" Crosslinked Polyurethane Based on a Reversible C–C Bond. *Adv. Funct. Mater.* **2018**, *28*, 1706050. [CrossRef]
13. Almutairi, M.D.; Aria, A.I.; Thakur, V.K.; Khan, M.A. Self-Healing Mechanisms for 3D-Printed Polymeric Structures: From Lab to Reality. *Polymers* **2020**, *12*, 1534. [CrossRef] [PubMed]
14. Kamei, K.; Khan, M.A. Investigating the Structural Dynamics and Crack Propagation Behavior under Uniform and Non-Uniform Temperature Conditions. *Materials* **2021**, *14*, 7071. [CrossRef] [PubMed]
15. Gattas, J.M.; You, Z. The Behaviour of Curved-Crease Foldcores under Low-Velocity Impact Loads. *Int. J. Solids Struct.* **2015**, *53*, 80–91. [CrossRef]
16. He, F.; Khan, M.; Aldosari, S. Interdependencies between Dynamic Response and Crack Growth in a 3D-Printed Acrylonitrile Butadiene Styrene (ABS) Cantilever Beam under Thermo-Mechanical Loads. *Polymers* **2022**, *14*, 982. [CrossRef] [PubMed]
17. Li, Y.; You, Z. Open-Section Origami Beams for Energy Absorption. *Int. J. Mech. Sci.* **2019**, *157–158*, 741–757. [CrossRef]
18. Richeton, J.; Ahzi, S.; Vecchio, K.S.; Jiang, F.C.; Adharapurapu, R.R. Influence of Temperature and Strain Rate on the Mechanical Behavior of Three Amorphous Polymers: Characterization and Modeling of the Compressive Yield Stress. *Int. J. Solids Struct.* **2006**, *43*, 2318–2335. [CrossRef]
19. Alshammari, Y.L.A.; He, F.; Khan, M.A. Modelling and Investigation of Crack Growth for 3d-Printed Acrylonitrile Butadiene Styrene (Abs) with Various Printing Parameters and Ambient Temperatures. *Polymers* **2021**, *13*, 3737. [CrossRef] [PubMed]
20. Yuan, L.; Dai, H.; Song, J.; Ma, J.; Chen, Y. The Behavior of a Functionally Graded Origami Structure Subjected to Quasi-Static Compression. *Mater. Des.* **2020**, *189*, 108494. [CrossRef]
21. Peraza Hernandez, E.A.; Hartl, D.J.; Lagoudas, D.C. Structural Mechanics and Design of Active Origami Structures. In *Active Origami*; Springer: Berlin/Heidelberg, Germany, 2019; pp. 331–409.
22. Kamei, K.; Khan, M.A.; Khan, K.A. Characterising Modal Behaviour of a Cantilever Beam at Different Heating Rates for Isothermal Conditions. *Appl. Sci.* **2021**, *11*, 4375. [CrossRef]
23. Banshiwal, J.K.; Tripathi, D.N. Self-Healing Polymer Composites for Structural Application. *Funct. Mater.* **2019**, *10*, 1–13. [CrossRef]
24. Yıldırım, G.; Khiavi, A.H.; Yeşilmen, S.; Şahmaran, M. Self-Healing Performance of Aged Cementitious Composites. *Cem. Concr. Compos.* **2018**, *87*, 172–186. [CrossRef]
25. Alazhari, M.; Sharma, T.; Heath, A.; Cooper, R.; Paine, K. Application of Expanded Perlite Encapsulated Bacteria and Growth Media for Self-Healing Concrete. *Constr. Build. Mater.* **2018**, *160*, 610–619. [CrossRef]
26. Teall, O.; Pilegis, M.; Davies, R.; Sweeney, J.; Jefferson, T.; Lark, R.; Gardner, D. A Shape Memory Polymer Concrete Crack Closure System Activated by Electrical Current. *Smart Mater. Struct.* **2018**, *27*, 075016. [CrossRef]
27. Pang, J.W.C.; Bond, I.P. A Hollow Fibre Reinforced Polymer Composite Encompassing Self-Healing and Enhanced Damage Visibility. *Compos. Sci. Technol.* **2005**, *65*, 1791–1799. [CrossRef]
28. Zai, B.A.; Khan, M.A.; Khan, S.Z.; Asif, M.; Khan, K.A.; Saquib, A.N.; Mansoor, A.; Shahzad, M.; Mujtaba, A. Prediction of Crack Depth and Fatigue Life of an Acrylonitrile Butadiene Styrene Cantilever Beam Using Dynamic Response. *J. Test. Eval.* **2020**, *48*, 20180674. [CrossRef]
29. Dhaliwal, G.S.; Dundar, M.A. Four Point Flexural Response of Acrylonitrile–Butadiene–Styrene. *J. Compos. Sci.* **2020**, *4*, 63. [CrossRef]

30. Lee, T.U.; You, Z.; Gattas, J.M. Elastica Surface Generation of Curved-Crease Origami. *Int. J. Solids Struct.* **2018**, *136–137*, 13–27. [CrossRef]
31. Damas, S.M.; Turner, C.J. The Material Testing of Nanoparticle Doped 3D Printed ABS Strain Gages for Resisitance and Stiffness. In Proceedings of the International Design Engineering Technical Conferences and Computers and Information in Engineering Conference IDETC/CIE2020, Virtual, 17–19 August 2020; pp. 1–9.
32. Tymrak, B.M.; Kreiger, M.; Pearce, J.M. Mechanical Properties of Components Fabricated with Open-Source 3-D Printers under Realistic Environmental Conditions. *Mater. Des.* **2014**, *58*, 242–246. [CrossRef]
33. Cantrell, J.; Rohde, S.; Damiani, D.; Gurnani, R.; Disandro, L.; Anton, J.; Young, A.; Jerez, A.; Steinbach, D.; Kroese, C.; et al. Experimental Characterization of the Mechanical Properties of 3D-Printed ABS and Polycarbonate Parts. *Rapid Prototyp. J.* **2017**, *23*, 811–824. [CrossRef]
34. Yu, Z.; Li, Y.; Zhao, Z.; Wang, C.; Yang, J.; Zhang, C.; Li, Z.; Wang, Y. Effect of Rubber Types on Synthesis, Morphology, and Properties of ABS Resins. *Polym. Eng. Sci.* **2009**, *49*, 2249–2256. [CrossRef]
35. Ritzen, L.; Montano, V.; Garcia, S.J. 3d Printing of a Self-Healing Thermoplastic Polyurethane through Fdm: From Polymer Slab to Mechanical Assessment. *Polymers* **2021**, *13*, 305. [CrossRef] [PubMed]
36. UL Prospector Generic Families of Plastic. Available online: https://www.ulprospector.com/plastics/en/generics (accessed on 15 September 2021).
37. Srivastava, V.K. A Review on Advances in Rapid Prototype 3D Printing of Multi-Functional Applications. *Sci. Technol.* **2017**, *7*, 4–24. [CrossRef]
38. Mitra, A.C.; Jagtap, A.; Kachare, S. Development and Validation of Experimental Setup for Flexural Formula of Cantilever Beam Using NI-LabVIEW. *Mater. Today Proc.* **2018**, *5*, 20326–20335. [CrossRef]
39. Hunt, C.; Kratz, J.; Partridge, I.K. Cure Path Dependency of Mode i Fracture Toughness in Thermoplastic Particle Interleaf Toughened Prepreg Laminates. *Compos. Part A Appl. Sci. Manuf.* **2016**, *87*, 109–114. [CrossRef]
40. Simon, I.; Banks-Sills, L.; Fourman, V. Mode I Delamination Propagation and R-Ratio Effects in Woven Composite DCB Specimens for a Multi-Directional Layup. *Int. J. Fatigue* **2017**, *96*, 237–251. [CrossRef]
41. Reis, P.N.B.; Ferreira, J.A.M.; Antunes, F.V.; Costa, J.D.M. Initial Crack Length on the Interlaminar Fracture of Woven Carbon/Epoxy Laminates. *Fibers Polym.* **2015**, *16*, 894–901. [CrossRef]
42. Brunner, A.J.; Blackman, B.R.K.; Davies, P. A Status Report on Delamination Resistance Testing of Polymer-Matrix Composites. *Eng. Fract. Mech.* **2008**, *75*, 2779–2794. [CrossRef]
43. Mishra, K.; Babu, L.K.; Vaidyanathan, R. Improvement of Fracture Toughness and Thermo-Mechanical Properties of Carbon Fiber/Epoxy Composites Using Polyhedral Oligomeric Silsesquioxane. *J. Compos. Mater.* **2020**, *54*, 1273–1280. [CrossRef]
44. Ait-Amir, B.; Pougnet, P.; El Hami, A. Meta-Model Development. *Embed. Mechatron. Syst.* **2015**, *2*, 151–179. [CrossRef]
45. Chen, L.; Ifju, P.G.; Sankar, B.V. A Novel Double Cantilever Beam Test for Stitched Composite Laminates. *J. Compos. Mater.* **2001**, *35*, 1137–1149. [CrossRef]
46. Kato, Y.; Minakuchi, S.; Ogihara, S.; Takeda, N.; Krull, B.; Patrick, J.; Hart, K.; White, S.; Sottos, N.; Yamagata, Y.; et al. Self-Healing Composites Structure Using Multiple through-Thickness Microvascular Channels. *Adv. Compos. Mater.* **2020**, *30*, 1–18. [CrossRef]

Article

# Piezoresistive Properties of 3D-Printed Polylactic Acid (PLA) Nanocomposites

Razieh Hashemi Sanatgar [1,2,3,*], Aurélie Cayla [2], Jinping Guan [3], Guoqiang Chen [3], Vincent Nierstrasz [1] and Christine Campagne [2]

1. Textile Materials Technology, Department of Textile Technology, Faculty of Textiles, Engineering and Business, University of Borås, SE-501 90 Borås, Sweden; vincent.nierstrasz@hb.se
2. ENSAIT, ULR 2461—GEMTEX—Génie et Matériaux Textiles, Université de Lille, F-59000 Lille, France; aurelie.cayla@ensait.fr (A.C.); christine.campagne@ensait.fr (C.C.)
3. College of Textile and Clothing Engineering, Soochow University, Suzhou 215006, China; guanjinping@suda.edu.cn (J.G.); chenguojiang@suda.edu.cn (G.C.)
* Correspondence: razieh.hashemi_sanatgar@hb.se

**Citation:** Hashemi Sanatgar, R.; Cayla, A.; Guan, J.; Chen, G.; Nierstrasz, V.; Campagne, C. Piezoresistive Properties of 3D-Printed Polylactic Acid (PLA) Nanocomposites. *Polymers* **2022**, *14*, 2981. https://doi.org/10.3390/polym14152981

Academic Editors: Hamid Reza Vanaei, Sofiane Khelladi and Abbas Tcharkhtchi

Received: 3 June 2022
Accepted: 8 July 2022
Published: 22 July 2022

**Publisher's Note:** MDPI stays neutral with regard to jurisdictional claims in published maps and institutional affiliations.

**Copyright:** © 2022 by the authors. Licensee MDPI, Basel, Switzerland. This article is an open access article distributed under the terms and conditions of the Creative Commons Attribution (CC BY) license (https://creativecommons.org/licenses/by/4.0/).

**Abstract:** An increasing interest is focused on the application of 3D printing for sensor manufacturing. Using 3D printing technology offers a new approach to the fabrication of sensors that are both geometrically and functionally complex. This work presents the analysis of the 3D-printed thermoplastic nanocomposites compress under the applied force. The response for the corresponding resistance changes versus applied load is obtained to evaluate the effectiveness of the printed layer as a pressure/force sensor. Multi-walled carbon nanotubes (MWNT) and high-structured carbon black (Ketjenblack) (KB) in the polylactic acid (PLA) matrix were extruded to develop 3D-printable filaments. The electrical and piezoresistive behaviors of the created 3D-printed layers were investigated. The percolation threshold of MWNT and KB 3D-printed layers are 1 wt.% and 4 wt.%, respectively. The PLA/1 wt.% MWNT 3D-printed layers with 1 mm thickness exhibit a negative pressure coefficient (NPC) characterized by a decrease of about one decade in resistance with increasing compressive loadings up to 18 N with a maximum strain up to about 16%. In the cyclic mode with a 1 N/min force rate, the PLA/1 wt.% MWNT 3D-printed layers showed good performance with the piezoresistive coefficient or gauge factor (G) of 7.6 obtained with the amplitude of the piezoresistive response ($A_r$) of about -0.8. KB composites could not show stable piezoresistive responses in a cyclic mode. However, under high force rate compression, the PLA/4 wt.% KB 3D-printed layers led to responses of large sensitivity ($A_r = -0.90$) and were exempt from noise with a high value of G = 47.6 in the first cycle, which is a highly efficient piezoresistive behavior.

**Keywords:** piezoresistive properties; 3D printing; fused deposition modelling (FDM); polylactic acid (PLA); multi-walled carbon nanotubes (MWNT); high-structured carbon black (KB)

## 1. Introduction

In recent years, 3D printing, also known as additive manufacturing (AM), has attracted significant attention from both industry and academia. In this technique, different methods such as material jetting [1], powder bed fusion [2], material extrusion [3], sheet lamination [4], directed energy deposition [5], photopolymerization [6,7], and binder jetting [8] are applied for the manufacturing of 3D items. These methods begin with a 3D model of the object, and then the special software digitizes and slices the object into the model layers. Afterward, the AM system prints 2D layers into a 3D build [9–12].

Three-dimensional printing is a novel method for the development of multifunctional components such as sensors with complex geometries and combined characteristics such as optical, chemical, electrical, and thermal, etc. [9]. It is possible to embed a sensor into a 3D-printed component or print the entire sensor consistently [13]. In the recent past, significant research has been carried out on the fabrication of 3D-printed sensors such as force [14], motion [15], optic [16], hearing [17], etc. by various 3D printing techniques with distinctive transduction mechanisms, applications and printing materials.

Fused deposition modeling (FDM) is one of the key AM methods, in which a thermoplastic filament is passed through a heated extrusion nozzle to be melted. Saari et al. [18] created a capacitive force sensor using the FDM method and ABS (Acrylonitrile butadiene styrene)-based materials consisting of a 3D-printed rigid frame with embedded wires in a spiral pattern imitating a flat plate capacitor and a thermoplastic elastomer dielectric spacer that compress under the applied force. An ear prosthesis fabricated by 3D printing of polyvinylidene fluoride (PVDF) [19] showed reliable responses under different conditions of pressure (0 to 16,350 Pa) and temperatures (2 to 90 °C) regarding the pyroelectric and piezoelectric properties. Krachunov [20] presented a novel method using 3D printing of ABS and polylactic acid (PLA) with silver coating for the design and manufacture of customized dry electrodes for Electroencephalography (EEG), which is a procedure that records brain activity in a non-invasive manner. The performance of the proposed electrodes is suitable for Brain–Computer Interface (BCI) applications, despite the presence of additional noise.

Applying electrically conductive polymer composites (CPCs) in FDM technology, some researchers have recently tried to develop sensors that are responsive to different stimuli such as chemicals including solvents, biological fluids, dopamine, serotonin, metals, vapors, mechanical flexing and liquid levels [21–24]. Kim et al. [25] 3D-printed thermoplastic polyurethane (TPU) and TPU/multiwalled carbon nanotubes (MWNT) (a structural part and a sensing part, respectively) to fabricate a 3D multiaxial force sensor that could detect the submillimeter scale deflection and its corresponding force on each axis.

TPU containing MWNT/graphene was used to develop flexible strain sensors [26,27]. The results demonstrated TPU nanocomposites as an excellent piezoresistive feedstock for 3D printing with the potential for wide-ranging applications in soft actuators, feedback from high-speed robotic applications and 3D-printed wearable devices.

Polylactic acid (PLA) has attracted researchers to apply this biodegradable thermoplastic as the matrix polymer in 3D-printed sensors. The total volatile organic compounds and ultrafine particles emitted while PLA printing is lesser in comparison to other polymers [28]. Three-dimensional printed PLA-carbon black could be effectively used as solvent [29] and capacitive sensors [30]. The tensile and impact strengths decreased after dipping them in solvents. The research showed that 3D-printed PLA containing nanographite/graphene is a promising economical electrochemical sensing platform; however, the performance of the 3D-printed devices is needed to be improved by increasing the percent of active material [31–33].

Printing in different geometries expands the utility of 3D-printed sensors in wearable forms and brings researchers closer to the desire of applying 3D printing for functional and smart textiles [34]. However, the existing high-sensitive pressure sensors in the medium- to high-pressure range could not be simply integrated into the garments without hindering the manual motion [35,36]. The sensors with a sensitivity in the medium pressure range (10–100 kPa) are required in gloves for monitoring hand stress during manual activity and object manipulation [37,38]. Foot pressure due to body weight as well as the applied force in using tools such as tennis rackets with repetitive motions are other examples of the medium pressure range [37]. Dios et al. investigated the piezoresistive performance of polymer-based nanocomposites in walking detection applications. Poly(vinylidene fluoride) (PVDF) in comparison with styrene-b-(ethylene-co-butylene)-b-styrene (SEBS) and thermoplastic polyurethane (TPU) is the most suitable polymer matrix in low deformation applications, whereas TPU and SEBS are suited for large deformation application due to their stretchability [39].

In fact, the piezoresistivity of 3D-printed PLA nanocomposites has not been investigated. Although the matrix is not flexible, the 3D-printed structure could bring functionality to the nanocomposites through possible complex geometries and the layer-by-layer structure which causes the inter-fillers and inter agglomerates gap to increase, leading the conductive nanocomposites to less dense and more sensitive to compression.

Therefore, in this research, the behavior of PLA 3D-printed nanocomposites under a load in the medium pressure range was investigated. Conductive filaments including multi-walled carbon nanotubes (MWNT) and high structured carbon blacks (Ketjenblack) (KB) in a PLA matrix were 3D printed and the electrical and compressive piezoresistive behavior of 3D-printed components were investigated. The piezoresistive behavior under compression was also studied in a cyclic mode in terms of filler type, filler content, and loading force rate.

## 2. Materials and Methods

### 2.1. Materials

As an electrically insulating thermoplastic matrix, a semi-crystalline polylactic acid (PLA) was purchased from NatureWorks, Minnetonka, MN, USA under the reference NatureWorks®-6202 D ($M_n$ = 58,300 g/mol; D-Isomer = 1.3%). Prior to compounding and extrusion, PLA pellets were dried at 60 °C for 12 h in oven to remove water.

The carbon black (KB) was obtained from AKZO NOBEL, Amersfoort, The Netherlands under the reference Ketjenblack® EC-600JD with the aggregate size of 10–50 nm, the apparent bulk density of 1–1.2 g/cm$^3$ and BET surface area of 1400 m$^2$/g. Multi-wall carbon nanotubes (MWNTs) were obtained from Nanocyl, Sambreville, Belgium under the reference Nanocyl®-7000 with a diameter of about 10 nm and lengths of 0.1–10 μm with a surface area of 250 m$^2$/g.

### 2.2. Nanocomposites Preparation

In first step, a Thermo Haake co-rotating intermeshing twin-screw extruder was used to disperse fillers (MWNT or KB) into PLA with a weight percentage of 10 wt.%. The screw size of Haake is 400 mm in length and an average diameter of 16 mm (L/D = 25). The pressure is about 20 bar. The rotational speed of the screw was set at 100 rpm and the temperature of the five heating zones of the extruder was set at 160, 175, 175, 170 and 160 °C. Upon exiting the extruder with an average speed of 1 m/min, the masterbatch was pelletized. In the following step, the pelletized masterbatch was diluted with PLA pellets to obtain the weight percentage of 0.5–5 wt.% for MWNT and 1.5–7 wt.% for KB in PLA. Before dilution, both pelletized masterbatch and PLA pellets were dried at 60 °C for 12 h. For cooling down the manufactured filaments, a bath of closed circulation of water at room temperature was applied. The developed 3D printer filaments were used to print the nanocomposite layers.

### 2.3. 3D Printing

The 3D printer used (a two-head WANHAO Duplicator 4/4x) supplied by Creative Tools AB (Halmstad, Sweden) with a nozzle diameter of 0.4 mm and maximum printing size of 22.5 × 14.5 × 15 cm$^3$. The 3D models were created in Rhinoceros software and exported as an STL (Standard Triangle Language that is the industry standard file type for 3D Printing), then transferred to Simplify3D software (Creative Tools AB) to be printed. Samples were 3D printed in different geometries including rectangular for electrical resistance measurement (1 × 12.75 × 60 mm$^3$) and circular for compression (1 mm thickness and 40 mm diameter) at 240° ± 2 °C. The raster angle was 0° with linear infill pattern (100%). The raster size was 0.3 mm in height and 0.4 mm in width. The printing speed was 3000 mm/min and the first layer speed was 50%.

### 2.4. Electrical Resistance Measurement

The electrical resistance of 3D-printed layers was measured using a two-point measurement method by a digital multimeter connected with alligator clips to rectangular 3D-printed layers (1 × 12.75 × 60 mm$^3$). Three measurements per CPC formulations (PLA/2, 3, 4, 5, 7 wt.% KB) and (PLA/0.5, 1, 1.5, 3, 5 wt.% MWNT) were carried out.

## 2.5. Piezoresistive Pressure Measurement

The piezoresistive properties of the 3D-printed samples under compression were measured by compression clamp (15 mm diameter) consisting of DMA Q800 and a multimeter/system switch (Keithley 3706A) controlled by the instrument web interface. Figure 1 describes the experimental setup applied to investigate the piezoresistive properties of the samples. The 3D-printed layers with a thickness of 1 mm and a diameter of 40 mm were clamped between two copper plates of 30 mm diameter as electrodes to investigate the piezoresistive behavior. Copper plates were connected to a multimeter to measure the nanocomposite layers' resistance. The samples and electrodes were clamped between a fixed part and the moving part providing the force (a Teflon tape was used for fixation). An initial preload of 2 N was applied to the sample in order to ensure full contact between the loading clamps and the sample surfaces. Then, compressive stress was used in the direction of resistance measurement to the sample. The geometry of the sample changes continuously due to the applied stress. Compressive loading was applied during the test at two different force rates (1 and 18 N/min) up to 18 N. DMA compression clamps yielded increasing pressure on the electrodes providing responses in the form of resistance.

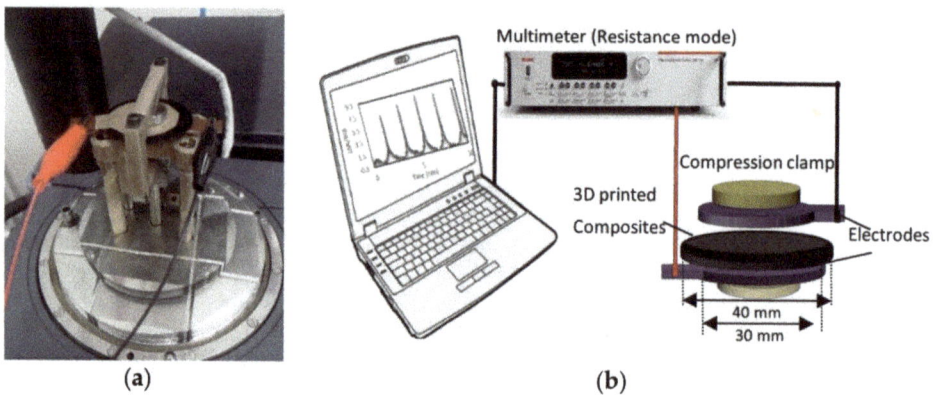

**Figure 1.** (a) Sample setup applied to investigate the piezoresistive properties of 3D-printed nanocomposite layers under compressive stress; (b) Schematic diagram of the positioning of the sample in clamps and electrodes.

The piezoresistive response ($A_r$) or relative difference of resistance amplitude of sensors was calculated according to Equation (1) [40]:

$$A_r = \frac{\Delta R}{R_0} = \frac{R - R_0}{R_0} \tag{1}$$

where $R$ represents the resistance of the composite under applied pressure and $R_0$ is the static resistance.

## 3. Results and Discussion

### 3.1. Electrical Characterization

The conductivity ($\sigma$) of the 3D-printed layers was calculated according to Equation (2):

$$\sigma = L/R \cdot A \tag{2}$$

where $L$ and $A$ are, respectively, the length (m) and the cross-sectional area (m$^2$) of the 3D-printed layers. $R$ is the electrical resistance ($\Omega$) and $\sigma$ is the electrical conductivity ($\Omega \cdot$m)$^{-1}$ or Siemens per meter (S/m).

The sudden transition from insulator to conductor, which is the indication of the percolation threshold happened in PLA/4 wt.% KB and PLA/1 wt.% MWNT (Figure 2).

The printed nanocomposites containing 2 and 3 wt.% KB as well as 0.5 wt.% MWNT were not conductive.

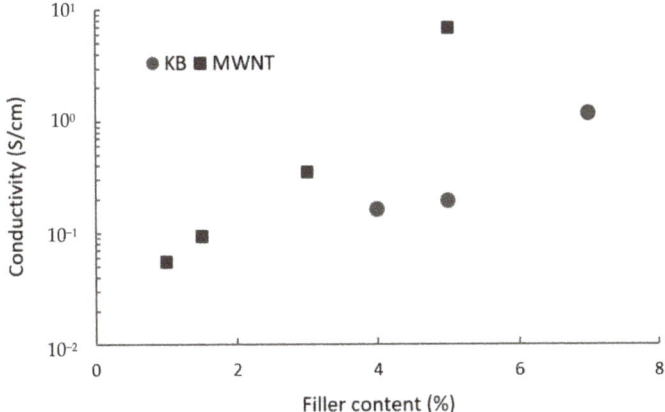

**Figure 2.** Electrical conductivity as a function of the filler content for 3D-printed layers of PLA nanocomposites containing MWNT and KB.

### 3.2. Compression Piezoresistive Properties

The piezoresistive behavior of 1% MWNT 3D-printed nanocomposite layers was investigated when subjected to compression stress ranging from 0.5 to 18 N. Figure 3 represents the piezoresistive source signals evolution and the related stress–strain diagrams. It is evident that the 1 wt.% MWNT 3D-printed composite layer shows a negative pressure coefficient (NPC) characterized by a decrease of about one decade in resistance with the compressive loadings increase up to 18 N with the maximum strain up to about 16%. Piezoresistive pressure sensors undergo a change in resistance under applied pressure that is assumed to be caused by the different compressibility of filler and polymeric matrix under an applied force. Fillers either separate or approach the applied compression and cause a positive or negative relationship between pressure and resistance depending on filler geometry and the magnitude of the pressure [41]. The 3D-printed layers approach by the applied compression causing to more effective connections between conductive nanofillers by decreasing the average inter-fillers distance and hence lower relative resistance, which describes the NPC effect detected in the 3D-printed nanocomposites. The layer-by-layer structure of the 3D-printed nanocomposites causes the inter-fillers and inter agglomerates gap to increase, leading the conductive nanocomposites being less dense and more sensitive to compression.

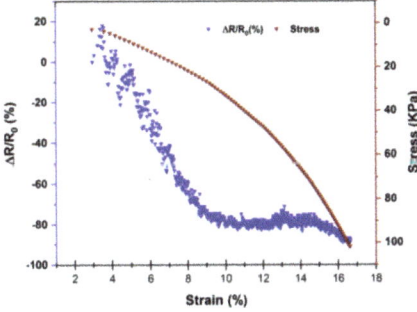

**Figure 3.** Piezoresistive responses of 1% MWNT nanocomposite 3D-printed layers under compressive loading.

## 3.3. Compression Piezoresistive Behavior in a Cyclic Mode

A resistance change of the 3D-printed nanocomposite layers was detected under cyclic loadings increasing from 10 to 100 kPa. Four cycles were carried out to check the reproducibility of the sample with different filler contents (a sample in percolation threshold and a sample with higher contents of fillers) in low force speed of 1 N/min and a 15 mm diameter compression clamp. Figure 4 illustrates the sensors' responses to applied stress and related strain.

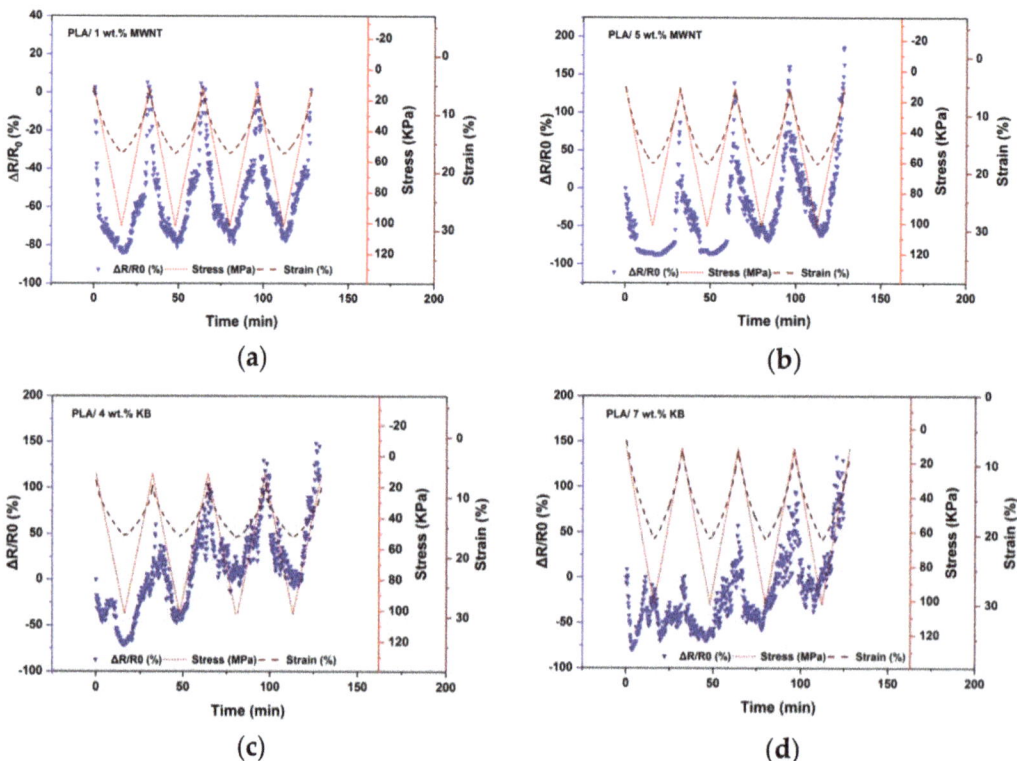

**Figure 4.** Comparison of 3D-printed nanocomposites piezoresistive responses: (**a**) PLA/1 wt.% MWNT, (**b**) PLA/5 wt.% MWNT, (**c**) PLA/4 wt.% KB and (**d**) PLA/7 wt.% KB.

As shown in Figure 4a, PLA/1 wt.% MWNT piezoresistive responses are synchronic with strain and stress and the resistance variation follows the deformation which turns back to its original value after unloading. However, the sensor responses of samples including 5 wt.% MWNT, 4 and 7 wt.% KB are not synchronic with the applied stress and strain. Figure 5 represents the resistance changes at the start (Sc) and end (Ec) of each cycle for all samples.

It is clear that except for the PLA/1 wt.% MWNT, the other 3D-printed layers have significant hysteresis behavior, which is because of the residual strain of the 3D-printed layer composites after the compression. The same behavior was reported in studies about the compression test of porous structures including carbon nanotubes [42,43]. The melt flow index of composites with higher filler contents is low [44], therefore, it is required to 3D print at a lower speed or use a higher nozzle temperature [45], which causes structures with an eventually larger hysteresis behavior under compression cycles. Moreover, in higher nanofiller contents than the percolation threshold, the dominant mechanism of conduction is percolation [46], therefore, the destruction of effective conductive paths in successive loading/unloading cycles is the dominant mechanism, especially when the related strain is also high (low force speed). Therefore, by successive loading/unloading cycles in higher filler contents, an increase in minimum and maximum sensitivity is observed.

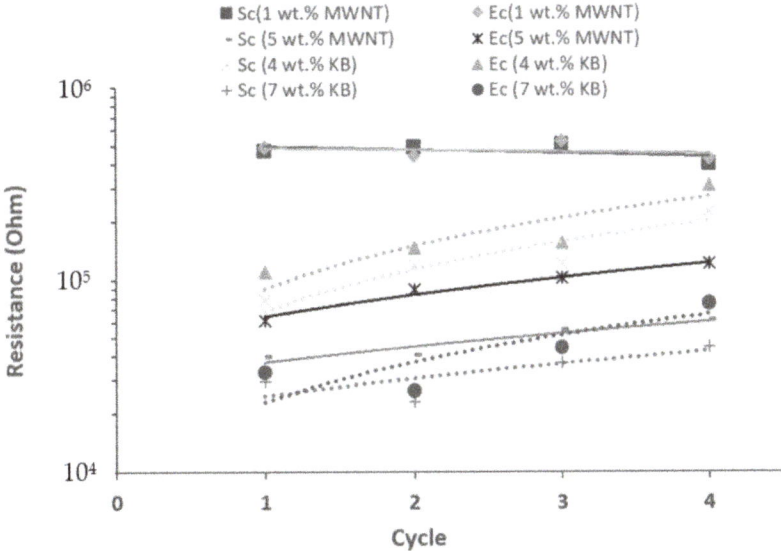

**Figure 5.** Resistance changes at the start and end of each cycle for different samples (Sc is the start of the cycle and Ec is the end of the cycle. The solid and dot linear trend lines represented MWNT and KB composites, respectively.

The hysteresis behavior of the PLA/1 wt.% MWNT layers is clearer in Figure 6, which shows that after a large hysteresis in the first cycle, the track of the three successive cycles of loading/unloading is nearly identical, with a minor deviation distinguished between the second and the fourth loops. This behavior is similar to the findings of Slobodian and Saha [47], where accordingly in the MWNT network, a ratcheting strain (mean value of the maximum and minimum strain in one cycle) takes place after the first compression cycle as a consequence of the primary deformation of the porous composition and blocked the reverse mobility of nanotubes in the middle of the dense networks. Through successive cycles of loading and unloading, the nanotubes' reorder becomes stable and the MWNT network gets to a steady stress–strain hysteresis loop order. This suggests that when the carbon nanotube network is well deformed, it can be applied as a sensing component of compression stress. In Figure 6, it can also be observed that the signal is linear with a slope difference below and over 30 kPa. Figure 6b depicts a schematic representation of a 1 wt.% MWNT 3D-printed layer sandwiched between two copper electrodes. Dashed lines between the MWNT individual particles and clusters represent quantum tunneling bridges which accordingly allow charge carriers to tunnel from one cluster to another without any physical contact in composite systems.

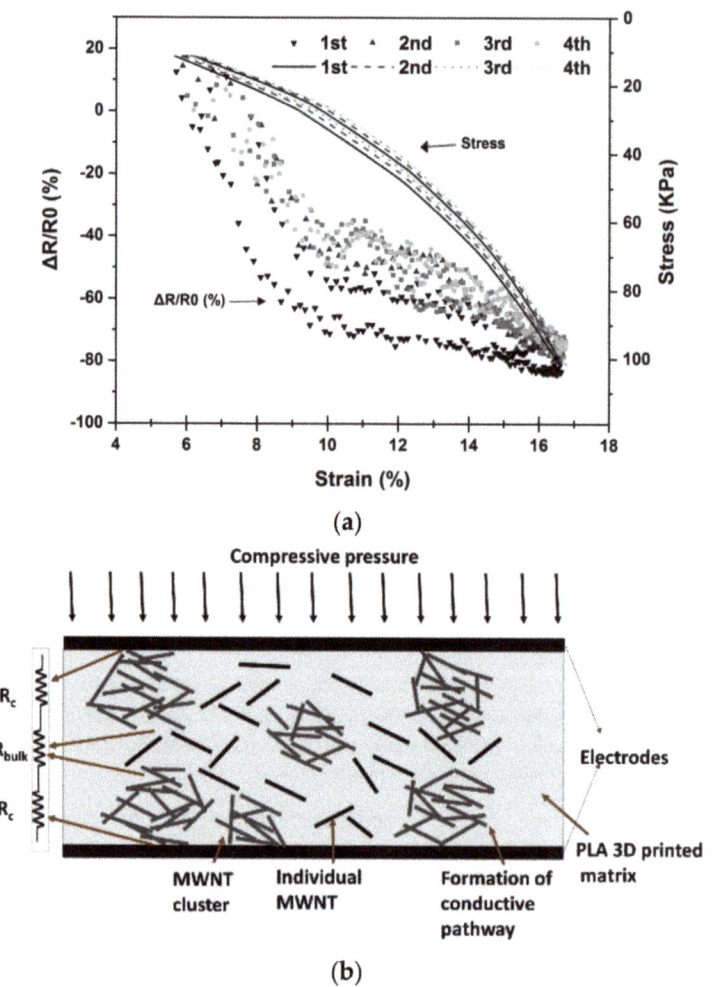

**Figure 6.** Piezoresistive behavior of 3D-printed PLA/1 wt.% MWNT nanocomposite: (**a**) Synchronism of $A_r$ with stress versus deformation. (**b**) Schematic diagram of the transduction mechanism of PLA/1 wt.% MWNT nanocomposite sandwiched between two metal electrodes towards compressive pressure. The electrical model of the FSR consists of a series of connections between the bulk (tunneling) resistance ($R_{bulk}$) and the contact resistance ($R_c$).

When the sample is subjected to external stress ($\sigma$), the inter-particle distance is reduced. According to the proposed model by Paredes-Madrid et al. [46], the total resistance across the Force Sensing Resistor (FSR) can be decomposed from Equation (3):

$$R_{FSR} = R_{bulk} + 2R_c \qquad (3)$$

where $R_{bulk}$ is the resistance of the CPC caused by the quantum tunneling phenomenon and $R_c$ is the contact resistance between the conductive particles and the metal electrodes. An FSR is created by the series connection between $R_{bulk}$ and $2R_c$ as shown in Figure 6b. However, three phenomena occur when incremental stress is applied to an FSR [46]: (1) the contact resistance of the existing paths is decreased according to power laws; (2) new contact paths are constructed to a greater extent contributing to a decrease in the contact resistance;

and (3) the average inter-particle distance is decreased, as a consequence decreasing the tunneling resistance, $R_{bulk}$. It seems that for the 1 wt.% MWNT 3D-printed layer, the contact resistance is decreased by forming new contact paths and decreasing the contact resistance of the existing paths under 30 KPa and 10% strain. However, over 30 KPa, the resistance decreases because of the diminishing of inter-particle distance and consequently decreasing the tunneling resistance. The piezoresistive coefficient, which is also called the gauge factor, G, can be graphically figured out from the slope of the curve in Figure 6 and calculated with Equation (4) [40]

$$G = \frac{A_r}{\varepsilon} \quad (4)$$

where $A_r = \frac{\Delta R}{R_0}$ (Equation (1)) is the piezoresistive response and $\varepsilon = \frac{\Delta L}{L_0}$ is the deformation of the sensor. For the 3D-printed PLA/1wt.% MWNT layers, the value of G = 7.6 was obtained with the amplitude of the piezoresistive response of about $A_r = -0.8$ (−80%).

To find out more about the sensitivity limitations of the developed FSR such as stress rate and related strain, the piezoresistive response of the 3D-printed composite layers under cyclic compressive stress with a high speed of 18 N/min was observed. Figure 7 shows the piezoresistive responses of the PLA/1 wt.% MWNT and PLA/4 wt.% KB samples, exposed to ten cycles of compressive stress from 10 up to 100 kPa.

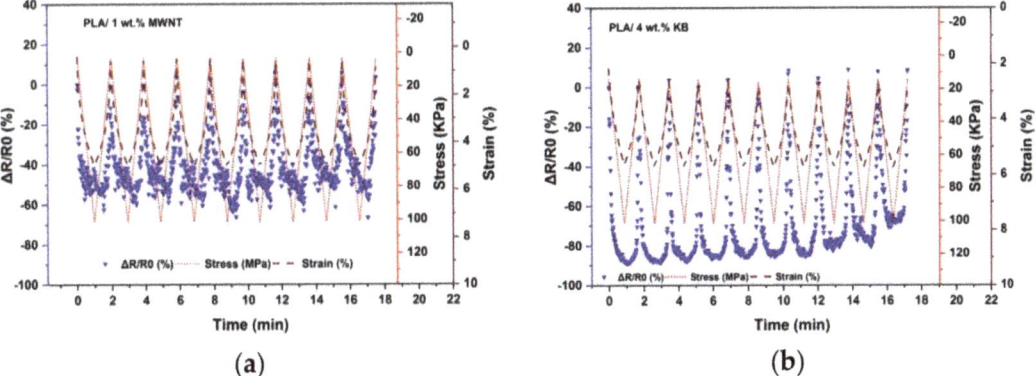

**Figure 7.** Comparison of 3D-printed nanocomposites piezoresistive responses in high force rate of 18 N/min (a) PLA/1 wt.% MWNT (b) PLA/4 wt.% KB.

Figure 7a shows that the piezoresistive response of the PLA/1 wt.% MWNT with an applied force rate of 18 N/min has a smaller amplitude ($A_r = -0.60$) and more noisy signals, but a higher value of G = 9.3 in comparison with low force rate of 1 N/min in Figure 4a. Figure 8 shows that applying a high force rate causes a smaller strain as there is insufficient time for the material to respond to stress with large-scale viscoelastic deformation or yielding [48].

In Figure 7b, the compression with a high force rate leads to responses of large sensitivity ($A_r = -0.90$) and exemption of noise for the PLA/4 wt.% KB 3D-printed layers. However, after almost seven cycles, the maximum sensitivity is not stable and starts to decrease. The high value of G = 47.6 in the first cycle shows the high piezoresistive properties of these layers if the cyclic functionality is not needed. The gauge factor decreases to G = 28 in the 10th cycle of stress. Therefore, the PLA/KB 3D-printed layers do not show stable piezoresistive behavior in a cyclic mode at low and high force rates.

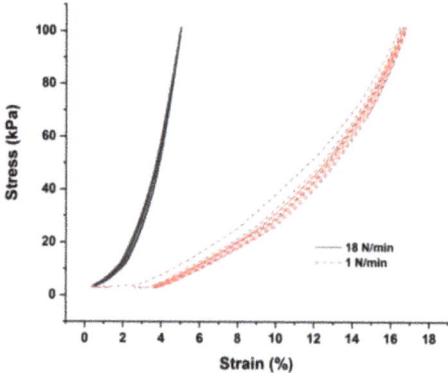

**Figure 8.** Stress–strain diagrams of 3D-printed PLA/1 wt.% MWNT with different force rates in a cyclic mode.

## 4. Conclusions

In this paper, fused deposition modeling 3D printing is used to develop CPC layers and investigate their electrical and piezoresistive behaviors. To this aim, multi-walled carbon nanotubes (MWNT) and high-structured carbon black (Ketjenblack) (KB) were incorporated into polylactic acid and 3D-printable filaments using the melt-mixing process were developed. The 3D-printed layers were created using fused deposition modeling. The percolation threshold of the MWNT and KB 3D-printed layers are 1 wt.% and 4 wt.%, respectively, by the two-point resistance measurement method. It was shown that it was possible to 3D print piezoresistive PLA nanocomposite layers from the MWNT and KB fillers and PLA matrix. The PLA/1 wt.% MWNT 3D-printed layers with a 1 mm thickness exhibit a negative pressure coefficient (NPC) characterized by a decrease of about one decade in resistance with increasing compressive loadings up to 18 N with a maximum strain up to about 16%. In the cyclic mode with a 1 N/min force rate, the PLA/1 wt.% MWNT 3D-printed layers showed good performance with a value of $G = 7.6$ obtained with the amplitude of the piezoresistive response of about $A_r = -0.8$ ($-80\%$). The response was linear in the range of pressure 10–100 kPa, with low noise and hysteresis that comes from the layer-by-layer architecture of the component and the tunneling effect of MWNT nanofillers in lower contents than the percolation threshold. At a high force rate of 18 N/min, the piezoresistive response of the PLA/1 wt.% MWNT has a smaller amplitude ($A_r = -0.60$) and more noisy signals but a value of $G = 9.3$. The KB composites could not show stable piezoresistive responses in a cyclic mode. However, the PLA/4 wt.% KB 3D-printed layers under high force rate compression lead to responses of large sensitivity ($A_r = -0.90$) and are exempt from noise with a high value of $G = 47.6$ in the first cycle. The results show that PLA/MWNT and PLA/KB can be considered good piezoresistive nanocomposites to be 3D printed where complex designs with functionality are needed for possible use in wearable electronics, soft robotics, and prosthetics, etc.

**Author Contributions:** Conceptualization, R.H.S., A.C., V.N. and C.C.; Data curation, R.H.S.; Formal analysis, R.H.S.; Funding acquisition, J.G., G.C., V.N. and C.C.; Investigation, R.H.S.; Methodology, R.H.S., A.C. and C.C.; Project administration, V.N. and C.C.; Resources, V.N. and C.C.; Supervision, A.C., J.G., G.C., V.N. and C.C.; Validation, R.H.S., A.C. and C.C.; Visualization, R.H.S.; Writing—original draft, R.H.S.; Writing—review and editing, R.H.S., A.C. and C.C. All authors have read and agreed to the published version of the manuscript.

**Funding:** This research was funded by [Erasmus Mundus Joint Doctorate Programme SMDTex—Sustainable Management and Design for Textile] grant number [n°2014-0683/001-001-EMJD] and the APC was funded by [University of Borås, Sweden].

**Institutional Review Board Statement:** Not applicable.

**Informed Consent Statement:** Not applicable.

**Data Availability Statement:** Not applicable.

**Conflicts of Interest:** The authors declare no conflict of interest.

## References

1. Tee, Y.L.; Tran, P.; Leary, M.; Pille, P.; Brandt, M. 3D Printing of polymer composites with material jetting: Mechanical and fractographic analysis. *Addit. Manuf.* **2020**, *36*, 101558. [CrossRef]
2. Mehrpouya, M.; Tuma, D.; Vaneker, T.; Afrasiabi, M.; Advanced, M.B.; Lab, M.; Zurich, E.; Gibson, I. Multimaterial powder bed fusion techniques. *Rapid Prototyp. J.* **2022**, *28*, 1–19. [CrossRef]
3. Boyle, B.M.; Xiong, P.T.; Mensch, T.E.; Werder, T.J.; Miyake, G.M. 3D printing using powder melt extrusion. *Addit. Manuf.* **2019**, *29*, 100811. [CrossRef] [PubMed]
4. Pohanka, M. Three-dimensional printing in analytical chemistry: Principles and applications. *Anal. Lett.* **2016**, *49*, 2865–2882. [CrossRef]
5. Ahn, D.-G. Directed energy deposition (DED) process: State of the art. *Int. J. Precis. Eng. Manuf. Technol.* **2021**, *8*, 703–742. [CrossRef]
6. Bagheri, A.; Jin, J. Photopolymerization in 3D Printing. *ACS Appl. Polym. Mater.* **2019**, *1*, 593–611. [CrossRef]
7. Prolongo, S.G.; Moriche, R.; Jiménez-Suárez, A.; Delgado, A.; Ureña, A. Printable self-heating coatings based on the use of carbon nanoreinforcements. *Polym. Compos.* **2020**, *41*, 271–278. [CrossRef]
8. Mostafaei, A.; Elliott, A.M.; Barnes, J.E.; Li, F.; Tan, W.; Cramer, C.L.; Nandwana, P.; Chmielus, M. Binder jet 3D printing—Process parameters, materials, properties, modeling, and challenges. *Prog. Mater. Sci.* **2021**, *119*, 100707. [CrossRef]
9. Xu, Y.; Wu, X.; Guo, X.; Kong, B.; Zhang, M.; Qian, X.; Mi, S.; Sun, W.; Lindquist, N.; Wittenberg, N.; et al. The boom in 3D-printed sensor technology. *Sensors* **2017**, *17*, 1166. [CrossRef]
10. Dudek, P. FDM 3D printing technology in manufacturing composite elements. *Arch. Metall. Mater.* **2013**, *58*, 10–13. [CrossRef]
11. Melnikova, R.; Ehrmann, A.; Finsterbusch, K. 3D printing of textile-based structures by Fused Deposition Modelling (FDM) with different polymer materials. *IOP Conf. Ser. Mater. Sci. Eng.* **2014**, *62*, 12018. [CrossRef]
12. Espalin, D.; Muse, D.W.; MacDonald, E.; Wicker, R.B. 3D Printing multifunctionality: Structures with electronics. *Int. J. Adv. Manuf. Technol.* **2014**, *72*, 963–978. [CrossRef]
13. MacDonald, E.; Wicker, R. Multiprocess 3D printing for increasing component functionality. *Science* **2016**, *353*, aaf2093. [CrossRef]
14. Kesner, S.B.; Howe, R.D. Design principles for rapid prototyping forces sensors using 3-D printing. *IEEE/ASME Trans. Mechatron.* **2011**, *16*, 866–870. [CrossRef] [PubMed]
15. Muth, J.T.; Vogt, D.M.; Truby, R.L.; Mengüç, Y.; Kolesky, D.B.; Wood, R.J.; Lewis, J.A. Embedded 3D printing of strain sensors within highly stretchable elastomers. *Adv. Mater.* **2014**, *26*, 6307–6312. [CrossRef]
16. Igrec, B.; Bosiljevac, M.; Sipus, Z.; Babic, D.; Rudan, S. Fiber optic vibration sensor for high-power electric machines realized using 3D printing technology. In Proceedings of the Photonic Instrumentation Engineering III, San Francisco, CA, USA, 17–18 February 2016; Volume 9754. [CrossRef]
17. Haque, R.I.; Ogam, E.; Loussert, C.; Benaben, P.; Boddaert, X. Fabrication of capacitive acoustic resonators combining 3D printing and 2D inkjet printing techniques. *Sensors* **2015**, *15*, 26018–26038. [CrossRef]
18. Saari, M.; Xia, B.; Cox, B.; Krueger, P.S.; Cohen, A.L.; Richer, E. Fabrication and analysis of a composite 3D Printed capacitive force sensor. *3D Print. Addit. Manuf.* **2016**, *3*, 137–141. [CrossRef]
19. Suaste-Gómez, E.; Rodríguez-Roldán, G.; Reyes-Cruz, H.; Terán-Jiménez, O. Developing an ear prosthesis fabricated in polyvinylidene fluoride by a 3D printer with sensory intrinsic properties of pressure and temperature. *Sensors* **2016**, *16*, 332. [CrossRef]
20. Krachunov, S.; Casson, A.J. 3D Printed Dry EEG Electrodes. *Sensors* **2016**, *16*, 1635. [CrossRef]
21. Hamzah, H.H.; Keattch, O.; Yeoman, M.S.; Covill, D.; Patel, B.A. Three-dimensional-printed electrochemical sensor for simultaneous dual monitoring of serotonin overflow and circular muscle contraction. *Anal. Chem.* **2019**, *91*, 12014–12020. [CrossRef]
22. Barbosa, J.R.; Amorim, P.H.O.; Mariana, M.C.; Dornellas, R.M.; Pereira, R.P.; Semaan, F.S. Evaluation of 3D printing parameters on the electrochemical performance of conductive polymeric components for chemical warfare agent sensing. In *Smart Innovation, Systems and Technologies*; Rocha, Á., Pereira, R., Eds.; Springer: Singapore, 2020. [CrossRef]
23. Kennedy, Z.C.; Christ, J.F.; Evans, K.A.; Arey, B.W.; Sweet, L.E.; Warner, M.G.; Erikson, R.L.; Barrett, C.A. 3D-printed poly(vinylidene fluoride)/carbon nanotube composites as a tunable, low-cost chemical vapour sensing platform. *Nanoscale* **2017**, *9*, 5458–5466. [CrossRef]
24. Leigh, S.J.; Bradley, R.J.; Purssell, C.P.; Billson, D.R.; Hutchins, D.A. A simple, low-cost conductive composite material for 3D printing of electronic sensors. *PLoS ONE* **2012**, *7*, e49365. [CrossRef] [PubMed]
25. Kim, K.; Park, J.; Suh, J.H.; Kim, M.; Jeong, Y.; Park, I. 3D printing of multiaxial force sensors using carbon nanotube (CNT)/thermoplastic polyurethane (TPU) filaments. *Sens. Actuators A Phys.* **2017**, *263*, 493–500. [CrossRef]
26. Christ, J.F.; Aliheidari, N.; Ameli, A.; Pötschke, P. 3D printed highly elastic strain sensors of multiwalled carbon nanotube/thermoplastic polyurethane nanocomposites. *Mater. Des.* **2017**, *131*, 394–401. [CrossRef]

27. Gul, J.Z.; Sajid, M.; Choi, K.H. 3D printed highly flexible strain sensor based on TPU-graphene composite for feedback from high speed robotic applications. *J. Mater. Chem. C* **2019**, *8*, 2597. [CrossRef]
28. Azimi, P.; Zhao, D.; Pouzet, C.; Crain, N.E.; Stephens, B. Emissions of ultrafine particles and volatile organic compounds from commercially available desktop three-dimensional printers with multiple filaments. *Environ. Sci. Technol.* **2016**, *50*, 1260–1268. [CrossRef]
29. Sathies, T.; Senthil, P.; Prakash, C. Application of 3D printed PLA-carbon black conductive polymer composite in solvent sensing. *Mater. Res. Express* **2019**, *6*, 115349. [CrossRef]
30. Thangavel, S.; Ponnusamy, S. Application of 3D printed polymer composite as capacitive sensor. *Sens. Rev.* **2019**, *40*, 54–61. [CrossRef]
31. Cardoso, R.M.; Silva, P.R.L.; Lima, A.P.; Rocha, D.P.; Oliveira, T.C.; do Prado, T.M.; Fava, E.L.; Fatibello-Filho, O.; Richter, E.M.; Muñoz, R.A.A. 3D-Printed graphene/polylactic acid electrode for bioanalysis: Biosensing of glucose and simultaneous determination of uric acid and nitrite in biological fluids. *Sens. Actuators B Chem.* **2020**, *307*, 127621. [CrossRef]
32. Kalinke, C.; Neumsteir, N.V.; de Oliveira Aparecido, G.; de Barros Ferraz, T.V.; Dos Santos, P.L.; Janegitz, B.C.; Bonacin, J.A. Comparison of activation processes for 3D printed PLA-graphene electrodes: Electrochemical properties and application for sensing of dopamine. *Analyst* **2020**, *145*, 1207–1218. [CrossRef]
33. Foster, C.W.; Elbardisy, H.M.; Down, M.P.; Keefe, E.M.; Smith, G.C.; Banks, C.E. Additively manufactured graphitic electrochemical sensing platforms. *Chem. Eng. J.* **2020**, *381*, 122343. [CrossRef]
34. Hashemi Sanatgar, R.; Campagne, C.; Nierstrasz, V. Investigation of the adhesion properties of direct 3D printing of polymers and nanocomposites on textiles: Effect of FDM printing process parameters. *Appl. Surf. Sci.* **2017**, *403*, 551–563. [CrossRef]
35. Chen, S.; Zhuo, B.; Guo, X. Large area one-step facile processing of microstructured elastomeric dielectric film for high sensitivity and durable sensing over wide pressure range. *ACS Appl. Mater. Interfaces* **2016**, *8*, 20364–20370. [CrossRef] [PubMed]
36. Trung, T.Q.; Lee, N.E. Flexible and stretchable physical sensor integrated platforms for wearable human-activity monitoring and personal healthcare. *Adv. Mater.* **2016**, *28*, 4338–4372. [CrossRef]
37. Tessarolo, M.; Possanzini, L.; Campari, E.G.; Bonfiglioli, R.; Violante, F.S.; Bonfiglio, A.; Fraboni, B. Adaptable pressure textile sensors based on a conductive polymer. *Flex. Print. Electron.* **2018**, *3*, 034001. [CrossRef]
38. Mannsfeld, S.C.B.; Tee, B.C.-K.; Stoltenberg, R.M.; Chen, C.V.H.-H.; Barman, S.; Muir, B.V.O.; Sokolov, A.N.; Reese, C.; Bao, Z. Highly sensitive flexible pressure sensors with microstructured rubber dielectric layers. *Nat. Mater.* **2010**, *9*, 859–864. [CrossRef]
39. Dios, J.R.; Garcia-Astrain, C.; Gonçalves, S.; Costa, P.; Lanceros-Méndez, S. Piezoresistive performance of polymer-based materials as a function of the matrix and nanofiller content to walking detection application. *Compos. Sci. Technol.* **2019**, *181*, 107678. [CrossRef]
40. Tung, T.T.; Robert, C.; Castro, M.; Feller, J.F.; Kim, T.Y.; Suh, K.S. Enhancing the sensitivity of graphene/polyurethane nanocomposite flexible piezo-resistive pressure sensors with magnetite nano-spacers. *Carbon* **2016**, *108*, 450–460. [CrossRef]
41. Toprakci, H.A.K. Piezoresistive Properties of Polyvinyl Chloride Composites. Ph.D. Thesis, North Carolina State University, Raleigh, NC, USA, 2012.
42. So, H.; Woo, J.; Kwon, J.; Yun, J.; Baik, S.; Seok, W. Carbon nanotube based pressure sensor for flexible electronics. *Mater. Res. Bull.* **2013**, *48*, 5036–5039. [CrossRef]
43. Slobodian, P.; Riha, P.; Lengalova, A.; Saha, P. Compressive stress-electrical conductivity characteristics of multiwall carbon nanotube networks. *J. Mater. Sci* **2011**, *46*, 3186–3190. [CrossRef]
44. Sanatgar, R.H.; Cayla, A.; Campagne, C.; Nierstrasz, V. Morphological and electrical characterization of conductive polylactic acid based nanocomposite before and after FDM 3D printing. *J. Appl. Polym. Sci.* **2019**, *136*, 47040. [CrossRef]
45. Gnanasekaran, K.; Heijmans, T.; van Bennekom, S.; Woldhuis, H.; Wijnia, S.; de With, G.; Friedrich, H. 3D printing of CNT- and graphene-based conductive polymer nanocomposites by fused deposition modeling. *Appl. Mater. Today* **2017**, *9*, 21–28. [CrossRef]
46. Paredes-Madrid, L.; Matute, A.; Bareño, J.O.; Vargas, C.A.P.; Velásquez, E.I.G. Underlying physics of conductive polymer composites and Force Sensing Resistors (FSRs). A study on creep response and dynamic loading. *Materials* **2017**, *10*, 1334. [CrossRef] [PubMed]
47. Slobodian, P.; Saha, P. Stress-strain hysteresis of a carbon nanotube network as polymer nanocomposite filler under cyclic deformation. *AIP Conf. Proc.* **2011**, *231*, 224–231. [CrossRef]
48. Hussein, M. Effects of strain rate and temperature on the mechanical behavior of carbon black reinforced elastomers based on butyl rubber and high molecular weight polyethylene. *Results Phys.* **2018**, *9*, 511–517. [CrossRef]

*Communication*

# Characterization of 3D Printed Metal-PLA Composite Scaffolds for Biomedical Applications

Irene Buj-Corral [1,*], Héctor Sanz-Fraile [2], Anna Ulldemolins [2], Aitor Tejo-Otero [1], Alejandro Domínguez-Fernández [1], Isaac Almendros [2,3] and Jorge Otero [2,3,*]

1. Department of Mechanical Engineering, School of Engineering of Barcelona (ETSEIB), Universitat Politècnica de Catalunya, Av. Diagonal 647, 08028 Barcelona, Spain; aitor.tejo@upc.edu (A.T.-O.); alejandro.dominguez-fernandez@upc.edu (A.D.-F.)
2. Unitat de Biofísica i Bioenginyeria, Facultat de Medicina i Ciències de la Salut, Universitat de Barcelona, 08036 Barcelona, Spain; hector.sanz.fraile@hotmail.com (H.S.-F.); anna.ulldemolins@ub.edu (A.U.); isaac.almendros@ub.edu (I.A.)
3. CIBER de Enfermedades Respiratorias, 28029 Madrid, Spain
\* Correspondence: irene.buj@upc.edu (I.B.-C.); jorge.otero@ub.edu (J.O.)

**Abstract:** Three-dimensional printing is revolutionizing the development of scaffolds due to their rapid-prototyping characteristics. One of the most used techniques is fused filament fabrication (FFF), which is fast and compatible with a wide range of polymers, such as PolyLactic Acid (PLA). Mechanical properties of the 3D printed polymeric scaffolds are often weak for certain applications. A potential solution is the development of composite materials. In the present work, metal-PLA composites have been tested as a material for 3D printing scaffolds. Three different materials were tested: copper-filled PLA, bronze-filled PLA, and steel-filled PLA. Disk-shaped samples were printed with linear infill patterns and line spacing of 0.6, 0.7, and 0.8 mm, respectively. The porosity of the samples was measured from cross-sectional images. Biocompatibility was assessed by culturing Human Bone Marrow-Derived Mesenchymal Stromal on the surface of the printed scaffolds. The results showed that, for identical line spacing value, the highest porosity corresponded to bronze-filled material and the lowest one to steel-filled material. Steel-filled PLA polymers showed good cytocompatibility without the need to coat the material with biomolecules. Moreover, human bone marrow-derived mesenchymal stromal cells differentiated towards osteoblasts when cultured on top of the developed scaffolds. Therefore, it can be concluded that steel-filled PLA bioprinted parts are valid scaffolds for bone tissue engineering.

**Keywords:** steel-filled PLA; FFF; scaffold; grid structure; cell culture

## 1. Introduction

Additive manufacturing (AM) is a group of techniques in which three-dimensional structures are manufactured layer-by-layer in an automated way. It offers several advantages over the traditional subtractive or forming techniques: (1) it allows manufacturing complex shapes and even porous structures, (2) cheaper parts are produced if low-cost machines are employed; (3) it implies material, waste, and energy savings. Within the AM field, there are seven different categories [1]: binder jetting (BJ) [2], directed energy deposition (DED) [3], material extrusion (includes FFF—Fused Filament Fabrication and DIW—Direct Ink Writing) [4,5], material jetting (MJ) [6], powder bed fusion (PBF) (includes SLM—Selective Laser Melting and SLS—Selective Laser Sintering) [7], sheet lamination [8], and vat photopolymerization (includes SLA—stereolithography and Digital Light Processing (DLP) printing, as well as volumetric 3D printing) [9,10].

Within the different AM techniques, FFF is one of the most widely used technologies for rapid prototyping within the biomedical field, as it presents several advantages in terms of costs and the range of materials that can be used [11]. Also known as FDM (Fused

Deposition Modelling), FFF uses a continuous filament of a thermoplastic material such as PolyLactic Acid (PLA) to build complex 3D structures in an automated way. One of the main disadvantages of FFF technology is the difficulty to ensure the correct bonding between layers [12]. The FFF 3D printing technique was patented in 1989 [13] and it bloomed up after its patent expired in 2009. After the technique became available to the general public, it has been used in different fields: automation, aeronautics, medicine, etc. Regarding the biomedical area, different applications could be highlighted, such as implants [14,15], 3D surgical planning prototypes [16,17], scaffolding [18,19], and regeneration of tissues [20].

There are numerous materials in the market for FFF 3D printing, for example acrylonitrile butadiene styrene (ABS) or Nylon, being PLA one of the most widely used, both alone and in combination with other materials such as wood, metals, or ceramics. Very little data have so far been published on systematic studies regarding the use of metal-filled filaments, since selecting compatible filler materials for the sake of improving the performance of polymeric composite materials is a difficult task [21]. In the present study, copper-, bronze-, and steel-filled PLA filaments are studied. Copper has an excellent heat and electric conductivity, it is easy to machine, bio-fouling resistant, and corrosion resistant [22]. Bronze alloy consists primarily of Cu, commonly with between 12 and 12.5% of Sn. It is a ductile alloy. Stainless steel is made of iron with typically a few tenths of carbon percentage, and with anti-corrosion elements such as Ni or Cr. It has high tensile strength, high corrosion resistance, and high biocompatibility. Therefore, it is used in a wide range of biomedical applications such as prostheses.

Regarding the mechanical properties of the metal-filled filaments, in some cases, increasing the metal content reduces the tensile strength and increases the thermal conductivity of the composite material studied. For example, Mohammadizadeh et al. [23] manufactured PLA filaments that contained copper, bronze, stainless steel, high carbon iron, and aluminum powders. They stated that the mechanical proper4ies of copper-filled-PLA were worse than those of PLA 3D printed parts. Additionally, they showed that the larger the layer height was, the lower the tensile strength, elastic modulus, and yield stress were. On the contrary, in different works, the mechanical properties were observed to increase when adding metals to the base polymer. Liu et al. [24] found that ceramic, copper, and aluminum-based PLA composite parts had similar or even superior mechanical properties when compared to bare PLA-made parts. Fafenrot et al. [25] developed polymer-metal materials 3D printed by FFF and concluded that the mechanical properties were similar to those of the PLA parts. On the other hand, there are other available options for the polymer matrix such as the use of ceramics. For instance, glass fiber-reinforced PLA can be employed in a wide range of applications, particularly in the biomedical, energy, and electronics industry [26]. In another example, Mahmoud et al. [27] studied the incorporation of two carbon fillers into the polypropylene: carbon nanotubes and synthetic graphite. The results showed that graphite-filled composites are more conductive than carbon nanotubes-filled composites. The flexural and tensile strength for both composites increased with the increase in the filler materials weight percentage. Later, the same authors [28] showed that flame-retardant MPP (melamine polyphosphate) had remarkable effects on the mechanical properties of the LLDPE (low-density polyethylene) composites. Five weight percentages of MPP were embedded into LLDPE, ranging from 5 to 30 wt%. It was concluded that the Young's modulus increased, and the tensile break strength and the tensile yield strength increased monotonically with the increase in MPP content.

The addition of metal components to the polymers used in FFF printing opens a new world in different fields such as bioengineering, but more knowledge needs to be obtained on the optimization of the production of these biomaterials for the fabrication of novel scaffolds. Although there are some studies about cell growth on 3D printed ceramic zirconia toughened alumina (ZTA) scaffolds [29], few and non-concluding studies have been done with metal-filled polymeric materials. Moreover, previous studies have not focused on the biological response of cells in metal-PLA 3D printed parts, as cells were cultured on metal-based scaffolds, such as titanium 3D printed bases [30]. On the other hand, several studies have evaluated the cytocompatibility of 3D printed iron-based scaffolds for bone regeneration [11,31,32] so we hypothesize that composite polymers incorporating metals will be appropriate for cell growth, since metals and alloys have been used extensively as bone substitutes [33,34]. These iron-based alloys have better mechanical properties than those based on lighter materials, such as magnesium.

Porosity is another key parameter that must be taken into consideration during the design and synthesis of a biomaterial [35]. The 3D printed porous materials should ideally fulfil conditions such as biocompatibility, noninflammatory response, tunable biodegradability, appropriate mechanic properties, defined pore structure, and, above all, promote a health improvement [36].

This work presents the characterization of three metal-reinforced PLA biomaterials for 3D printing biomedical scaffolds regarding porosity, surface roughness and cell culture. For that purpose, first the surface of the parts was analyzed, and the line spacing was measured. Surface roughness was then measured on the upper surface of the specimens. Regarding biological characterization, human-derived bone marrow mesenchymal stromal cells (hBM-MSC) were cultured on the composite scaffolds to assess their biocompatibility and the effect of the 3D printing scaffolds on determining cell fate, specifically in osteogenic differentiation.

## 2. Materials and Methods

### 2.1. Materials

The materials used in the present study were metal-filled PLA filaments of 2.85 mm diameter manufactured by ColorFabb (Belfeld, Netherlands). The specific materials used were: (1) steel-filled PLA, (2) bronze-filled PLA, and (3) copper-filled PLA. As stated by the manufacturer, these materials were developed for aesthetic purposes and need a polishing treatment after being printed if a brilliant appearance is to be required. All reagents were purchased from Sigma-Aldrich (Saint Louis, MO, USA), unless specified otherwise.

### 2.2. 3D Printing Process

Parts were additively manufactured by using a Sigma R19 3D printer (BCN3D Technologies, Gavà, Spain). The 3D printing parameters are presented in Table 1. Figure 1 shows a scheme of an FFF 3D printer. Cura BCN3D software was used to generate the G-code that is required to print the parts.

**Table 1.** 3D printing parameters.

| Variable | Value |
|---|---|
| Infill pattern | Linear |
| Layer height (mm) | 0.15 |
| Nozzle diameter (mm) | 0.4 |
| Print speed (mm/s) | 7 |
| Extrusion multiplier (%) | 100 |
| Temperature (°C) | 190 |

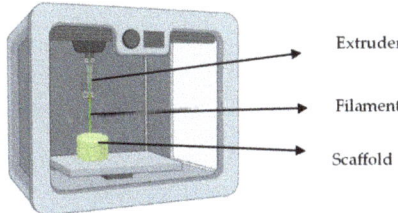

**Figure 1.** FFF 3D printer scheme.

Disk-shaped samples of 6 mm in diameter and 2 mm in height were manufactured. The linear infill pattern was selected, using three different line spacing values: 0.6 mm, 0.7 mm, and 0.8 mm (Figure 2). The shell width was set to 0.4 mm and the bottom width was set to 1.2 mm. No top layer was used.

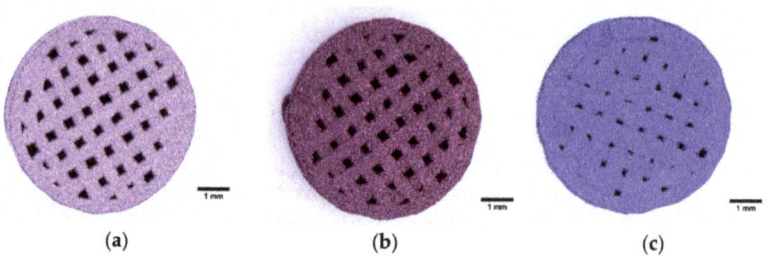

(a)                    (b)                    (c)

**Figure 2.** Disk-shaped 3D printed samples with line spacing of 0.7 mm of (a) bronze-filled PLA, (b) copper-filled PLA, (c) steel-filled PLA. Scale bars correspond to 1 mm.

*2.3. Pictures*

Pictures of the 3D printed scaffolds were obtained by using a Leica S8AP0 binocular magnifier (Leica Camera AG, Wetzlar, Germany) with 8× (Figure 2) and 16× (Figure 3) magnification, respectively.

*2.4. Porosity*

The porosity of the disk samples was quantified from the images of the cross section of the scaffolds, assuming that the length of the pores corresponds to the length of the sample and using Equation (1):

$$Pt = \frac{Vp}{Vt} \qquad (1)$$

where $Pt$ is the porosity, $Vp$ the pore volume, and $Vt$ the total volume of the scaffold. The software used for cross-sectional images quantification was ImageJ.

*2.5. Roughness*

Roughness was measured with a Talysurf 2 contact roughness meter from Taylor Hobson Ltd., Leicester, UK. A diamond tip was used with tip angle of 90° and tip radius of 2 µm. The measuring force was 0.8 mN and speed was 0.5 mm/s. A Gaussian filter was employed. A cut-off value of 0.8 mm was used according to ISO 4288 [37]. Total sampling length was 4.8 mm (6 × 0.8 mm). Roughness was measured on the upper surface of the disks, along the generatrices of the filaments in the two perpendicular directions, in order to assess if there were differences regarding their surface finish. As an example, the steel-filled samples were measured with line spacing 0.6 and 0.7 mm. Two different samples were measured for each line spacing value.

The roughness parameters that were analyzed in this present study are:

- Arithmetical mean roughness value or arithmetical mean of the absolute values of the profile deviations from the mean line of the roughness profile (Ra) (Equation (2)), which is one of the most commonly employed parameters in industry;

$$Ra = \frac{1}{L} \int_0^L |Z(x)| dx \qquad (2)$$

- Mean roughness depth or average maximum peak to valley of five consecutive sampling lengths of the profile within a sampling length (Rz);
- Kurtosis (Rku), which is a measure of the sharpness of the profile (Equation (3)); and:

$$\text{Rku} = \frac{1}{R_q^4}\left[\frac{1}{L}\int_0^L Z^4(x)\,dx\right] \quad (3)$$

— Skewness (Rsk), which measures the symmetry of the profile (Equation (4)).

$$\text{Rsk} = \frac{1}{R_q^3}\left[\frac{1}{L}\int_0^L Z^3(x)\,dx\right] \quad (4)$$

These parameters are defined in the UNE-EN-ISO 4287:1999 standard [38].

*2.6. Human Bone Marrow-Derived Mesenchymal Stromal Cells Culture on the Developed Scaffolds*

Primary human Bone Marrow-Derived Mesenchymal Stromal cells (hBM-MSCs, ATCC PCS-500-012, ATCC, Manassas, VA, USA) were expanded following the manufacturers' instructions. Cells from passage 3–6 were used for all the experiments presented herein.

Printed parts were coated by incubating with rat tail-derived type I collagen at a concentration of 0.1 mg/mL for 30 min at 37 °C. hBM-MSCs were cultured on the scaffolds at a seeding density of $4 \times 10^4$ cells/cm$^2$ for 24 and 72 h. Cells seeded on uncoated parts were cultured in parallel. At the defined time points, cells were fixed with paraformaldehyde (PFA) 4% for further immunohistochemical imaging.

In a subsequent set of experiments, hBM-MSCs were cultured with a seeding density of $9.4 \times 10^4$ cells/cm$^2$ for 1, 4, and 7 days, respectively, on the different collagen-coated/uncoated samples and then fixed with PFA 4%.

Finally, to evaluate the impact of the developed scaffolds on the cell fate, hBM-MSCs were cultured with αMEM (12509069, Gibco, Waltham, MA, USA) 10% FBS (Gibco) on the steel-filled PLA scaffolds with 0.6, 0.7, and 0.8 mm line spacing, respectively, with a seeding density of $1.3 \times 10^4$ cells/cm$^2$. Control cells were cultured in parallel on conventional culture plates. After 21 days, the cells were fixed with PFA 4% for further analysis.

*2.7. Immunohistochemical Analysis*

After PFA fixation, samples were permeabilized with Triton 0.1%, blocked with 10% FBS solution, and incubated overnight at 4 °C with primary antibodies (anti-hOsteocalcein 967801, R&D Systems, Minneapolis, MN, USA) and subsequently, incubated for 2 h at 37 °C with the secondary Alexa 488 anti-rabbit antibody for differentiation studies. For morphology analysis, nuclei were stained with NucBlue (Thermo Scientific, Waltham, MA, USA) and actin cytoskeleton with phalloidin (Thermo Scientific, Waltham, MA, USA). Images were acquired with a Nikon D-Eclipse Ci confocal microscope (Nikon, Tokyo, Japan) with 10× and 20× Plan Apo objectives (Nikon).

## 3. Results and Discussion

*3.1. 3D Printed Samples*

Figure 3 shows the pictures of bronze-filled, copper-filled, and steel-filled samples manufactured with 0.6, 0.7, and 0.8 mm line spacing, respectively. Achieving 0.6 mm line spacing was more difficult than 0.8 mm, because of smaller pores. Despite that, appropriate scaffolds were achieved for the three different materials.

Additionally, scaffolds for biomedical applications should have a porous architecture. This porosity provides the necessary environment for promoting cell migration, proliferation, etc. [39].

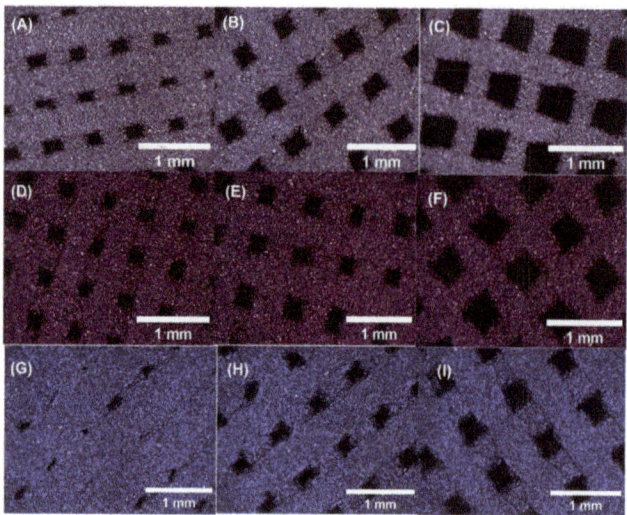

**Figure 3.** Surface of the samples with 16× magnification: (**A**) bronze-filled 0.6 mm, (**B**) bronze-filled 0.7 mm, (**C**) bronze-filled 0.8 mm, (**D**) copper-filled 0.6 mm. (**E**) copper-filled 0.7 mm, (**F**) copper-filled 0.7 mm, (**G**) steel-filled 0.6 mm, (**H**) steel-filled 0.7 mm, (**I**) steel-filled 0.8 mm. The scale bars correspond to 1 mm.

### 3.2. Porosity

As shown in Figure 4, the higher the line spacing, the higher the porosity of the 3D printed scaffolds is. Among the different scaffolds, for a certain line spacing value, most porous scaffolds are the bronze-filled ones, followed by the copper-filled ones, although they were manufactured with the same 3D printing conditions:

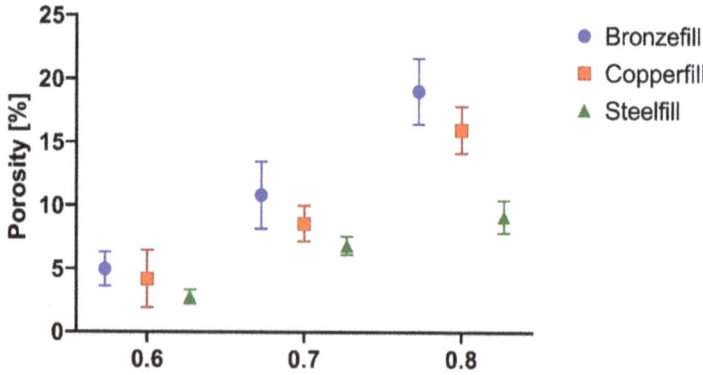

**Figure 4.** Porosity of the 3D printed scaffolds: bronze-filled, copper-filled and steel-filled. $N = 3$.

### 3.3. Roughness

Table 2 presents the roughness results for both the internal and the external surfaces of the 3D printed samples. One measurement was performed on each surface.

Table 2. Roughness on the external and internal layer of 3D printed steel-filled samples.

| Line Spacing | Sample | External Roughness | | | | Internal Roughness | | | |
|---|---|---|---|---|---|---|---|---|---|
| | | Ra (µm) | Rz (µm) | Rku | Rsk | Ra (µm) | Rz (µm) | Rku | Rsk |
| 0.6 mm | 1 | 25.36 | 128.90 | 3.12 | −1.00 | 7.99 | 38.67 | 2.91 | 0.24 |
| | 2 | 25.60 | 129.11 | 3.14 | −1.03 | 7.97 | 37.16 | 3.29 | 0.44 |
| 0.7 mm | 1 | 32.38 | 142.63 | 2.54 | −0.73 | 12.13 | 52.91 | 2.60 | −0.09 |
| | 2 | 35.04 | 158.71 | 2.73 | −0.80 | 15.68 | 78.31 | 2.89 | 0.04 |

Figure 5 depicts the roughness profiles of samples with a line spacing of 0.6 and 0.7 mm, on the external (first) and internal (second) layers, respectively, starting from the top of the part. Higher Ra values were obtained on the external (first) layer (Table 2 and Figure 5a,c) than on the internal (second) layer (Table 2 and Figure 5b,d). On the external layer, slightly higher Ra values (up to 35.04 µm) were found for line spacing 0.7 mm than for line spacing 0.6 mm (up to 25.60 µm). The Rz parameter shows a similar trend than Ra.

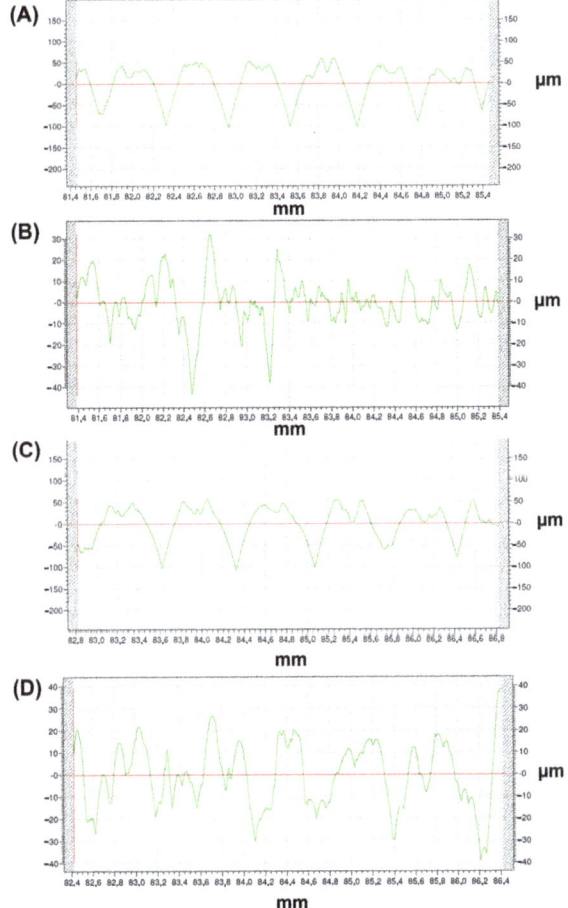

Figure 5. Roughness profiles of the steel-filled 3D printed samples. (A) 0.6 mm external roughness (first layer). (B) 0.6 mm internal roughness (second layer). (C) 0.7 mm external roughness (first layer). (D) 0.7 mm internal roughness (second layer).

Rku values around 3 were found in all cases, corresponding to a normal distribution of the roughness heights in each profile. On the external layer, slightly negative Rsk values were obtained, corresponding to longer valleys than crests. On the contrary, on the second layer Rsk values are close to 0, corresponding to symmetric profiles.

The external or first layers show more regular roughness profiles (Figure 5a,c) than the internal or second layers (Figure 5b,d).

### 3.4. Human Bone Marrow-Derived Mesenchymal Stromal Cells Cultured on the Developed Scaffolds

hBM-MSCs showed good adhesion to both collagen-coated and untreated 3D printed steel-filled PLA samples. Cells cultured for both 24 and 72 h were well-adhered to the 3D printed PLA composites (Figure 6a,b), with no observed differences between both conditions.

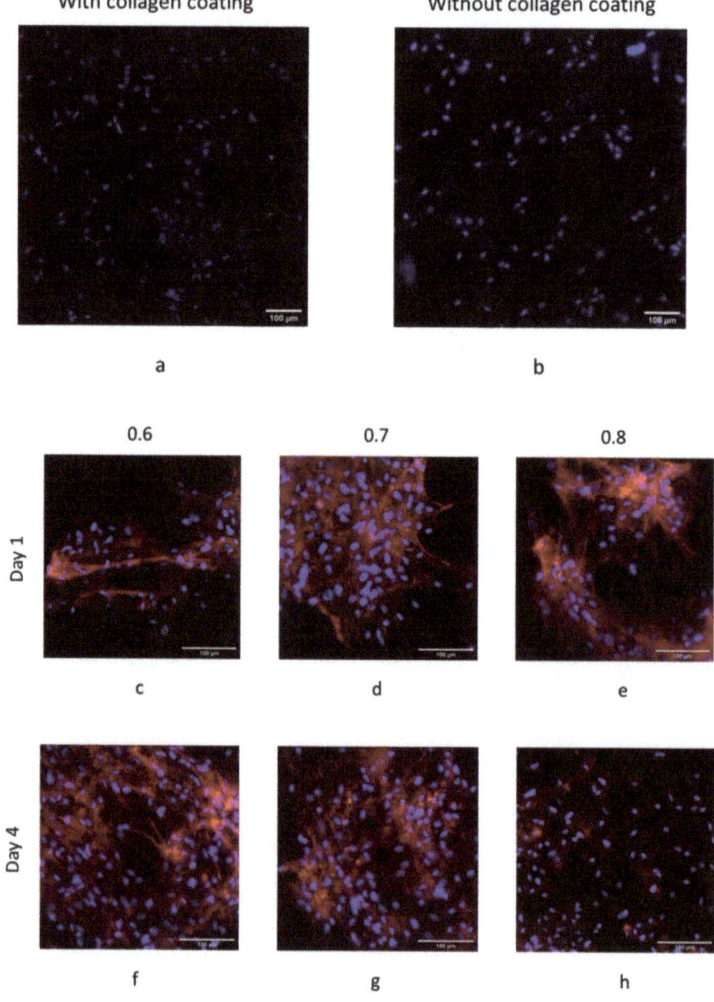

Figure 6. *Cont.*

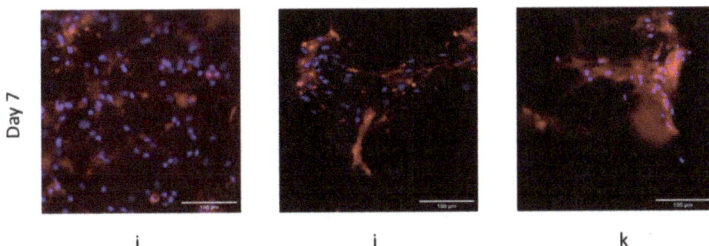

**Figure 6.** (**a**,**b**) Adhesion test to PLA with and without collagen coating. (**c**–**e**) Viability test at day 1 of cell culture. (**f**–**h**) Viability test at day 4 of cell culture. (**i**–**k**) Viability test at day 7 of cell culture.

Steel-filled PLA samples showed a very good cytocompatibility, especially on 0.6 and 0.7 line spacing (Figure 6c,d,f,g,i,j). On the contrary, copper-filled and bronze-filled PLA presented higher cytotoxicity since there were no cells adhered to the scaffolds after 24 h of culture (data not shown).

Steel-filled scaffolds showed high biocompatibility, unlike copper-filled and bronze-filled materials. This is in concordance with Kuroda et al. [39]. Additionally, the best biological behavior was found with the lowest porosity achieved (0.6 mm line spacing). This is in accordance with data presented by Chen et al. [40], who concluded that samples with 30% porosity exhibit the best biocompatibility, which were the lowest porosity scaffolds of their research.

The different scaffolds manufactured by means of FFF showed to have different behavior. As mentioned, bronze-filled as well as copper-filled scaffolds presented high cytotoxicity since there were no cells adhered to the scaffolds after 24 h of culture.

It is interesting to highlight that there were no differences observed when a specific protein coating was used in the parts. Cells form specific adhesions to the collagen protein while they are expected to form unspecific adhesions to uncoated materials. From the experiments presented herein, it can be concluded that steel-filled PLA promotes the formation of unspecific adhesion in MSCs, while this is not happening with copper- or bronze-filled polymers.

*3.5. hBM-MSCs Differentiated towards Osteoblasts When Cultured on the Developed Scaffolds*

hBM-MSCs cultured on the steel-filled 3D printed scaffolds (0.6 and 0.7 mm line spacing) without osteogenic supplements showed the presence of osteocalcin after 21 days of culture (Figure 6). Moreover, cells exhibited a broad spreading area compared with those cultured under conventional culture conditions (Figure 7).

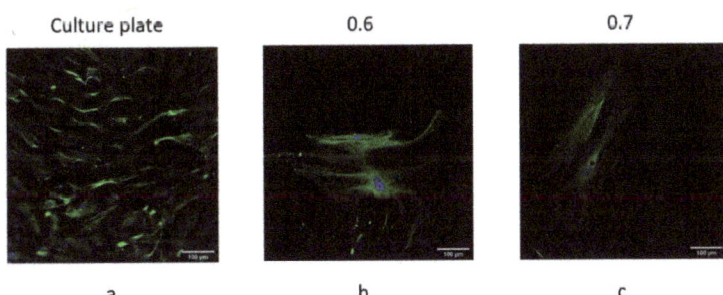

**Figure 7.** Osteocalcin detection by immunofluorescence in hBM-MSCs (at day 21) cultured on (**a**) conventional culture conditions, steel-filled PLA scaffolds of (**b**) 0.6 mm and (**c**) 0.7 mm line spacing. Osteocalcin (green) and nucleus (blue). Scale bars correspond to 100 μm.

The images show the classical spindle-like shape of MSCs, so it seems that the steel-filled 3D printed structures are a suitable scaffold for cell culturing. The enlarged phenotype is similar to differentiated osteoblasts when cultured on rigid substrates [41–43].

## 4. Conclusions

In the present work, results are presented for the cell growth of stem cells on metal-filled PLA composites that were printed with a grid structure by means of the FFF technique. Three different composites were tested: bronze, copper, and stainless steel, respectively. The main conclusions are as follows:

- Given a certain line spacing, higher porosity was observed for the copper-filled scaffolds than for the bronze-filled scaffolds and the steel-filled scaffolds, although they were 3D printed with similar printing conditions;
- Steel-filled composite showed important cell growth, both with and without protein coating, so it is promoting the formation of unspecific adhesions in MSCs;
- Neither bronze-filled nor copper-filled composites favored cell growth, so they cannot be considered to be biocompatible;
- When considering steel-filled composite, line spacing of 0.6 and 0.7 mm provided the best results, while line spacing of 0.8 mm is not recommended.

In future work, the effect of the use of other infill patterns on both cell growth and the mechanical strength of the structures will be addressed.

**Author Contributions:** Conceptualization, I.B.-C. and J.O.; methodology, A.U., H.S.-F., A.D.-F. and A.T.-O.; software, A.D.-F. and H.S.-F.; validation, I.B.-C. and J.O.; formal analysis, A.U., H.S.-F., I.A. and A.T.-O.; investigation, A.U., H.S.-F., I.A. and A.T.-O.; resources, I.A., I.B.-C. and J.O.; data curation, A.T.-O.; writing—original draft preparation, all; writing—review and editing, all; visualization, I.B.-C., A.D.-F. and J.O.; supervision, I.B.-C. and J.O.; project administration, I.B.-C. and J.O.; funding acquisition, I.B.-C., I.A. and J.O. All authors have read and agreed to the published version of the manuscript.

**Funding:** This research was co-funded by the European Union Regional Development Fund within the framework of the ERDF Operational Program of Catalonia 2014–2020 with a grant of 50% of total cost eligible, project BASE3D, grant number 001-P-001646. Research was also funded in part by the Spanish Ministry of Science, Innovation and Universities, grants numbers PID2019-108958RB-I00/AEI/10.13039/501100011033 and PGC2018-097323-A-I00.

**Institutional Review Board Statement:** Not applicable.

**Informed Consent Statement:** Not applicable.

**Data Availability Statement:** Data supporting the findings of this study are available from the corresponding authors upon reasonable request.

**Acknowledgments:** The authors thank Ramón Casado for his help with the experimental tests.

**Conflicts of Interest:** The authors declare no conflict of interest.

## References

1. *ISO/ASTM 52900:2021*; Additive Manufacturing—General Principles—Fundamentals and Vocabulary. International Organization for Standardization: Geneva, Switzerland, 2021; pp. 1–28.
2. Ziaee, M.; Crane, N.B. Binder jetting: A review of process, materials, and methods. *Addit. Manuf.* **2019**, *28*, 781–801. [CrossRef]
3. Ryu, D.J.; Sonn, C.-H.; Hong, D.H.; Kwon, K.B.; Park, S.J.; Ban, H.Y.; Kwak, T.Y.; Lim, D.; Wang, J.H. Titanium Porous Coating Using 3D Direct Energy Deposition (DED) Printing for Cementless TKA Implants: Does It Induce Chronic Inflammation? *Materials* **2020**, *13*, 472. [CrossRef] [PubMed]
4. Lewis, J.A.; Smay, J.E.; Stuecker, J.; Cesarano, J. Direct Ink Writing of Three-Dimensional Ceramic Structures. *J. Am. Ceram. Soc.* **2006**, *89*, 3599–3609. [CrossRef]
5. Tejo-Otero, A.; Colly, A.; Courtial, E.-J.; Fenollosa-Artés, F.; Buj-Corral, I.; Marquette, C.A. Soft-tissue-mimicking using silicones for the manufacturing of soft phantoms by fresh 3D printing. *Rapid Prototyp. J.* **2021**, *28*, 285–296. [CrossRef]
6. Tee, Y.L.; Tran, P.; Leary, M.; Pille, P.; Brandt, M. 3D Printing of polymer composites with material jetting: Mechanical and fractographic analysis. *Addit. Manuf.* **2020**, *36*, 101558. [CrossRef]

7. Murr, L.E. Metallurgy principles applied to powder bed fusion 3D printing/additive manufacturing of personalized and optimized metal and alloy biomedical implants: An overview. *Integr. Med. Res.* **2019**, *9*, 1087–1103. [CrossRef]
8. Bhatt, P.M.; Kabir, A.M.; Peralta, M.; Bruck, H.A.; Gupta, S.K. A robotic cell for performing sheet lamination-based additive manufacturing. *Addit. Manuf.* **2019**, *27*, 278–289. [CrossRef]
9. Zhu, Y.; Ramadani, E.; Egap, E. Thiol ligand capped quantum dot as an efficient and oxygen tolerance photoinitiator for aqueous phase radical polymerization and 3D printing under visible light. *Polym. Chem.* **2021**, *12*, 5106–5116. [CrossRef]
10. Shi, X.; Zhang, J.; Corrigan, N.A.; Boyer, C. Controlling mechanical properties of 3D printed polymer composites through photoinduced reversible addition–fragmentation chain transfer (RAFT) polymerization. *Polym. Chem.* **2021**, *13*, 44–57. [CrossRef]
11. Bozkurt, Y.; Karayel, E. 3D printing technology; methods, biomedical applications, future opportunities and trends. *J. Mater. Res. Technol.* **2021**, *14*, 1430–1450. [CrossRef]
12. Vanaei, H.R.; Khelladi, S.; Deligant, M.; Shirinbayan, M.; Tcharkhtchi, A. Numerical Prediction for Temperature Profile of Parts Manufactured Using Fused Filament Fabrication. *J. Manuf. Process.* **2022**, *76*, 548–558. [CrossRef]
13. Deckard, C.R. Method and Apparatus for Producing Parts by Selective Sintering. U.S. Patent No. 4,863,538, 5 September 1989.
14. Buj-Corral, I.; Bagheri, A.; Petit-Rojo, O. 3D Printing of Porous Scaffolds with Controlled Porosity and Pore Size Values. *Materials* **2018**, *11*, 1532. [CrossRef]
15. Bagheri, A.; Buj-Corral, I.; Ferrer, M.; Pastor, M.M.; Roure, F. Determination of the Elasticity Modulus of 3D-Printed Octet-Truss Structures for Use in Porous Prosthesis Implants. *Materials* **2018**, *11*, 2420. [CrossRef]
16. Tejo-Otero, A.; Buj-Corral, I.; Fenollosa-Artés, F. 3D Printing in Medicine for Preoperative Surgical Planning: A Review. *Ann. Biomed. Eng.* **2020**, *48*, 536–555. [CrossRef]
17. Tejo-Otero, A.; Lustig-Gainza, P.; Fenollosa-Artés, F.; Valls, A.; Krauel, L.; Buj-Corral, I. 3D printed soft surgical planning prototype for a biliary tract rhabdomyosarcoma. *J. Mech. Behav. Biomed. Mater.* **2020**, *109*, 103844. [CrossRef]
18. Polonio-Alcalá, E.; Rabionet, M.; Guerra, A.J.; Yeste, M.; Ciurana, J.; Puig, T. Screening of Additive Manufactured Scaffolds Designs for Triple Negative Breast Cancer 3D Cell Culture and Stem-like Expansion. *Int. J. Mol. Sci.* **2018**, *19*, 3148. [CrossRef]
19. Polonio-Alcalá, E.; Rabionet, M.; Gallardo, X.; Angelats, D.; Ciurana, J.; Ruiz-Martínez, S.; Puig, T. PLA Electrospun Scaffolds for Three-Dimensional Triple-Negative Breast Cancer Cell Culture. *Polymers* **2019**, *11*, 916. [CrossRef]
20. Nejad, Z.M.; Zamanian, A.; Saeidifar, M.; Vanaei, H.R.; Amoli, M.S. 3D Bioprinting of Polycaprolactone-Based Scaffolds for Pulp-Dentin Regeneration: Investigation of Physicochemical and Biological Behavior. *Polymers* **2021**, *13*, 4442. [CrossRef]
21. Yousry, M.; Zaghloul, M.; Mahmoud, M.; Zaghloul, Y.; Mahmoud, M.; Zaghloul, Y. Developments in polyester composite materials—An in-depth review on natural fibres and nano fillers. *Compos. Struct.* **2021**, *278*, 114698.
22. Fuseini, M.; Mahmoud, M.; Zaghloul, Y.; El-shazly, A.H. Evaluation of synthesized polyaniline nanofibres as corrosion protection film coating on copper substrate by electrophoretic deposition. *J. Mater. Sci.* **2022**, *57*, 6085–6101. [CrossRef]
23. Mohammadizadeh, M.; Lu, H.; Fidan, I.; Tantawi, K.; Gupta, A. Mechanical and Thermal Analyses of Metal-PLA. *Inventions* **2020**, *5*, 44. [CrossRef]
24. Liu, Z.; Lei, Q.; Xing, S. Mechanical characteristics of wood, ceramic, metal and carbon fiber-based PLA composites fabricated by FDM. *J. Mater. Res. Technol.* **2019**, *8*, 3743–3753. [CrossRef]
25. Fafenrot, S.; Grimmelsmann, N.; Wortmann, M.; Ehrmann, A. Three-Dimensional (3D) Printing of Polymer-Metal Hybrid Materials by Fused Deposition Modeling. *Materials* **2017**, *10*, 1199. [CrossRef]
26. Mahmoud, M.; Zaghloul, Y.; Mohamed, Y.S. Fatigue and tensile behaviors of fiber-reinforced thermosetting composites embedded with nanoparticles. *J. Compos. Mater.* **2019**, *53*, 709–718.
27. Mahmoud, M.; Zaghloul, Y.; Yousry, M.; Zaghloul, M.; Mahmoud, M.; Zaghloul, Y. Experimental and modeling analysis of mechanical-electrical behaviors of polypropylene composites filled with graphite and MWCNT fillers. *Polym. Test.* **2017**, *63*, 467–474.
28. Mahmoud, M.; Mahmoud, Y. Mechanical properties of linear low-density polyethylene fi re-retarded with melamine polyphosphate. *J. Appl. Polym. Sci.* **2018**, *135*, 46770.
29. Stanciuc, A.-M.; Sprecher, C.M.; Adrien, J.; Roiban, L.I.; Alini, M.; Gremillard, L.; Peroglio, M. Robocast zirconia-toughened alumina scaffolds: Processing, structural characterisation and interaction with human primary osteoblasts. *J. Eur. Ceram. Soc.* **2018**, *38*, 845–853. [CrossRef]
30. Rifai, A.; Tran, N.; Reineck, P.; Elbourne, A.; Mayes, E.L.H.; Sarker, A.; Dekiwadia, C.; Ivanova, E.P.; Crawford, R.J.; Ohshima, T.; et al. Engineering the Interface: Nanodiamond Coating on 3D-Printed Titanium Promotes Mammalian Cell Growth and Inhibits *Staphylococcus aureus* Colonization. *ACS Appl. Mater. Interfaces* **2019**, *11*, 24588–24597. [CrossRef]
31. Ma, H.; Li, T.; Huan, Z.; Zhang, M.; Yang, Z.; Wang, J.; Chang, J.; Wu, C. 3D printing of high-strength bioscaffolds for the synergistic treatment of bone cancer. *NPG Asia Mater.* **2018**, *10*, 31–44. [CrossRef]
32. Yang, C.; Huan, Z.; Wang, X.; Wu, C.; Chang, J. 3D Printed Fe Scaffolds with HA Nanocoating for Bone Regeneration. *ACS Biomater. Sci. Eng.* **2018**, *4*, 608–616. [CrossRef]
33. Hermawan, H. Updates on the research and development of absorbable metals for biomedical applications. *Prog. Biomater.* **2018**, *7*, 93–110. [CrossRef] [PubMed]
34. Han, H.-S.; Loffredo, S.; Jun, I.; Edwards, J.; Kim, Y.-C.; Seok, H.-K.; Witte, F.; Mantovani, D.; Glyn-Jones, S. Current status and outlook on the clinical translation of biodegradable metals. *Mater. Today* **2018**, *23*, 57–71. [CrossRef]

35. Ahumada, M.; Jacques, E.; Calderon, C.; Martínez-Gómez, F. Porosity in biomaterials: A key factor in the development of applied materials in biomedicine. *Handb. Ecomater.* **2019**, *5*, 3503–3522.
36. Chai, Q.; Jiao, Y.; Yu, X. Hydrogels for Biomedical Applications: Their Characteristics and the Mechanisms behind Them. *Gels* **2017**, *3*, 6. [CrossRef] [PubMed]
37. *ISO 4288:1996*; Geometrical Product Specifications (GPS)—Surface Texture: Profile Method—Rules and Procedures for the Assessment of Surface Texture. International Organization for Standardization: Geneva, Switzerland, 1996.
38. *ISO 4287:1997*; Geometrical Product Specifications (GPS)—Surface Texture: Profile Method—Terms, Definitions and Surface Texture Parameters. International Organization for Standardization: Geneva, Switzerland, 1997.
39. Zimina, A.; Senatov, F.; Choudhary, R.; Kolesnikov, E.; Anisimova, N.; Kiselevskiy, M.; Orlova, P.; Strukova, N.; Generalova, M.; Manskikh, V.; et al. Biocompatibility and Physico-Chemical Properties of Highly Porous PLA/HA Scaffolds for Bone Reconstruction. *Polymers* **2020**, *12*, 2938. [CrossRef]
40. Chen, Y.; Frith, J.E.; Dehghan-Manshadi, A.; Attar, H.; Kent, D.; Soro, N.D.M.; Bermingham, M.J.; Dargusch, M.S. Mechanical properties and biocompatibility of porous titanium scaffolds for bone tissue engineering. *J. Mech. Behav. Biomed. Mater.* **2017**, *75*, 169–174. [CrossRef]
41. Matta, C.; Szűcs-Somogyi, C.; Kon, E.; Robinson, D.; Neufeld, T.; Altschuler, N.; Berta, A.; Hangody, L.; Veréb, Z.; Zákány, R. Osteogenic differentiation of human bone marrow-derived mesenchymal stem cells is enhanced by an aragonite scaffold. *Differentiation* **2019**, *107*, 24–34. [CrossRef]
42. Mazzoni, E.; Mazziotta, C.; Iaquinta, M.R.; Lanzillotti, C.; Fortini, F.; D'Agostino, A.; Trevisiol, L.; Nocini, R.; Barbanti-Brodano, G.; Mescola, A.; et al. Enhanced Osteogenic Differentiation of Human Bone Marrow-Derived Mesenchymal Stem Cells by a Hybrid Hydroxylapatite/Collagen Scaffold. *Front. Cell Dev. Biol.* **2021**, *8*, 610570. [CrossRef]
43. Zhang, T.; Lin, S.; Shao, X.; Zhang, Q.; Xue, C.; Zhang, S.; Lin, Y.; Zhu, B.; Cai, X. Effect of matrix stiffness on osteoblast functionalization. *Cell Prolif.* **2017**, *50*, e12338. [CrossRef]

Article

# Modeling Impact Mechanics of 3D Helicoidally Architected Polymer Composites Enabled by Additive Manufacturing for Lightweight Silicon Photovoltaics Technology

Arief Suriadi Budiman [1,2,3,4,*], Rahul Sahay [3,5,*], Komal Agarwal [3], Rayya Fajarna [4], Fergyanto E. Gunawan [4], Avinash Baji [6] and Nagarajan Raghavan [5]

1. Oregon Renewable Energy Center (OREC), Klamath Falls, OR 97601, USA
2. Department of Manufacturing and Mechanical Engineering and Technology, Oregon Institute of Technology, Klamath Falls, OR 97601, USA
3. Xtreme Materials Lab, Engineering Product Development, Singapore University of Technology and Design (SUTD), Singapore 487372, Singapore; komal_agarwal@alumni.sutd.edu.sg
4. Industrial Engineering Department, BINUS Graduate Program—Master of Industrial Engineering, Bina Nusantara University, Jakarta 11480, Indonesia; rayya.fajarna@binus.ac.id (R.F.); fgunawan@binus.edu (F.E.G.)
5. Nano-Macro Reliability Lab, Engineering Product Development Pillar, Singapore University of Technology and Design (SUTD), Singapore 487372, Singapore; nagarajan@sutd.edu.sg
6. Department of Engineering, La Trobe University, Melbourne, VIC 3086, Australia; a.baji@latrobe.edu.au
* Correspondence: suriadi@alumni.stanford.edu (A.S.B.); rahul@sutd.edu.sg (R.S.)

**Abstract:** When silicon solar cells are used in the novel lightweight photovoltaic (PV) modules using a sandwich design with polycarbonate sheets on both the front and back sides of the cells, they are much more prone to impact loading, which may be prevalent in four-season countries during wintertime. Yet, the lightweight PV modules have recently become an increasingly important development, especially for certain segments of the renewable energy markets all over the world—such as exhibition halls, factories, supermarkets, farms, etc.—including in countries with harsh hailstorms during winter. Even in the standard PV module design using glass as the front sheet, the silicon cells inside remain fragile and may be prone to impact loading. This impact loading has been widely known to lead to cracks in the silicon solar cells that over an extended period of time may significantly degrade performance (output power). In our group's previous work, a 3D helicoidally architected fiber-based polymer composite (enabled by an electrospinning-based additive manufacturing methodology) was found to exhibit excellent impact resistance—absorbing much of the energy from the impact load—such that the silicon solar cells encapsulated on both sides by this material breaks only at significantly higher impact load/energy, compared to when a standard, commercial PV encapsulant material was used. In the present study, we aim to use numerical simulation and modeling to enhance our understanding of the stress distribution and evolution during impact loading on such helicoidally arranged fiber-based composite materials, and thus the damage evolution and mechanisms. This could further aid the implementation of the lightweight PV technology for the unique market needs, especially in countries with extreme winter seasons.

**Keywords:** 3D helicoidal architecture; fiber-based polymer composite; impact resistance; lightweight photovoltaics (PV); numerical modeling

## 1. Introduction

Lightweight photovoltaic (PV) modules are becoming increasingly important to certain sectors of the renewable energy market. Lightweight PV technology could potentially help address global climate and sustainability issues by being deployed in countries where electricity infrastructure is often lacking in very remote, poor locations separated by oceans [1–4]. For such a region, centralized energy sources may not be the best solution. By reducing the cost of transporting and installing solar PV systems, lightweight PV could play a critical part in building the kind of self-sufficient power infrastructure that is desperately desired in distant, rustic areas of developing countries. In order to be transported to remote locations with limited road and mobility infrastructure, PV power infrastructure must be lightweight. Lightweight PV modules are desired both for use in urban structures in advanced countries [5–8] and for easy set up in distant and impoverished locations in unusual areas of the developing world.

Furthermore, many large structures, such as exhibition halls, industrial plants, supermarkets, farms, etc., have a large footprint with few supporting pillars, resulting in a rooftop with low load-supporting capacity. Such rooftops require lightweight PV modules. If not, the cost of reinforcing such building before installing the heavy glass-based PV modules would make the renewable energy venture (building plus PV power infrastructure) inefficient and unappealing to potential business interests [5,9,10]. Lightweight PV has been extensively and comprehensively researched elsewhere as part of building-integrated PV and for predominantly urban building applications [5–8,11].

In addition, as silicon will be the predominant PV technology for the foreseeable future [5,6], we need to aid lightweight silicon-based PV modules. Structural strength, and impact resistance of lightweight PV modules, especially against heavy winds and hailstorms in four-season countries in Europe and North America [12,13], is one of the most important technological concerns in the development of lightweight PV modules. The front panels of a PV module can be made strong enough not to break under impact loads, such as hailstorms; nevertheless, the energy is transferred straight to the underlying material—first to the encapsulation (usually ethylene vinyl acetate), which would simply give way. The energy was then passed to the brittle silicon cells, which are particularly susceptible to such point impact loading, causing cracks to develop (nucleate) and/or propagate [7,13,14]. As a result, electrical performance degrades gradually or dramatically, which can lead to hotspots and potentially dangerous situations (such as fires, etc.).

Even though the idea of lightweight PV modules is tempting, nevertheless, it is not currently a practical option owing to concerns about structural stiffness and reliability [5–7,12,13]. Numerous commercial lightweight PV systems (even those that meet IEC/UL standards) have limited operational time [6,7]. Despite the fact that silicon cells remain internally delicate and extremely vulnerable to certain impact loads, recent studies with substantial material advances and ingenious design have allowed great improvements in the impact resistance of numerous polymer-based substrates used as front panels (as an alternative of glass) in conventional PV modules [13–16].

Natural structural materials, such as those found in mantis shrimp and nacre, have been demonstrated to provide superior mechanical and, particularly, impact resistant properties [17–19]. For example, the dactyl club of mantis shrimps has a 3D construction with a helicoidal shape that can disperse energy through quasi-plastic compression responses, providing a barrier to the spread of microcracks throughout recurring impacts [20–23]. Recent publications from our own research have stated higher impact resistance of such materials [24,25]. Further, the comparison in terms of mechanical properties of the 3D helicoidally aligned layered materials vs. layered materials without rotation offsets (or in other words, unidirectional layered materials) has been shown experimentally in our own previous publication by Agarwal et al. [26]. In addition, such comparisons have also been reported experimentally by other research groups using various materials (glass filament epoxy [27], carbon epoxy [27,28], and fiber sizes [27–29]). Agarwal et al. [26] reported for the smaller scale (using the custom electrospinning set up as an additive manufacturing methodology) of the other groups' studies, in terms of fiber diameters. Furthermore, comparison in terms of mechanical properties of the 3D helicoidally aligned layered materials vs. a sample of same thickness as the layer stack (or in other words, bulk sample) has also been demonstrated experimentally in our own previous publication [26], in addition to other previous reports [27–29], chiefly from Kisailus et al. [29]. Again, our group's studies—started by Agarwal et al. [26] to the more recent Sahay et al. [25] and Budiman et al. [30]—simply further pursued this line of investigation into the smaller scale fibers (using the custom electrospinning set up as an additive manufacturing methodology) and for the potential application of the unique materials for silicon-based PV module technology. The general outcome of such studies was that the layered structure of such materials, which consists of helicoidally aligned 3D fibers compared to typical layered structure/bulk material, would effectively absorb the impact energy and transfer very slight energy to the fragile silicon solar cells. This would allow novel lightweight PV module design with improved impact resistance and structural reliability (based on polymer materials for the front and back panel) particularly against cracks in the silicon cell.

The objective of this work is to provide a numerical analysis and modeling to predict how stress changes during impact loading in 3D-architectured layered polymer systems with helicoidally oriented fibers. This model demonstrates the fundamental feasibility of the proposed concept, namely, the use of 3D-architected layered polymer assemblies of helicoidally oriented fibers to protect silicon solar cells against the nucleation and proliferation of cracks caused by impact loads (e.g., from hailstorms) in the design of lightweight PV modules. We are expanding our approaches to enable this unique material for use in lightweight PV technologies. We are building on our earlier studies on novel materials [24–26,31] as well as numerical modeling of stresses in PV module design [32–37]. Furthermore, the design of lightweight PV modules would allow the integration of PV into curved or contoured surfaces, resulting in a more appealing design for integrating PV into urban structures.

## 2. Methodology

### 2.1. Material

The material used in making the multilayered composite plate is PVDF-HFP fibers, as was used in the experimental impact testing [30]. PVDF-HFP is polyvinyl alcohol (PVA) with MW = 98,000, polyvinylidene fluoride-co-hexafluropropylene (PVDF-HFP) with MW = 400,000, acetone, and dimethylacetamide (DMAc) were obtained from Merck, Singapore, as reported in our previous study [38]. More complete information about the materials used in the experimental impact testing can be found in [24].

The multilayered composite plate was modeled with fiber alignment in the layers changed from 90° (i.e., grid pattern) to 45° and then later to 15° to simulate the experiments (as illustrated in Figure 1), in which the impact resistance increase was observed [30]. Thus, in the present study, three multilayered composite plates were modeled:

A. First composite plate (Composite Plate A) consists of layers with fiber alignments of [0°, 90°, 180°, 270°, 360°].
B. Second composite plate (Composite Plate B) consists of layers with fiber alignments of [0°, 45°, 90°, 135°, 180°, 225°, 270°, 315°, 360°], as illustrated in Figure 1b.
C. Third composite plate (Composite Plate C) consists of layers with fiber alignments of [0°, 15°, 30°, 45°, 60°, 75°, 90°, 105°, 120°, 135°, 150°, 165°, 180°, 195°, ... , 360°]

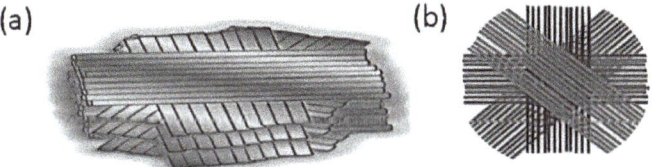

**Figure 1.** Schematic illustration of the multilayered composite (**a**) consisting of layers with 3D-architected helicoidally aligned fibers; (**b**) with fiber alignments of [0°, 45°, 90°, 135°, 180°, 225°, 270°, 315°, 360°] described in our previous report [24].

The dimensions of the composite plate in the Finite Element Model (FEM) used are 100 mm × 100 mm × 0.4 mm (length × width × thickness). Impact loading uses a steel ball with a radius of 30 mm. The mass of the ball is 0.1 kg with an impact velocity of 1 mm/ms. These parameters scale with the experimental impact testing as reported in [30]. Mechanical properties of the PVDF-HFP fiber were obtained from our previous publication [24], which reported in detail the materials used in the experimental impact testing in Ref. [30] as well as in the present numerical simulation study (See Table 1).

**Table 1.** Mechanical properties of PVDF-HFP fiber composite material (in each layer, with 0° and 90° represent longitudinal and transverse direction, respectively, of the fibers) used in simulation modeling [24].

| Properties | Simbol | Unit | PVDF-HFP Fiber |
|---|---|---|---|
| Young modulus 0° | E1 | MPa | 70 |
| Young modulus 90° | E2 | | 30 |
| Poisson ratio | V12 | | 0.1 |
| Ultimate tensile strength 0° | Xt | MPa | 60 |
| Ultimate compression strength 0° | Xc | | 57 |
| Ultimate tensile strength 90° | Yt | MPa | 30 |
| Ultimate compression strength 90° | Yc | | 27 |
| Ultimate tensile strain 0° | ext | % | 85 |
| Ultimate compression strain 0° | exc | | 80 |
| Ultimate tensile strain 90° | eyt | % | 45 |
| Ultimate compression strain 90° | eyc | | 35 |
| Density | ρ | g/cm$^3$ | 1.6 |

*2.2. Electrospinning-Based Additive Manufacturing (Es-AM)*

Only recently has electrospinning-based additive manufacturing (Es-AM) technology enabled such intricate 3D designs [25,31]. Near-field electrospinning (NFES) has been used to fabricate helicoidally oriented fiber layers as an additive manufacturing approach [7,8]. Agarwal et al. [24] describe in detail the fabrication of helicoidally oriented fiber layers with different angular orientations. NFES typically creates helicoidally oriented fiber layers by depositing one-dimensional fibers at precise locations in a controlled manner and then stacking the fibers layer by layer with angular offsets to create a 3D helicoidally arranged synthetic structural composite (HA-SSC) (see Figure 1). Typically, in the case of 90° HA-SSC90, the fiber layers were deposited with 90° angular offsets starting from 0°, 90°, 180°, 270°, to 360°—as mimicked in this numerical simulation study as Composite Plate A. Similarly, in HA-SSC45 with 45° angular offsets, the fiber layers were deposited at 0°, 45°, 90°, 135°, 180°, 225°, 270°, 315°, 360°—as mimicked in this numerical simulation study with Composite Plate B, and schematically illustrated further in Figure 1b. In HA-SSC15 with 15° angular offsets, the fiber layers were deposited at 0°, 15°, 30°, 45°, 60°, 75°, 90°, 105°, 120°, 135°, 150°, 165°, 180°, 195°, ... , 360°—as mimicked in the present study as Composite Plate C. The HA-SSCs samples were between 230 and 250 μm thick. In the experimental impact tests published in [30], we used only HA-SSC15 and HA-SSC45—with fiber orientation at rotation angles of 15° and 45°, respectively—from the different variants of the HA-SSC samples we reported in [24]. See [30] for more information on the impact testing of HA-SSCs.

This composite fabricated and reported in [24] is not yet ready for incorporation into PV power infrastructure. An optically transparent material is required for practical PV application with similar transmission of sunlight (in terms of intensity and range of suitable wavelengths). Nevertheless, as elucidated in the Introduction and the Materials section, the emphasis of the study is to establish the viability of the idea of improved impact resistance through the 3D-architected impact-resistant encapsulant, not its complete industrial incorporation into PV module design.

## 2.3. Finite Element Modeling of the Impact Loading on Multilayered Composite with 3D Helicoidal Architecture

Finite element (FE) models allow us to estimate the mechanical stress that develops during impact loading of the multilayered fiber-based composite materials with the 3D helicoidal arrangement to predict the damage mechanisms associated with the stress evolution during impact loading. However, to keep computational complexities to a minimum while still gaining the fundamental deformation mechanics, we approximate the fiber-based composite layers with laminate geometry that has anisotropic mechanical properties in a certain direction in each layer. We used thin film geometry in the FE model [38]. We used commercially available general-purpose LS-DYNA (4.6.19, LSTC (Ansys, Inc.), Canonsburg, PA, USA) FE software to obtain the stresses induced in the multilaminate structures to understand the evolution of stresses during impact loading. The FE model uses shell elements of a conventional thin film. The thin film sample was modeled using regular 8-noded quadrilateral elements (CPEG8). A generalized plane strain condition was assumed.

Figure 2 shows the multilayered composite plate, which in the LS-DYNA software was modeled using 4N plat on shape meshes. The size of the mesh is 2 mm. All the edges of plates are simply supported. The plates are subjected to impact loading on the middle of a surface in the form of a ball dropping onto the plate—mimicking the ball-dropping impact test as described in our previous report on a similar multilayered composite with 3D helicoidal architecture [30]. The impact loading was restricted to the z-direction. Data on displacement in the z-direction is collected on each node element of the middle layer of the multilayered composite plate.

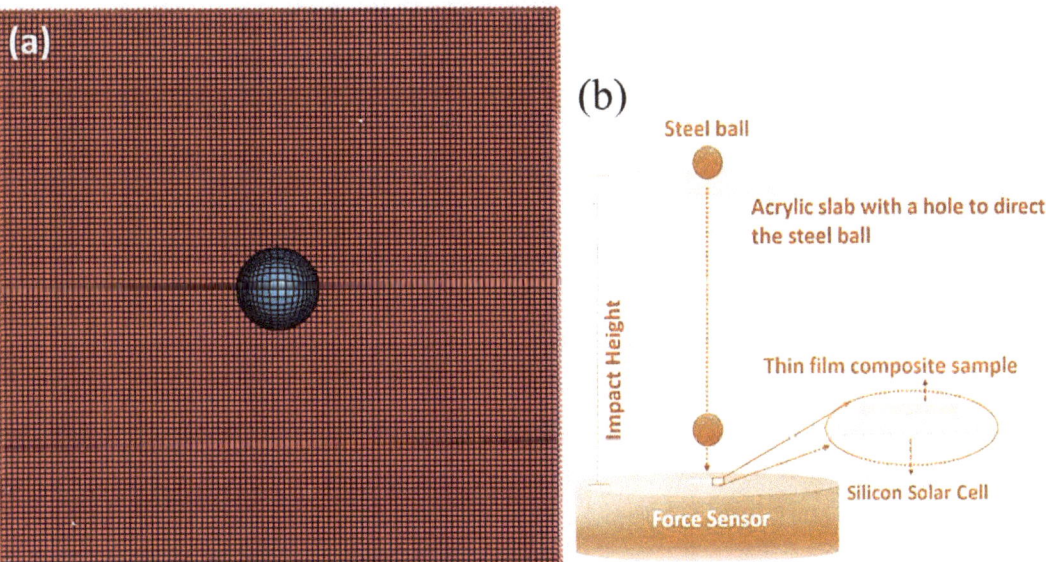

Figure 2. Schematic of the 2D plane-strain FE model (a) with thin film shell elements of multilaminate structures under impact loading mimicking the ball-dropping impact test (b) as described in our previous report [30].

The mechanics of the thin film plate during the impact loading here were modeled with the surrogate approach [39]. The strain in the film here due to the impact loading was surrogated by the scaling approach (size of the ball with respect to size of the multilayered composite geometry are equivalent between the model and the experiment) to keep the complexity to a minimum, while still obtaining important insights about stress evolution and damage distribution mechanisms during the impact loading. Thus, the relative magnitude of the stresses with respect to time during impact loading represents the evolution of energy distribution during the ball-dropping impact testing, with actual silicon solar cells under the 3D-architected layered polymer structures consisting of helicoidally aligned fibers, as reported in [30]. Uniform deformation over the volume of each layer of the multilayered composite is assumed during the process—leaving only the asymmetric deformation due to the anisotropy of each of the laminate due to different rotational angle of fiber alignment in each layer of the 3D helicoidal architected polymer composite. Fully elastic behavior of each of the laminate was assumed, which is reasonable given the scaling approach and the actual experimental results [30]. Interfaces between laminates were modeled as surface-to-surface tie constraint [39].

## 3. Results and Discussion

### 3.1. Results of the Finite Element Simulation: Impact Contact between the Steel Ball and the Plate

The position of the steel ball (modeled as sphere) with respect to the composite plate for various time instances during the impact contact is shown in Figure 3 (from the side of the model for Composite Plate A). The steel ball impinges the plate at an initial velocity of 1 mm/ms. The simulation shows the sphere has started contact with the plate surface at the time of t = 1.4399 ms. The simulation then shows in Figure 3 that the steel ball has exerted sufficient force to deform the plate laterally at t = 1.8899 ms. Consequently, at time t = 5.5788 ms, the plate has reached its maximum deformation. Lastly, at the time instant of t = 9.9811 ms, the steel ball bounced back and lost contact with the plate. Although the above time evolution was shown for Composite Plate A, similar occurrences represent the impact contact evolution between the steel ball and the plate in the other composite plates studied in the present study (Composite Plates B and C).

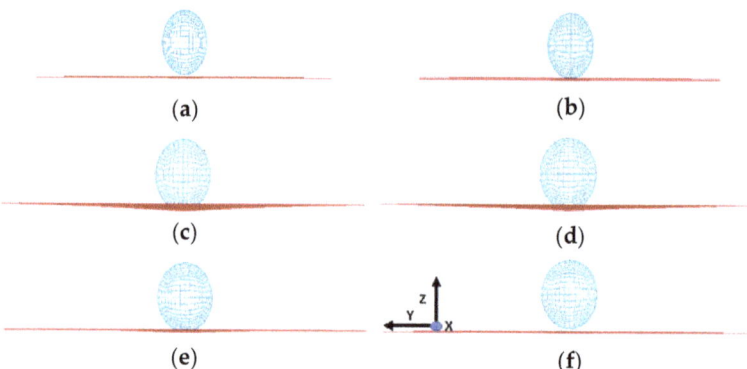

**Figure 3.** FE simulation showing impact contact between the sphere and the composite plate, (a) t = 0.4388 ms, (b) t = 1.4399 ms, (c) t = 1.8899 ms, (d) t = 5.5788 ms, (e) t = 7.8338 ms and (f) t = 9.9811 ms.

### 3.2. Results of the Finite Element Simulation: Deformation of the Composite Plate

The distribution of the Max Principal Stress (or the first principal stress, S1) with respect to time during the impact loading is shown in Figures 4–6 for Composite Plates A, B, and C, respectively. Initially, when the sphere first impinged on the plate, Figures 4–6 all show that stress concentration at the plate center. This is expected, as it is where the impact of the sphere on the plate occurred. The stress that is initiated at the plate center spreads out quickly to the whole plate and is absorbed by the composite plate structures at different rates.

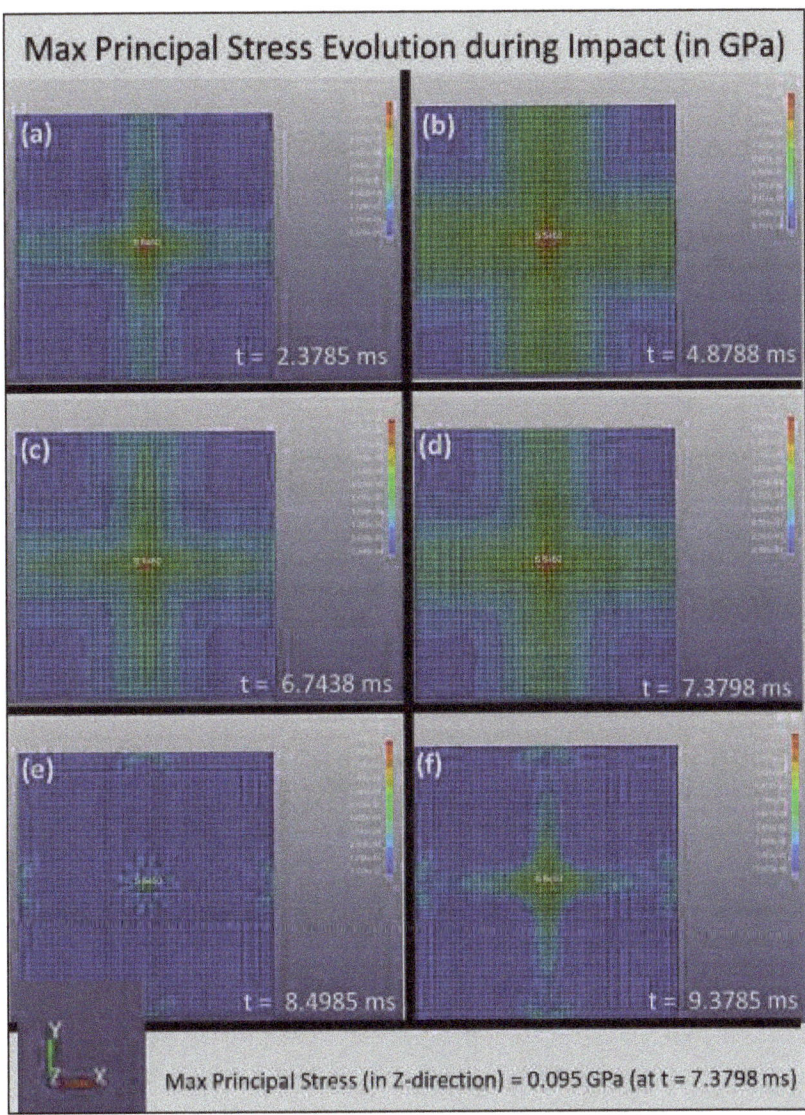

**Figure 4.** The distribution of the maximum principal stress, S1, with respect to time for Composite Plate A during impact loading, (**a**) 2.3785 ms, (**b**) 4.8788 ms, (**c**) 6.7438 ms, (**d**) 7.3798 ms, (**e**) 8.4985 ms and (**f**) 9.3785 ms.

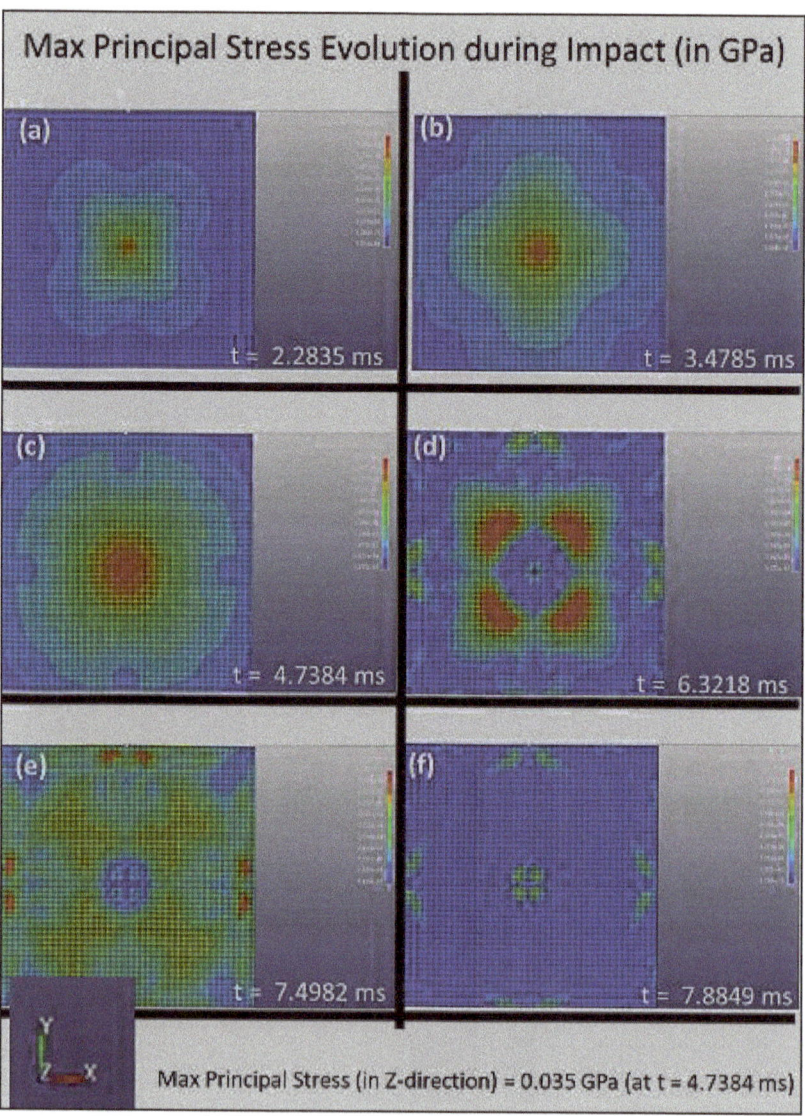

**Figure 5.** The distribution of the maximum principal stress, S1, with respect to time for Composite Plate B during impact loading, (**a**) 2.2835 ms, (**b**) 3.4785 ms, (**c**) 4.7384 ms, (**d**) 6.3218 ms, (**e**) 7.4982 ms and (**f**) 7.8849 ms.

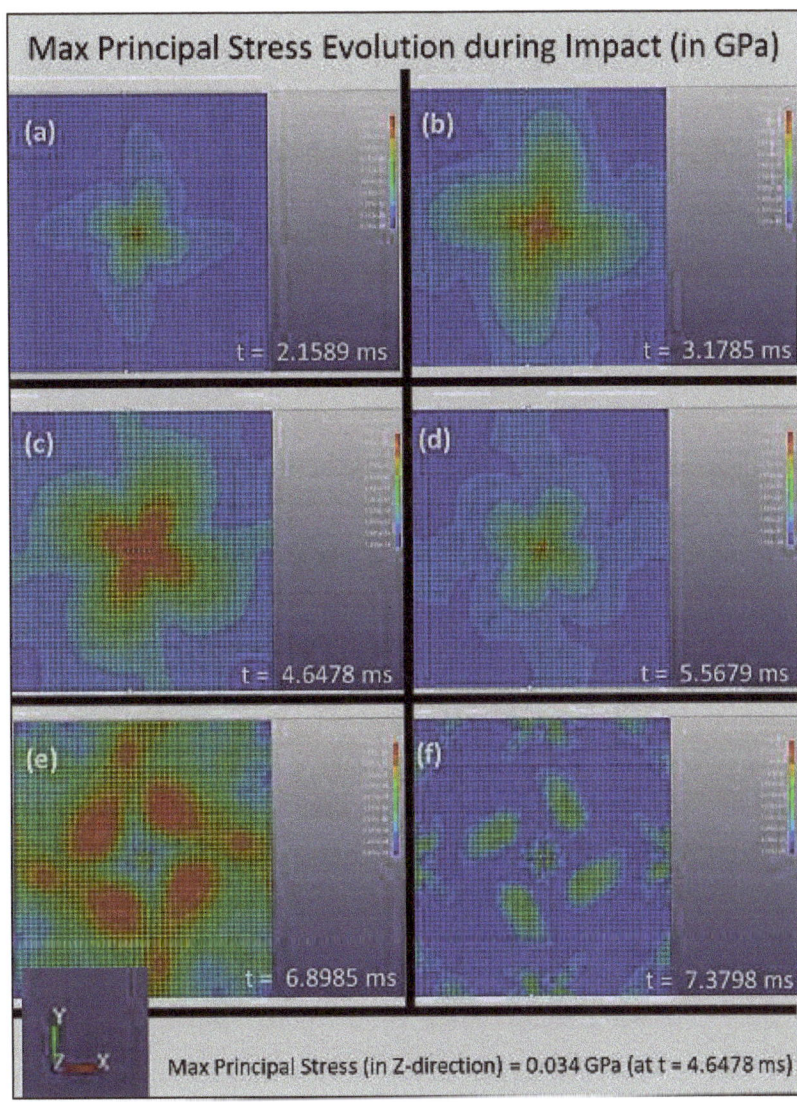

**Figure 6.** The distribution of the maximum principal stress, S1, with respect to time for Composite Plate C during impact loading, (**a**) 2.1589 ms, (**b**) 3.1785 ms, (**c**) 4.6478 ms (**d**) 5.5679 ms, (**e**) 6.8985 ms and (**f**) 7.3798 ms.

Figure 4 clearly shows that the stress distribution during impact loading follows the four-symmetry that is created by the grid pattern of Composite Plate A, which consists of layers with fiber alignments of [0°, 90°, 180°, 270°, 360°]. Stress concentrated at the center of the plate and reaches the maximum principal stress (which occurred in all composite plates in the present study in the z-direction) of 0.095 GPa at t = 7.3798 ms. Over time, the stress reduces and finally reaches the zero-stress state after about 9.5 ms.

Figure 5 clearly shows that the stress distribution during impact loading follows the double four-fold symmetry that is created by the grid pattern of Composite Plate B, which consists of layers with fiber alignments of [0°, 45°, 90°, 135°, 180°, 225°, 270°, 315°, 360°], as illustrated in Figure 2b. Figure 5 also shows that the stress concentrated at the center of the plate and reaches the maximum principal stress (in the z-direction) of 0.035 GPa at t = 4.7384 ms. Over time, the stress reduces and finally reaches the zero-stress state after about 8 ms. These parameters clearly demonstrate a much higher dissipation rate of damage in Composite Plate B, as compared to Composite Plate A. The maximum principal stress shown here in Figure 5 (for Composite Plate B) is lower, and the rate in which the stress is distributed and subsequently reduced to zero upon the same impact loading (impact load and loading rate) is much higher. This suggests that Composite Plate B with 3D architecture consisting of layers with fiber alignments of [0°, 45°, 90°, 135°, 180°, 225°, 270°, 315°, 360°] is much more effective and efficient in absorbing and dissipating impact energy and damage compared to Composite Plate A.

Figure 6 further suggests that Composite Plate C with 3D architecture consisting of layers with fiber alignments of [0°, 15°, 30°, 45°, 60°, 75°, 90°, 105°, 120°, 135°, 150°, 165°, 180°, 195°, ... , 360°] is also much more effective and efficient in absorbing and dissipating impact energy and damage compared to Composite Plate A. It also exhibits an incremental increase in the absorption and dissipation rate of impact energy and damage compared to that of Composite Plate B. Figure 6 shows that the stress concentrated at the center of the plate and reached the maximum principal stress (in the z-direction) of 0.034 GPa at t = 4.6478 ms. Over time, the stress reduces and finally reaches the zero-stress state after about 7.5 ms. These parameters clearly suggest that the rotational angle of the 3D helicoidal architecture influences the impact resistance of the materials. The smaller rotational angles clearly play a key role in increasing the dissipation rate of the impact damage, although the increase could be moderated after some levels of rotational angles. This is further substantiated by the stress vs. time response data shown in Figure 7.

Figure 7 clearly indicates that the Composite Plates B and C have a much shorter time response in terms of absorbing the impact energy (and thus damage dissipation in the event of plastic deformation and fracture events). Furthermore, the much lower stress intensity in Figure 7b,c demonstrates that the smaller rotational angle steps are effective in quickly and efficiently distributing the stress concentration. Both Composite Plates B and C exhibit stress levels getting back to pre-impact level (zero stress state) within under 7.5 ms, compared to just under 10 ms for Composite Plate A. Composite Plates B and C also show maximal principal stress intensities of 0.035 GPa and 0.034 GPa, respectively. This is much lower stress level compared to that of Composite Plate A, which reaches 0.095 GPa (almost three times). It is evident through the stress evolution during the impact loading simulation that the smaller rotational angles play a key role in increasing the dissipation rate of the impact damage, although the increase could be moderated after some levels of rotational angles. The increase in the absorption rate of impact damage and dissipation rate of stress intensity with reduction of rotational angle in the 3D helicoidal architecture seems to taper off after 45°. The FE simulation was conducted with the surrogate approach, thus while the absolute magnitudes still need to be further verified with experimental study using the same scales as in the modeling, the relative comparison between the effects of rotational angles in increasing the rate of absorption of damage and dissipation of stress intensity suggests self-consistency and excellent agreement with the experimental impact loading as reported in our earlier publication [30] and as will be further elaborated in the following section.

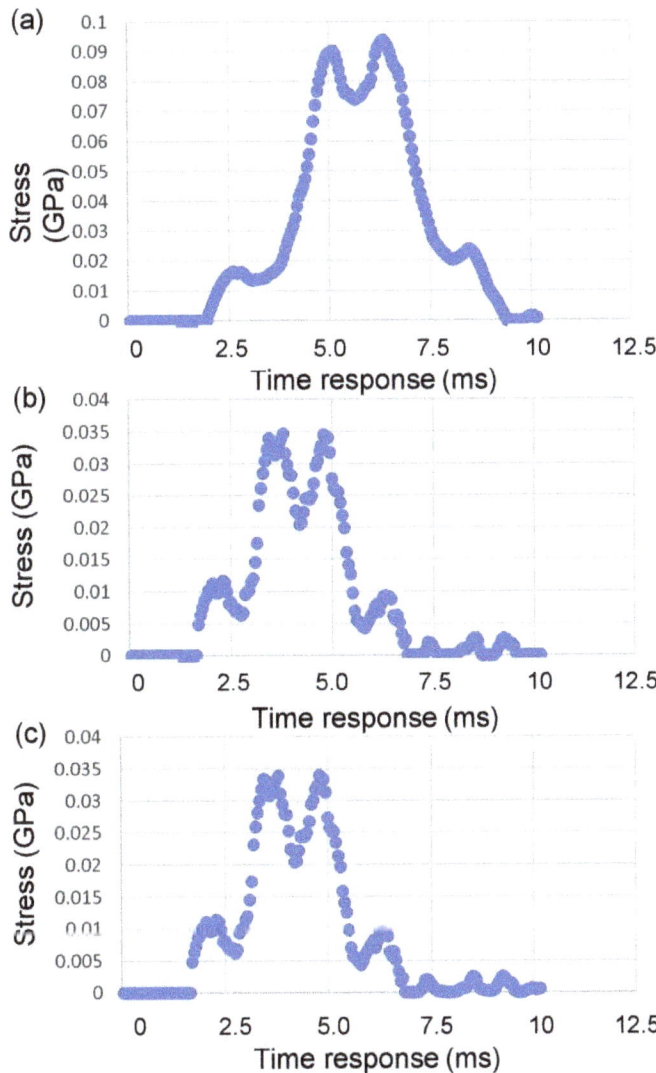

**Figure 7.** The time response of the distribution of maximum principal stress, S1, for (**a**) Composite Plate A, (**b**) Composite Plate B, and (**c**) Composite Plate C, during impact loading.

*3.3. Comparison with Impact Test of Photovoltaic (PV) Cells*

The experimental study on the impact-loading test of the 3D helicoidally architected polymer composite materials (with different rotational angles) has been reported in our recent publication [30]. In it, we use fragile silicon solar cells, which are highly susceptible to impact load underneath the polymer composite materials, such that the polymer composite materials act as a protection layer for the fragile silicon solar cells [30]. A customized impact testing setup (as illustrated in Figure 2b) was used to determine the impact resistance of such solar cells when protected by the samples [40]. More complete information about the impact test using the steel ball dropping method can be obtained in [30].

The fracture height indicates the height at which we began observing fracture of the silicon solar cells under the polymer composite materials. The nominal height of the bare Si solar cell group was found to be 25 (±5) cm, which was the real height at which all silicon cells fracture in the at least six times we repeated the tests (we performed the ball drop tests to more than six samples at this height—up to 11 samples). The uncertainty level of the experiment was ±5 cm. Consequently, the fracture data indicated the heights at which the fracture started. Heights greater than the abovementioned value would evidently fracture the silicon cells.

The fracture heights when we had typical encapsulant: ethylene vinyl acetate, on top of the silicon solar cells, was 50 ± 4 cm. The fracture heights are 69 ± 2 cm and 82 ± 4 cm for the Composite Plate B: HA-SSC15 and Composite Plate C: HA-SSC45, respectively [24,25]. Therefore, it was evident that the HA-SSC materials were the better protector of the silicon solar cells against impact loading, compared to the nominal EVA. Both HA-SSC composites (HA-SSC15 and HA-SSC45, respectively, Composite Plates B and C) enable significantly higher fracture heights and related specific potential energies (well beyond experimental uncertainties) before the underlying silicon solar cells fracture or trigger/spread catastrophic fracture events in ball drop tests. Because of their helicoidally oriented fiber-reinforced layered structures, both HA-SSC would efficiently disperse the impact energy and deflect the crack laterally by following the helicoidal progression of fiber directions, rather than immediately fracturing in a straight line through the thickness of the HA-SSCs [24,25,41]. Therefore, a smaller amount of impact energy is transferred to the underlying silicon solar cell.

The fractography images described in Ref [30] show abovementioned efficient mechanism for load dissipation as well as effective damage/energy absorption. In addition, both HA-SSC permit multiple crack lines on the silicon cells, signifying high dissipated impact energy and load transfer to the side (as evidenced by crack lines at different angles on the silicon cell surfaces). In typical monocrystalline silicon wafers, cracks seem to track the favored crystallographic directions of <110> linked to the weakest crystallographic planes of {111}, as has been extensively stated in the literature for both PV and other silicon-based devices [32–37,42,43].

The crack regularly encounters variations in the modulus of the fibers and the matrix material. The crack in HA-SSC travels along one fibrous layer, encounters a modulus deviation due to the matrix material present, penetrates further into the matrix, then reaches another fibrous layer in a dissimilar direction and plane, and finally diverges from its original route to follow a dissimilar path. The crack proliferates in multiple planes and orientations, rotates and twists inside and outside the fiber and matrix phases, and places greater stress on the fibers short of catastrophic failure. Due to the helicoidal network of the fibers in the composites, more energy is required for the propagation of the crack, so only a limited amount of impact energy/damage is transferred to the underlying silicon solar cells. According to Budiman et al. [30], the Composite Plate B: HA-SSC45 appears to be the most successful at consistently deflecting fracture/ impact damage along different angular orientations. The fracture heights show that the solar cells fracture at 82 ± 4 cm, which is the highest value among the samples investigated in this study. However, the study [35] paints a somewhat different depiction for the Composite Plate C: HA-SSC15. The breakage of the solar cells under HA-SSC15 happens at a lower height (69 ± 2 cm), resulting in a lower specific potential energy (and thus a lower impact energy/damage absorption rate).

In the present study, our FE simulation shows that the absorption rate of impact damage and dissipation rate of stress intensity increases with reduction of rotational angle in the 3D helicoidal architecture, although the effect seems to taper off after 45°. The FE simulation here shows that Composite Plate C: HA-SSC15 exhibits faster absorption rate of impact damage and stress level reduction (back to pre-impact state), although by a much smaller margin compared to that between Composites Plate A and B (from rotational angle of 90° to 45°). While this may not be in general agreement with our previous study reported in Budiman et al. [30], the FE simulation findings in the present study do agree very well with another earlier experimental report from our group, as described in Agarwal et al. [24]. The difference between these two studies is silicon solar cells were used under the HA-SSC in the impact loading experiments in Budiman et al. [30], while glass coverslips were used in Agarwal et al. [24]. The HA-SSCs used were, however, the same [30].

We believe that the variance (in the ability to absorb impact damage between HA-SSC15 and HA-SSC45) is due to the monocrystalline silicon solar cells that were among the samples of HA-SSC, because we followed the same method of the ball-dropping test as [24,25], together with the identical size of the steel ball. Compared to the glass slip, the monocrystalline silicon samples exhibit a crystallographic dependency of the mechanical properties, such as the preferential occurrence of fractures. Consequently, the whole relationship amongst the helicoidal orientation in HA-SSC and fracture of monocrystalline silicon solar cells needs to be further explored, which is outside the scope of the current FE simulation. Nevertheless, FE simulation in the present study takes into consideration the fact that the material put under the HA-SSC was a silicon monocrystalline solar cell (with its anisotropy in the mechanical properties), and not a glass coverslip (which has mostly isotropic mechanical properties).

*3.4. Enabling Next-Gen Lightweight Photovoltaic (PV) Module Technology*

It is clear from the experimental results [24,30] and the FE simulation shown in the present study that the HA-SSCs are excellent at absorbing and dispersing impact energy/damage, protecting the delicate solar cells underneath from point impact loads to which silicon PV is particularly susceptible. The objective of this paper is to complement the previously published experimental data with FE simulation study on the basic viability of using HA-SSCs for PV encapsulation to shield the sensitive solar cells, particularly in the design of lightweight PV modules that are predominantly susceptible to impact loadings, such as hailstorms, as outlined in IEC 61215/61646 clause 10.17. Although many current studies have revealed that a variety of polymer-based materials [12–16] can be used to improve the fracture and impact resistance of lightweight PV modules; nevertheless, the encapsulation materials employed in these studies were all EVA—albeit with different thicknesses or slight variations, such as low curing temperature [44].

Due to the economic impact and production maturity of the entire lightweight PV module, EVA was clearly selected as the encapsulation material of choice in previous studies. Our previous experimental results, as well as the numerical simulation results reported in this manuscript, provide initial indication from the standpoint of basic technological viability for the use of other types of novel polymer films (i.e., not EVA) with unique 3D architectures to allow strong and impact resilient lightweight PV module designs.

Nevertheless, there are many potential technological challenges to be resolved that prevent this unique idea from being fully realized. First, the HA-SSCs created were not transparent. Transparent protective layers on silicon cells in PV modules are an absolute must. It should be emphasized that the tests [30] and modeling of FE in the current study were performed to confirm the feasibility of integrating HA-SSC into PV modules. After our FE simulation results demonstrated the basic, complete feasibility using the electrospinning-based additive manufacturing (AM) method, we were able to identify additional polymers that we could fabricate in transparent systems, such as nylon [45] and poly(methyl methacrylate) (PMMA), or PMMA-based composites [46,47]. Furthermore, not only the interfacial adhesion with the front panel but also with the solar cell itself must be adequate [44]. The interfacial adhesion was not modeled in the current work and is the next step in the study. To maintain the 3D design throughout the lamination process, these novel materials may need further development [44,18,19]. All these may pave the way for future research of innovative polymeric composites/materials with 3D architecture to improve the use of lightweight PV modules and technologies.

## 4. Conclusions and Future Perspectives

In this FE simulation study, fiber-based 3D composites with helicoidally architecture were revealed to have outstanding impact energy/damage absorption and high-energy dissipation rate. Therefore, when applied to solar cells, they offer better shield against impact loads. Composite Plates C and B permit considerably higher fracture heights of $69 \pm 2$ and $82 \pm 4$, respectively, in comparison to $25 \pm 5$ for unprotected solar cells, and $50 \pm 4$ for EVA-protected solar cells, as observed during the ball-drop experiment. These helicoidally aligned synthetic composites (HA-SSCs) were fabricated using an electrospinning-based additive manufacturing (AM) technology that has only recently become possible at our group (Xtreme Materials Laboratory) in the past few years. During the ball-drop impact simulation using the Finite Element (FE) method, the HA-SSC composite materials (Composite Plates A, B, and C) showed consistent increase in damage absorption rate and stress level dissipation mechanism with reduced rotational angle of the HA-SSC materials. This is in excellent agreement with reports in the literature using isotropic materials (such as amorphous glass), although the agreement may be limited for anisotropic materials (such as monocrystalline silicon). The current FE simulation results add to the growing body of data that the innovative HA-SSC could be used for PV encapsulation to allow the design and fabrication of lightweight PV modules. The FE simulation results, as reported in the present manuscript, indicate impact protection increased with the reduction in the azimuthal angle to $45°$, after which the impact protection remained more or less constant. This is a unique insight which could have important implications for the silicon-based PV technology industry, as well as for other societally important applications of the HA-SSC materials, such as for lighter army combat vests and sports gear (helmets, etc.).

**Author Contributions:** Conceptualization, A.S.B. and R.S.; methodology, R.S., K.A. and A.B.; software, A.S.B. and F.E.G., validation, A.S.B., R.S. and K.A.; formal analysis, A.S.B.; investigation, R.S. and K.A.; resources, N.R.; data curation, R.S. and K.A.; writing—original draft preparation, R.S., R.F. and A.S.B.; writing—review and editing, A.S.B., R.F., F.E.G. and R.S.; visualization, R.S., K.A. and R.F.; supervision, A.S.B., N.R. and A.B.; project administration, N.R. and A.B.; funding acquisition, N.R. and A.B. All authors have read and agreed to the published version of the manuscript.

**Funding:** The authors would like to acknowledge the funding from the Ministry of Education (MOE) Academic Research Funds MOE2017-T2-2-175 titled "Materials with Tunable Impact Resistance via Integrated Additive Manufacturing as well as MOE2019-T2-1-197 titled "Monte Carlo Design and Optimization of Multicomponent Polymer Nano-composites". This work is supported by the La Trobe University Leadership RFA Grant, La Trobe University Start-up Grant and Collaboration and Research Engagement (CaRE) Grant offered by the School of Engineering and Mathematical Sciences (SEMS), La Trobe University. The authors also acknowledge the receipt of funding support from Temasek Labs@SUTD Singapore, through its SEED grant program for the project IGDSS1501011 and SMART (Singapore-MIT Alliance for Research and Technology) through its Ignition grant program for the project SMART ING-000067 ENG IGN. N.R. would like to acknowledge the funding from the Ministry of Education (MOE) Academic Research Fund MOE2019-T2-1-197 titled "Monte Carlo Design and Optimization of Multicomponent Polymer Nanocomposites" as well as support from EDB-IPP Surplus Funds Grant No. RGSUR08 for payment of article processing charges (APC).

**Institutional Review Board Statement:** Not applicable.

**Informed Consent Statement:** Not applicable.

**Data Availability Statement:** The data presented in this study are available upon request from the corresponding author.

**Acknowledgments:** The authors gratefully acknowledge the critical support and infrastructure provided by the Singapore University of Technology and Design (SUTD). S. Anbazhagan, and A.S. Budiman also gratefully acknowledge the funding and support from the National Research Foundation (NRF)/Economic Development Board (EDB) of Singapore for the project under EIRP Grant "(NRF EWT-EIRP002-017)—Enabling Thin Silicon Technologies for Next Generation, Lower Cost Solar PV Systems". The authors gratefully acknowledge PT Impack Pratama Indonesia for providing materials, exploratory supports, and technical discussions related to possible implementations of the materials development for the applications in building-integrated PV (photovoltaics) technology/design. ASB gratefully acknowledges the support and infrastructure provided by the Oregon Institute of Technology (OIT), and especially the Oregon Renewable Energy Center (OREC) at OregonTech.

**Conflicts of Interest:** The authors declare no conflict of interest.

## References

1. Handara, V.; Illya, G.; Tippabhotla, S.K.; Shivakumar, R.; Budiman, A.S. Novel and Innovative Solar Photovoltaics Systems Design for Tropical and Near-Ocean Regions—An Overview and Research Directions. *Proc. Eng.* **2016**, *139*, 22–31. [CrossRef]
2. Illya, G.; Handara, V.A.; Yujing, L.; Shivakumar, R.; Budiman, A.S. Backsheet Degradation under Salt Damp Heat Environments-Enabling Novel and Innovative Solar Photovoltaic Systems Design for Tropical Regions and Sea Close Areas. *Proc. Eng.* **2016**, *139*, 7–14. [CrossRef]
3. Illya, G.; Handara, V.; Siahandan, M.; Nathania, A.; Budiman, A.S. Mechanical Studies of Solar Photovoltaics (PV) Backsheets under Salt Damp Heat Environments. *Proc. Eng.* **2017**, *215*, 238–245. [CrossRef]
4. Handara, V.A.; Radchenko, I.; Tippabhotla, S.K.; Narayanan, K.; Illya, G.; Kunz, M.; Tamura, N.; Budiman, A.S. Probing stress and fracture mechanism in encapsulated thin silicon solar cells by synchrotron X-ray microdiffraction. *Sol. Energy Mater. Sol. Cells* **2017**, *162*, 30–40. [CrossRef]
5. Martins, A.C.; Chapuis, V.; Virtuani, A.; Li, H.Y.; Perret-Aebi, L.E.; Ballif, C. Thermo-mechanical stability of lightweight glass-free photovoltaic modules based on a composite substrate. *Sol. Energy Mater. Sol. Cells* **2018**, *187*, 82–90. [CrossRef]
6. Martins, A.C.; Chapuis, V.; Sculati-Meillaud, F.; Virtuani, A.; Ballif, C. Light and durable: Composite structures for building-integrated photovoltaic modules. *Prog. Photovolt. Res. Appl.* **2018**, *26*, 718–729. [CrossRef]
7. Martins, A.C.; Chapuis, V.; Virtuani, A.; Ballif, C. Robust Glass-Free Lightweight Photovoltaic Modules with Improved Resistance to Mechanical Loads and Impact. *IEEE J. Photovolt.* **2019**, *9*, 245–251. [CrossRef]
8. Ballif, C.; Perret-Aebi, L.E.; Lufkin, S.; Rey, E. Integrated thinking for photovoltaics in buildings. *Nat. Energy* **2018**, *3*, 438–442. [CrossRef]
9. Berger, K.; Cueli, A.B.; Boddaert, S.; Buono, M.D.; Delisle, V.; Fedorova, A.; Frontini, F.; Hendrick, P.; Inoue, S.; Ishi, H.; et al. International Definitions of BIPV. In *EA 597 Photovoltaic Power Systems Programme*; Report IEA-PVPS T15-04; IEA: Paris, France, 2018; Available online: https://iea-pvps.org/key-topics/international-definitions-of-bipv/ (accessed on 14 December 2021).
10. Zhang, F.; Deng, H.; Margolis, R.; Su, J. Analysis of distributed-generation photovoltaic deployment, installation time and cost, market barriers, and policies in China. *Energy Policy* **2015**, *81*, 43–55. [CrossRef]
11. Kajisa, T.; Miyauchi, H.; Mizuhara, K.; Hayashi, K.; Tokimitsu, T.; Inoue, M.; Hara, K.; Mauda, A. Novel lighter weight crystalline silicon photovoltaic module using acrylic-film as a cover sheet. *Jpn. J. Appl. Phys.* **2014**, *53*, 092302. [CrossRef]
12. Dhere, N.G. *Flexible Packaging for PV Modules Reliability of Photovoltaic Cells, Modules, Components, and Systems*; International Society for Optics and Photonics: Bellingham, WA, USA, 2008; Volume 7048, p. 70480.
13. Wright, A.; Lee, E.J. Impact Resistant Lightweight Photovoltaic Modules. U.S. Patent WO 2016/077402, 19 May 2016.
14. Martins, A.C.O.; Chapuis, V.; Virtuani, A.; Perret-Abie, L.E.; Ballif, C. Hail Resistance of Composite-Based Glass-Free Lightweight Modules for Building Integrated Photovoltaics Applications. In Proceedings of the 33rd European Photovoltaic Solar Energy Conference and Exhibition, Amsterdam, The Netherlands, 25–29 September 2017; pp. 2604–2608.
15. Gaume, J.; Quesnel, F.; Guillerez, S.; LeQuang, N.; Williatte, S.; Goaer, G. Solight: A new lightweight PV module complying IEC standards. In Proceedings of the 33rd European Photovoltaic Solar Energy Conference and Exhibition, Amsterdam, The Netherlands, 27 September 2017.
16. Boulanger, A.; Gaume, J.; Quesnel, F.; Ruols, P.; Rouby, F. Operasol: A light photovoltaic panel with integrated connectors. In Proceedings of the 33rd European Photovoltaic Solar Energy Conference and Exhibition, Amsterdam, The Netherlands, 25–29 September 2017.
17. Yang, W.; Chen, I.H.; Gludovatz, B.; Zimmermann, E.A.; Ritchie, R.O.; Meyers, M.A. Natural flexible dermal armor. *Adv. Mater.* **2013**, *25*, 31–48. [CrossRef]
18. Naleway, S.E.; Porter, M.M.; McKittrick, J.; Meyers, M.A. Structural Design Elements in Biological Materials: Application to Bioinspiration. *Adv. Mater.* **2015**, *27*, 5455–5476. [CrossRef] [PubMed]
19. Wegst, U.G.K.; Bai, H.; Saiz, E.; Tomsia, A.P.; Ritchie, R.O. Bioinspired structural materials. *Nature* **2015**, *14*, 23–36. [CrossRef] [PubMed]
20. Patek, S.N.; Caldwell, R.L. Extreme impact and cavitation forces of a biological hammer: Strike forces of the peacock mantis shrimp Odontodactylus scyllarus. *J. Exp. Biol.* **2005**, *208*, 3655–3664. [CrossRef] [PubMed]
21. Cronin, T.W.; Marshall, N.J.; Quinn, C.A.; King, C.A. Ultraviolet Photoreception in Mantis Shrimp. *Vis. Res.* **1994**, *34*, 44–1452. [CrossRef]
22. Amini, S.; Tadayon, M.; Idapalapati, S.; Miserez, A. The role of quasi-plasticity in the extreme contact damage tolerance of the stomatopod dactyl club. *Nature* **2015**, *14*, 943–950. [CrossRef]
23. Fratzl, P.; Weinkamer, R. Nature's hierarchical materials. *Prog. Mater. Sci.* **2007**, *52*, 1263–1334. [CrossRef]

24. Agarwal, K.; Sahay, R.; Baji, A.; Budiman, A.S. Impact Resistant and Tough Helicoidally Aligned Ribbon Reinforced Composite with Tunable Mechanical Properties via Integrated Additive Manufacturing Methodologies. *ACS Appl. Polym. Mater.* **2020**, *2*, 2491–3504. [CrossRef]
25. Sahay, R.; Agarwal, K.; Subramani, A.; Raghavan, N.; Budiman, A.S.; Baji, A. Helicoidally arranged polyacrylonitrile fiber-reinforced strong and impact-resistant thin polyvinyl alcohol film enabled by electrospinning-based additive manufacturing. *Polymers* **2020**, *12*, 2376. [CrossRef]
26. Agarwal, K.; Zhou, Y.; Ali, H.P.A.; Radchenko, I.; Baji, A.; Budiman, A.S. Additive manufacturing enabled by electrospinning for tougher bio-inspired materials. *Adv. Mater. Sci.* **2018**, *2018*, 1–9. [CrossRef]
27. Chen, B.; Peng, X.; Cai, C.; Niu, H.; Wu, X. Helicoidal microstructure of Scarabaei cuticle and biomimetic research. *Mater. Sci. Eng. A* **2006**, *423*, 237–242. [CrossRef]
28. Apichattrabrut, T.; Ravi-Chandar, K. Helicoidal composites. *Mech. Adv.* **2006**, *13*, 61–76. [CrossRef]
29. Grunenfelder, L.K.; Suksangpanya, N.; Salinas, C.; Milliro, G.; Yaraghi, N.; Herrera, S.; Lutterodt, K.E.; Nutt, S.R.; Zavattieri, P.; Kisailus, D. Bio-inspired impact-resistant composites. *Acta Biomater.* **2014**, *10*, 3997–4008. [CrossRef] [PubMed]
30. Budiman, A.S.; Sahay, R.; Argawal, K.; Illya, G.; Widjaja, R.G.; Baji, A.; Raghavan, N. Impact-Resistant and Tough 3D Helicoidally Architected Polymer Composites Enabling Next-Generation Lightweight Silicon Photovoltaics Module Design and Technology. *Polymers* **2021**, *13*, 3315. [CrossRef] [PubMed]
31. Agarwal, K.; Sahay, R.; Baji, A.; Budiman, A.S. Biomimetic tough helicoidally structured material through novel electrospinning based additive manufacturing. *MRS Adv.* **2019**, *4*, 2345–2354. [CrossRef]
32. Song, W.J.R.; Tippabhotla, S.K.; Tay, A.A.O.; Budiman, A.S. Effect of interconnect geometry on the evolution of stresses in a solar photovoltaic laminate during and after lamination. *Sol. Energy Mater. Sol. Cells* **2018**, *187*, 241–248. [CrossRef]
33. Budiman, A.S.; Illya, G.; Handara, V.; Caldwell, W.A.; Bonelli, C.; Kunz, M.; Tamura, N.; Verstraeten, D. Enabling Thin Silicon Technologies for Next Generation c-Si Solar PV Renewable Energy Systems using Synchrotron X-ray Microdiffraction as Stress and Crack Mechanism Probe. *Sol. Energy Mater. Sol. Cells* **2014**, *130*, 303–308. [CrossRef]
34. Tippabhotla, S.K.; Radchenko, I.; Song, W.J.R.; Illya, G.; Handara, V.; Kunz, M.; Tamura, N.; Tay, A.A.O.; Budiman, A.S. From Cells to Laminate: Probing and Modeling Residual Stress Evolution in Thin Silicon Photovoltaic Modules using Synchrotron X-ray Micro-Diffraction Experiments and Finite Element Simulations. *Prog. Photovolt.* **2017**, *25*, 791–809. [CrossRef]
35. Tippabhotla, S.K.; Song, W.J.R.; Tay, A.A.O.; Budiman, A.S. Effect of encapsulants on the thermomechanical residual stress in the back-contact silicon solar cells of photovoltaic modules—A constrained local curvature model. *Sol. Energy* **2019**, *182*, 134–147. [CrossRef]
36. Tian, T.; Morusupalli, R.; Shin, H.; Son, H.; Byun, K.; Joo, Y.; Caramto, R.; Smith, L.; Shen, Y.; Kunz, M.; et al. On the Mechanical Stresses of Cu Through-Silicon Via (TSV) Samples Fabricated by SK Hynix vs. SEMATECH—Enabling Robust and Reliable 3-D Interconnect/Integrated Circuit (IC) Technology. *Proc. Eng.* **2016**, *139*, 101–111. [CrossRef]
37. Ali, I.; Tippabhotla, S.K.; Radchenko, I.; Al-Obeidi, A.; Stan, C.V.; Tamura, N.; Budiman, A.S. Probing Stress States in Silicon Nanowires during Electrochemical Lithiation using In Situ Synchrotron X-ray Microdiffraction. *Front. Energy Res.* **2018**, *6*, 19. [CrossRef]
38. Mariyadi, B.; Gunawan, F.E.; Budiman, A.S. New Indicator for Health Monitoring of Structures Made of Fiber-Reinforced Composite Materials Under Low Impact Loading. *Turk. J. Comput. Math.* **2021**, *12*, 14.
39. Harito, C.; Wijaya, J.; Fajarna, R.; Tippabhotla, S.K.; Thomas, O.; Gunawan, F.E.; Budiman, A.S. Mechanical behaviours and modelling of nanostructure anode in Lithium-Ion Battery (LIB)—Enabling low-cost 3D nanostructures in future LIB for sustainable energy storage. In Proceedings of the 3rd Internation Conference on Biospheric Harmony Advanced Research (ICOBAR), Jakarta, Indonesia, 24–25 June 2021.
40. Safri, S.N.A.; Sultan, M.T.H.; Jawaid, M.; Jayakrishna, K. Impact behaviour of hybrid composites for structural applications: A review. *Compos. B. Eng.* **2018**, *133*, 112–121. [CrossRef]
41. Chen, R.; Liu, J.; Yang, C. Transparent Impact-Resistant Composite Films with Bioinspired Hierarchical Structure. *ACS Appl. Mater. Interfaces* **2019**, *11*, 23616–23622. [CrossRef] [PubMed]
42. Tippabhotla, S.K.; Radchenko, I.; Stan, C.V.; Tamura, N.; Budiman, A.S. Stress evolution in silicon nanowires during electrochemical lithiation using in situ synchrotron X-ray microdiffraction. *J. Mater. Res.* **2019**, *34*, 1–10. [CrossRef]
43. Budiman, A.S.; Shin, H.A.S.; Kim, B.J.; Hwang, S.H.; Son, H.Y.; Suh, M.S.; Chung, Q.H.; Byun, K.Y.; Tamura, N.; Kunz, M.; et al. Measurement of stresses in Cu and Si around through-silicon via by synchrotron X-ray microdiffraction for 3-dimensional integrated circuits. *Microelectron. Reliab.* **2012**, *52*, 530–533. [CrossRef]
44. Budiman, A.S.; Illya, G.; Anbazhagan, S.; Tippabhotla, S.K.; Song, W.J.R.; Sahay, R.; Tay, A.A.O. Enabling lightweight PC-PC Photovoltaics Module Technology—Enhancing Integration of Silicon Solar Cells into Aesthetic Design for Greener Building and Urban Structures. *Sol. Energy* **2021**, *227*, 38–45. [CrossRef]
45. Kim, I.C.; Kim, T.H.; Lee, S.H.; Kim, B.S. Extremely Foldable and Highly Transparent Nanofiber-Based Electrodes for Liquid Crystal Smart Devices. *Sci. Rep.* **2018**, *8*, 11517. [CrossRef] [PubMed]
46. Matei, E.; Busuioc, C.; Evanghelidis, A.; Zgura, I.; Enculescu, M.; Beregoi, M.; Enculescu, I. Hierarchical Functionalization of Electrospun Fibers by Electrodeposition of Zinc Oxide Nanostructures. *Appl. Surf. Sci.* **2018**, *458*, 555–563. [CrossRef]
47. Wu, M.; Wu, Y.; Liu, Z.; Liu, H. Optically transparent poly(methyl methacrylate) composite films reinforced with electrospun polyacrylonitrile nanofibers. *J. Compos. Mater.* **2012**, *46*, 2731–2738. [CrossRef]

48. Ali, H.P.A.; Radchenko, I.; Li, N.; Budiman, A.S. The roles of interfaces and other microstructural features in Cu/Nb nanolayers as revealed by in situ beam bending experiments inside an scanning electron microscope (SEM). *Mater. Sci. Eng. A* **2018**, *738*, 253–263. [CrossRef]
49. Shivakumar, R.; Tippabhotla, S.K.; Handara, V.A.; Illya, G.; Tay, A.A.O.; Novoa, F.; Dauskardt, R.H.; Budiman, A.S. Fracture Mechanics and Testing of Interface Adhesion Strength in Multilayered Structures-Application in Advanced Solar PV Materials and Technology. In Proceedings of the 8th International Conference on Materials for Advanced Technologies, Singapore, 28 June–3 July 2015; pp. 47–55. [CrossRef]

Article

# Clinical Application of 3D-Printed Patient-Specific Polycaprolactone/Beta Tricalcium Phosphate Scaffold for Complex Zygomatico-Maxillary Defects

Woo-Shik Jeong [1], Young-Chul Kim [1], Jae-Cheong Min [1], Ho-Jin Park [1], Eun-Ju Lee [2,3], Jin-Hyung Shim [2,3] and Jong-Woo Choi [1,*]

[1] Department of Plastic and Reconstructive Surgery, Asan Medical Center, University of Ulsan College of Medicine, 88 Olympicro 43 gil, Songpa-gu, Seoul 05505, Korea; woosjeong.ps@gmail.com (W.-S.J.); youngchulkk@naver.com (Y.-C.K.); psdoc87@gmail.com (J.-C.M.); leptonfamily@gmail.com (H.-J.P.)

[2] Research Institute, T&R Biofab Co., Ltd., Seongnam-si 13487, Korea; ejlee@tnrbiofab.com (E.-J.L.); happyshim@tnrbiofab.com (J.-H.S.)

[3] Department of Mechanical Engineering, Korea Polytechnic University, Siheung-si 15073, Korea

* Correspondence: pschoi@amc.seoul.kr; Tel.: +82-2-3010-3604

**Abstract:** (1) Background: In the present study, we evaluated the efficacy of a 3D-printed, patient-specific polycaprolactone/beta tricalcium phosphate (PCL/β-TCP) scaffold in the treatment of complex zygomatico-maxillary defects. (2) Methods: We evaluated eight patients who underwent immediate or delayed maxillary reconstruction with patient-specific PCL implants between December 2019 and June 2021. The efficacy of these techniques was assessed using the volume and density analysis of computed tomography data obtained before surgery and six months after surgery. (3) Results: Patients underwent maxillary reconstruction with the 3D-printed PCL/β-TCP scaffold based on various reconstructive techniques, including bone graft, fasciocutaneous free flaps, and fat graft. In the volume analysis, satisfactory volume conformity was achieved between the preoperative simulation and actual implant volume with a mean volume conformity of 79.71%, ranging from 70.89% to 86.31%. The ratio of de novo bone formation to total implant volume (bone volume fraction) was satisfactory with a mean bone fraction volume of 23.34%, ranging from 7.81% to 66.21%. Mean tissue density in the region of interest was 188.84 HU, ranging from 151.48 HU to 291.74 HU. (4) Conclusions: The combined use of the PCL/β-TCP scaffold with virtual surgical simulation and 3D printing techniques may replace traditional non-absorbable implants in the future owing to its accuracy and biocompatible properties.

**Keywords:** polycaprolactone; tricalcium phosphate; PCL/β-TCP; 3D printing; maxillary defect

## 1. Introduction

The management of a maxillary defect is complicated when surgeons must replace the original 3D structure of the bone and carry out functional midfacial restoration in the periorbital and perioral region. Vascularized bone flaps have been the standard option in the field of mandibular reconstruction [1]. They provide a rigid and durable structure that allows adjuvant radiation treatment, a skin paddle for additional soft tissue defects, space for dental implant placement, and reasonable adaptation to remnant bony structures. However, no single flap can provide sufficient volume or support in larger or complex defects, especially when orbital adnexae and dental components are involved.

In complex maxillary treatments, alloplastic material has been combined with autogenous reconstruction. Titanium mesh has been widely applied because it is easy to use and biocompatible, allowing the ingrowth of connective tissue through the implant. Moreover, it can be molded into the complex maxillary structure [2]. However, it can lead to implant exposure or palpability due to the breakdown of the mucocutaneous lining. Deformative change can also occur during scar contracture and adjuvant radiation treatment [3,4]. To address these limitations, biodegradable or bioabsorbable materials have gained popularity; they are rigid and biocompatible, induce bone regeneration, and confer a lower chance of foreign body reaction [5–7].

Combined with computer-aided techniques, such as virtual surgical planning, various alloplastic materials have improved the accuracy of maxillofacial reconstruction [8–10]. 3D printing technology, combined with preoperative planning and modeling, enables more effective patient-specific treatment. In addition, biodegradable printing materials can now be used in a customized fashion to reconstruct complicated craniomaxillofacial defects with acceptable outcomes. Among these various biodegradable materials, PCL (polycaprolactone) has been used as guided bone regeneration (GBR) membrane owing to its favorable mechanical properties and biocompatibility with a slower degradation rate [11,12]. The beta-tricalcium phosphate (β-TCP), a bioceramic material, has been used in the field of bone tissue engineering owing to its chemical properties resembling bone minerals and excellent osteoconductivity [13–15]. The use of PCL blended with beta-tricalcium phosphate (β-TCP) was reported as a promising GBR membrane to promote new bone formation, with an initial stability comparable to cortical bone [11,16–18]. Traditionally, promising results were achieved in terms of osteogenic activity when a PCL scaffold was blended with 20% TCP [19–21]. To the best of our knowledge, there are few studies using a 3D-printed PCL/β-TCP scaffold in complex zygomatico-maxillary defects. The present work aims to evaluate the new bone formation and 3D conformity using a computed tomographic data and clinical outcomes in zygomatico-maxillary reconstruction with a 3D-printed PCL/β-TCP scaffold.

## 2. Materials and Methods

We evaluated a prospective series of eight patients with complex zygomatico-maxillary defects who underwent reconstruction with 3D-printed PCL implants between December 2019 and June 2021. The inclusion criteria were as follows: (1) unilateral zygomatic maxillary defect with or without orbital floor involvement, (2) maxillary defect resulting from cancer ablation, benign tumor resection, trauma, or degenerative change of the hemiface such as Parry–Romberg syndrome, (3) requirement of immediate or delayed reconstruction due to maxillary defect, and (4) follow-up period of at least six months. The exclusion criteria were as follows: (1) bilateral defect, (2) critical infectious disease or immune deficiency, (3) current or anticipated chemotherapy or immune suppression therapy, and (4) pregnancy or possibility of pregnancy.

Demographic information regarding sex, age, underlying disease, cause of defect, onset of reconstruction (immediate or delayed), type of maxillary defect, and postoperative complications were reviewed. The maxillary defects were categorized based on the amount of vertical and horizontal maxillary defect, as suggested by Brown et al. [22]. Surgical details regarding reconstructive options, incisional approach, application of bone forming material, implant fixation method, and revisional operation were described. Each patient underwent computed tomography (CT) scans with a slice thickness of 0.6 mm at three time periods, including before surgery and six months after.

This study was conducted according to the Declaration of Helsinki and approved by the independent Ethics Committee/Institutional Review Board of the Asan Medical Center (approval number: 2021-1292), with written informed consent obtained from all patients

### 2.1. 3D Simulation and 3D Printing of Patient-Specific Implants

The patient-specific implants were designed using 3D modeling software (Materialise Mimics; Materialise NV, Leuven, Belgium). The anticipated maxillary defect was marked on a stereolithography model of the skull, and the contralateral normal orbit was flipped to obtain the ideal normal contours of the defect. A patient-specific implant was designed over the region of interest, fabricated and then refined, with smoothing of the contour

(Figure 1). All processes were performed under close communication between modeling experts and plastic surgeons.

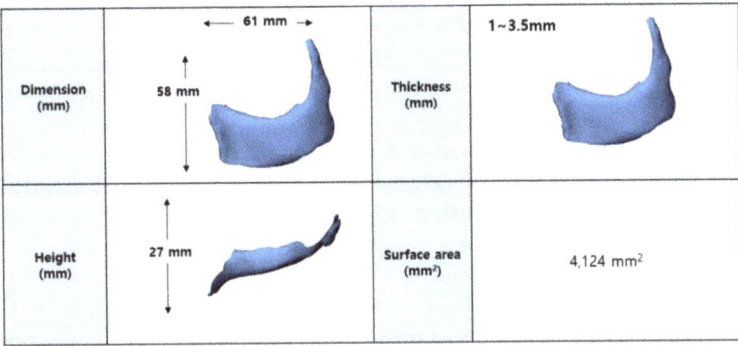

**Figure 1.** Design of patient-specific PCL/β-TCP scaffold in patient #1. The implant was three-dimensionally designed using 3D modeling software based on the mirror imaging of a contralateral normal zygomatico-maxillary structure.

The PCL (Evonik Industries, Essen, Germany) and β-TCP (Foster corporation, Putnam, CT, USA) were mixed in a ratio of 8:2. After the PCL was melted by heating for 15 min at 110 °C, powdered β-TCP was added, which was then blended for 10 min. The PCL/β-TCP mixture was 3D printed using a multi-head deposition system using computer-aided manufacturing software. It had a rectangular pore architecture with a porosity of 50% and a pore size of 500 μm, as determined by 3D modeling software (3-Matic Research 9.0, Materialise, Leuven, Belgium). The scaffolds were freeze-dried at −85 °C for 24 h, then sterilized under a 450 W UV lamp for 4 h. All manufacture process was managed by a facility with Good Manufacturing Practice certification (T&R Biofab Co. Ltd., Seoul, Korea). The image of the 3D printed scaffold is depicted in Figure 2.

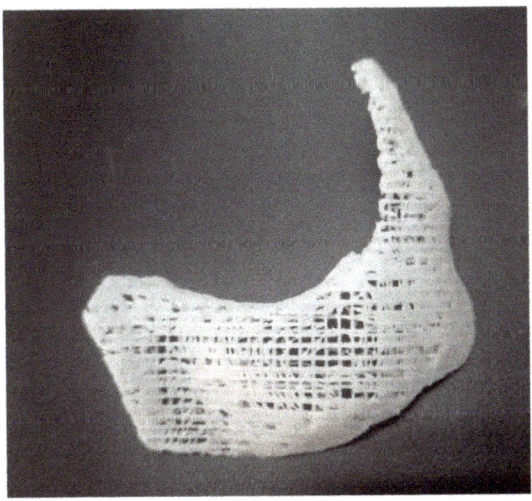

**Figure 2.** Photograph of 3D printed PCL/β-TCP scaffold with line with 500 μm, and 50% of porosity.

## 2.2. Surgical Procedure

Patients presented with a wide range of maxillary bone and soft tissue defects of various etiologies. The bone defect area was exposed as much as possible so that the implant could be inserted. The bone defect was covered with the 3D-printed implant, with or without osteocutaneous free flaps. The PCL implant was embedded in a betadine solution for 10 min before insertion. If necessary, it was easily molded using a No. 15 blade or scissors, depending on the actual defect. After the PCL implant was inserted into the defect, it was fixed to the adjacent bony structure using mini-plates and 6–8 mm titanium screws. Additional free flaps were indicated if the alloplastic implant necessitated soft tissue envelop to cover the defect. Immediate adverse reactions related to the implant, such as allergic reactions, were checked during surgery.

## 2.3. Volume and Density Analysis Based on CT Data

A CT scan was performed before surgery and six months after to evaluate volumetric and density change. The DICOM data were translated into a stereolithography model in 3D modeling software (Mimics; Materialise Software Solutions, Leuven, Belgium) to simulate a postoperative image using a volume rendering technique. The region of interest was defined before surgery along the contour of the simulated implant object, as well as six months after surgery along the outer surface of the inserted implant. Two images were superimposed based on anatomical landmarks, including the anterior nasal spine, nasion, gonion, and menton. Overlapping between the simulated implant volume and postsurgical implant volume was calculated using the Boolean operation. The volume conformity was defined as the percentage of overlapping volume between the simulated and postsurgical images (Figure 3).

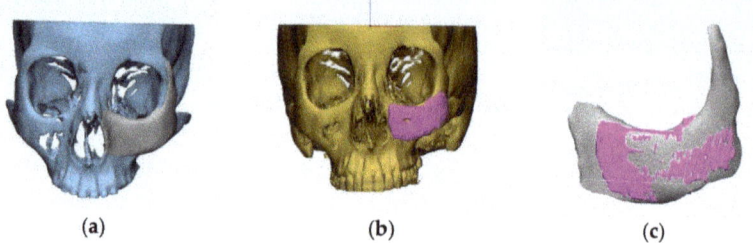

**Figure 3.** Volumetric analysis between preoperative planned model and actual surgical result. (a) Preoperatively designed STL model. (b) Actual surgical result volume-rendered as STL model. (c) Two images were superimposed based on anatomic landmarks. Overlapping between the simulated implant volume and postsurgical implant volume was calculated using the Boolean operation.

To identify de novo bone formation, the CT images were subjected to radiodensity analysis using a 3D modeling software (Mimics, Materialise Software Solutions, Leuven, Belgium); the radiodensity was measured in Hounsfield units (HU) in the region of interest. The applied threshold to measure the bone mineral density of newly regenerated bone was 200 HU. The bone volume fraction was defined as the volume ratio of de novo bone to the total implant within the region of interest (Figure 4). In addition, the mean tissue density of the region of interest was investigated at different time periods, including before surgery and six months after surgery.

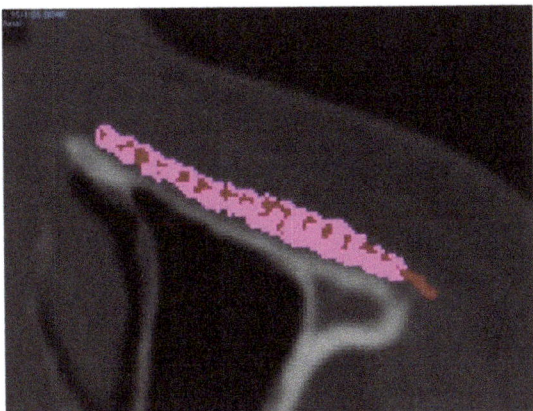

**Figure 4.** The bone volume fraction was defined as a ratio of the de novo bone volume to the total implant volume within the region of interest. Red area notes the region where the tissue density was measured over 200HU, while the purple area denotes the PCL/β-TCP scaffold.

*2.4. Tensile Test of the Scaffold*

Tensile testing was performed using a single column universal testing machine (Instron, Norwood, MA, USA). The dimension of the scaffold sample was standardized to 10 × 40 × 1 (mm), and porosity was 50%. The number of the sample for the test was 7. The Young's modulus was calculated by the linear curve of the stress–stain curve.

## 3. Results

Eight patients were included in this study, presenting a wide range of maxillary defects of various etiologies. The causes of the defects were as follows: intraosseous hemangioma in two patients, immediate reconstruction following cancer ablation in three patients, and Romberg disease, traumatic facial deformity, and fibrous dysplasia in one patient each. Five of the eight patients underwent immediate reconstruction following tumor ablation, while three underwent delayed reconstruction. There was a case of wound dehiscence caused by partial flap necrosis, which required wound coverage by a local flap. Detailed information regarding demographics are depicted in Table 1.

**Table 1.** Demographics of patients.

| | Sex | Age | Cause of Defect | Location | Type of Defect | Onset of Reconstruction | Postoperative Complication | Underlying Disease |
|---|---|---|---|---|---|---|---|---|
| Patient #1 | F | 21 | Intraosseous hemangioma | Rt. | N.A. | 24-month delayed | None | None |
| Patient #2 | M | 19 | Romberg disease | Rt. | N.A | Delayed | None | None |
| Patient #3 | M | 51 | Intraosseous hemangioma | Lt. | V | Immediate | None | None |
| Patient #4 | F | 50 | Traumatic facial deformity | Lt. | N.A | 60-month delayed | None | None |
| Patient #5 | M | 21 | Fibrous dysplasia | Lt. | IIIb | Immediate | None | None |
| Patient #6 | F | 43 | Radiation necrosis following nasal cavity cancer ablation | Lt. | IIIb | Immediate | Wound dehiscence due to delayed wound healing | Diabetes |
| Patient #7 | F | 44 | Radiation necrosis following maxillary sinus cancer ablation | Lt. | IIIb | Immediate | None | Hypertension |
| Patient #8 | M | 42 | Maxillary sinus cancer | Rt. | V | Immediate | None | None |

N.A.: Not applicable.

Regarding surgical details, in four of the eight patients, the 3D-printed implant was inserted through a perioral and conjunctival incision. The other four patients underwent concurrent free flap or free bone grafts. In patients who had undergone cancer ablation, a head and neck surgeon used lateral rhinotomy and a Weber–Ferguson incision. A bone-forming substance was used in three patients: a demineralized bone matrix (DMB) in two patients and a demineralized calcium phosphate bone substitute in one patient. Revisional operation was required in four patients who underwent a secondary fat graft and one patient who underwent local wound coverage to treat partial flap necrosis (Table 2).

Table 2. Surgical details.

| | Reconstructive Option | Incisional Approach | Application of Bone Regeneration Material | Implant Fixation | Revisional Operation |
|---|---|---|---|---|---|
| Patient #1 | Fat graft | Gingivobuccal and transconjunctival | None | HA-PLLA resorbable plate and screws | Secondary fat graft |
| Patient #2 | Fat graft | Gingivobuccal and transconjunctival | Resorbable calcium phosphate bone substitute | Titanium miniplate and screws | Secondary fat graft |
| Patient #3 | Fat graft | Gingivobuccal and transconjunctival | DBM | Titanium miniplate and screws | Secondary fat graft |
| Patient #4 | Fat graft | Gingivobuccal and transconjunctival | None | Titanium miniplate and screws | Secondary fat graft |
| Patient #5 | Iliac bone graft | Gingivobuccal and transconjunctival | DBM | Wire steel | None |
| Patient #6 | RFFF, Iliac bone graft | Weber-Ferguson approach | None | Titanium miniplate and screws | Local wound coverage |
| Patient #7 | ALT FF, RFFF | Lateral rhinotomy and subcillary approach | None | Titanium miniplate and screws | None |
| Patient #8 | None | Lateral rhinotomy and subcillary approach | None | Wire steel | None |

HA-PLLA: Hydroxyapatite/poly-l-lactide; DBM: Demineralized bone matrix; ALT FF: Anterolateral thigh free flap; RFFF: Radial forearm free flap.

The result of the volume analysis was as follows. The mean preoperatively planned implant volume was 11.32 mm$^3$, ranging from 2.16 mm$^3$ to 30.37 mm$^3$. The mean postoperatively actual implant volume was 10.21 mm$^3$, ranging from 1.84 mm$^3$ to 28.22 mm$^3$. After the superimposition of two images, the mean volume conformity was 79.71%, ranging from 70.89% to 86.31%. Postoperatively, the de novo formation of bone was calculated and the mean was 2.15 mm$^3$, ranging from 0.22 mm$^3$ to 7.15.mm$^3$. The bone volume fraction was obtained as the ratio of de novo bone volume and postoperative implant volume, with a mean of 23.34%, ranging from 7.81% to 66.21%. Mean tissue density in the region of interest was 188.84 HU, ranging from 151.48 HU to 291.74 HU (Table 3).

Table 3. Volume and density analysis.

| | Preoperatively Planned Implant Volume (mm$^3$) | Postoperative Actual Implant Volume (mm$^3$) | Conforming Volume after Superimposition (mm$^3$) | Volume Conformity (%) | Postoperative Newly Generated Bone Volume (mm$^3$) | Bone Volume Fraction (%) | Postoperative Mean Tissue Density (HU) |
|---|---|---|---|---|---|---|---|
| Patient #1 | 11.82 | 10.55 | 9.62 | 81.39 | 1.25 | 11.87 | 165.55 |
| Patient #2 | 8.76 | 8.42 | 7.51 | 85.77 | 3.15 | 37.41 | 184.22 |
| Patient #3 | 3.72 | 3.22 | 2.64 | 70.89 | 0.25 | 7.81 | 223.00 |
| Patient #4 | 2.16 | 1.84 | 1.66 | 76.76 | 1.22 | 66.21 | 291.74 |
| Patient #5 | 30.37 | 28.22 | 26.22 | 86.31 | 7.15 | 25.34 | 184.55 |
| Patient #6 | 15.88 | 13.51 | 11.53 | 72.59 | 2.13 | 15.73 | 168.44 |
| Patient #7 | 2.74 | 2.49 | 2.16 | 79.05 | 0.22 | 8.80 | 151.48 |
| Patient #8 | 15.09 | 13.42 | 12.82 | 84.96 | 1.82 | 13.54 | 182.51 |

HU: Hounsfield unit.

In the mechanical property test, the Young's modulus of the standardized scaffold with 50% porosity was 162.7 ± 12.8 MPa (Table 4).

**Table 4.** Experimental result of mechanical property test.

| Scaffold Dimension (mm) | Porosity (%) | Young's Modulus | Number of Sample |
| --- | --- | --- | --- |
| 10 × 40 × 1 | 50 | 162.7 ± 12.8 MPa | 7 |

*3.1. Case Presentation*

Representative cases with clinical pictures are described in this section.

3.1.1. Case 1

Patient #1 was 21-year-old female who underwent delayed reconstruction 24 months after ablation of intraosseous hemangioma. The maxillary bone defect was exposed using the gingivobuccal and transconjunctival approaches. A 3D-printed PCL/β-TCP scaffold was fitted into the defect, and the patient required no further resection of the bony structures. The implant was fixed using a resorbable plate and screws made of HA-PLLA (hydroxyapatite/poly-l-lactide (Figures 5 and 6).

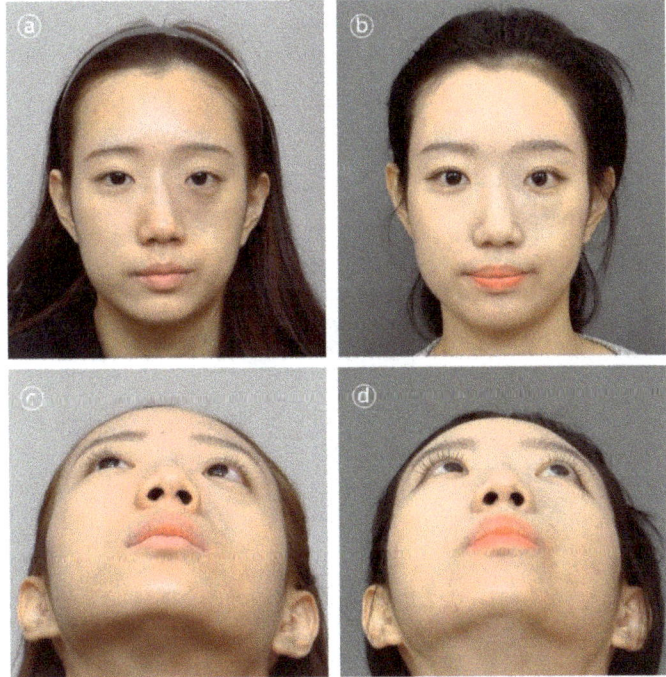

**Figure 5.** Clinical photographs in patient #1. Contour and symmetry of left cheek region was restored. (**a,b**) Pre- and postoperative 6-month frontal view photographs. (**c,d**) Pre- and postoperative 6-month basal view photographs.

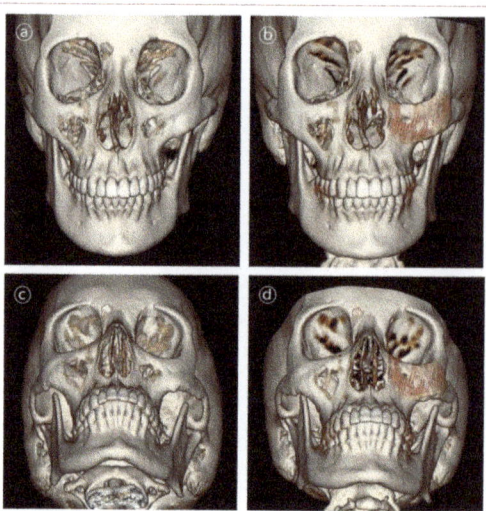

**Figure 6.** 3D CT images in patient #1. Contour and symmetry of left zygomatico-maxillary region was restored with de novo bone formation. (**a,b**) Pre- and postoperative 6-month frontal view CT images. (**c,d**) Pre- and postoperative 6-month basal view CT images.

3.1.2. Case 2

Patient #5 was 21-year-old male who underwent immediate reconstruction following the en bloc resection of maxillary fibrous dysplasia, defined as a type V defect. The patient underwent reconstruction with the 3D-printed PCL/β-TCP scaffold through a conventional gingivobuccal and transconjunctival incisions. The 3D-printed implant was fixated with wire steel. There was no complication in the long-term follow-up (Figures 7 and 8).

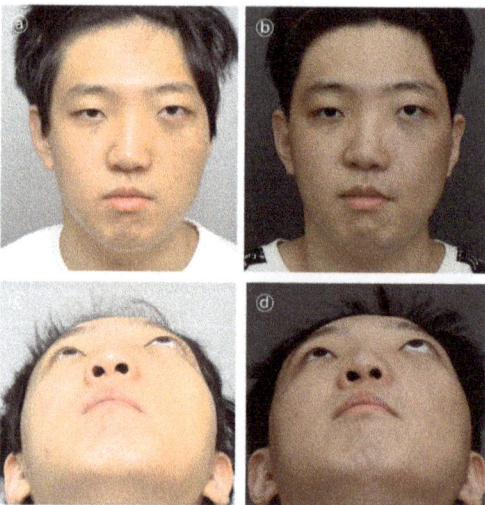

**Figure 7.** Clinical photographs in patient #5. Contour and symmetry of left cheek region was improved. (**a,b**) Pre- and postoperative 6-month frontal view photographs. (**c,d**) Pre- and postoperative 6-month basal view photographs.

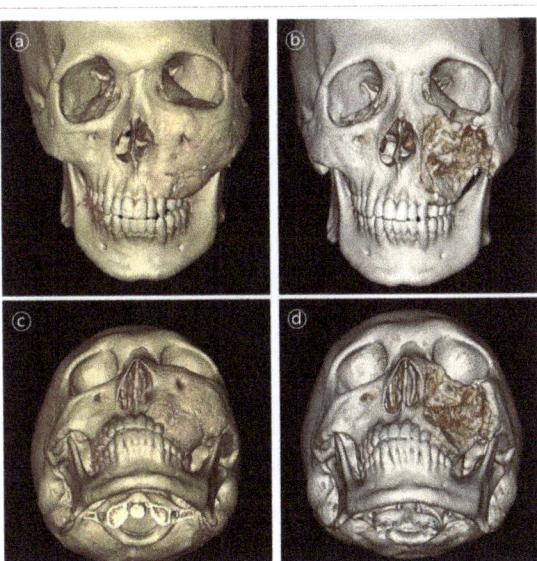

**Figure 8.** 3D CT images in patient #5. Contour and symmetry of left zygomatico-maxillary region was improved. (**a,b**) Pre- and postoperative 6-month frontal view CT images. (**c,d**) Pre- and postoperative 6-month basal view CT images.

## 4. Discussion

Polycaprolactone (PCL) is one of the polymers prepared by ring opening polymerization of ε-caprolactone using a variety of catalysts. It safely degrades into carbon dioxide and water over 2–3 years and provides a suitable scaffold for guided bone regeneration [23,24]. The PCL/β-TCP scaffolds used in this study had a 3D shape, moderate rigidity, and relatively high elasticity and were manufactured with a patient-specific design. This property allows surgeons to manipulate and mold the implants using a blade or scissors. In our mechanical property test, Young's modulus of the scaffold with 50% porosity was 162.7 ± 12.8 MPa, which is a similar level to that of the human mandibular trabecular bone (6.9 to 199.5 MPa) [25]. It was strong enough to maintain a three-dimensional shape when applied to clinical practice, and also had an adequate elasticity to be carved using tools available in the operating room. However, this might be insufficient to mimic the compressive strength and modulus of cortical bone itself [26,27]. Thus, the characteristics of PCL/β-TCP should be carefully considered depending on the amount of bony defect and surrounding soft tissue. The scaffold might be insufficient to be applied alone in the reconstruction of the whole zygomatico-maxillary complex. However, it was sufficient to bear the tension and compression during biomechanics of the upper jaw as when indicated as an onlay graft onto the bony surface or interpositional graft between the bony gaps. Overall, we did not find any bony instability or occlusal complication during the follow-up period. We suggested that the loading force should be distributed to the underlying bony strut through secure fixation with titanium screws and to overlap with the surrounding bony structure.

The PCL scaffold has been widely used in craniofacial reconstruction of various forms, including mesh, membrane, plate, and 3D implants [28–31]. Several authors have used PCL mesh in rhinoplasty to replace autogenous cartilage grafts [32]. They have reported that PCL mesh with a 3D structure was a safe and effective material and that it could maintain volume without any foreign body reaction [28]. However, unlike our study, PCL implants in the previous literature have only been applied to 2D reconstruction. Recently, Han et al. used 3D PCL implants in three cases of maxillary reconstruction following cancer ablation.

All patients showed favorable outcomes. No signs of infection were observed in any of the three patients, and the existing native tissue was successfully fused with filling of the pores. So far, there has been few reports on the combined use of PCL and β-TCP as a 3D scaffold in clinical cases. We applied a patient-specific PCL/β-TCP scaffold to treat various maxillary defects with a range of etiologies, including facial asymmetry due to Romberg's disease and ablation of fibrous dysplasia and hemangioma. Notably, we performed a more structured analysis in our cases, measuring volume conformity and bone density.

Regarding the volume conformity, suboptimal results were obtained in two of the eight cases who underwent immediate reconstruction following maxillary sinus cancer ablation. Although we designed the implants with a 3D shape following the resection plan, the design did not always fit the actual resection margin. This resulted in less conformity between the preoperative simulated and postoperative actual implant volumes. However, experienced head and neck surgeons were fully capable of adjusting the shape of the implants because the material had elastic properties.

Meanwhile, our study reported a case of implant exposure in a patient who had undergone radiation treatment. We reasoned that the wound dehiscence had resulted from delayed wound healing in the irradiated field, especially in the naso-orbital region, rather than from the implant itself. It follows that the implant should be covered with a durable and thick flap, especially when patients have undergone previous radiation, and that meticulous debridement of remaining unhealthy tissue should be carried out to avoid wound complications.

Another complication of the biomaterial that should be considered is the possibility of an allergic reaction. Some rare complications have been reported with the use of biodegradable material due to the wide range of foreign body reactions [33–35]. Although there was no allergic reaction reported in our cases, the use of PCL might lead to serious foreign body reactions. Some researchers reported on long-term, late-onset inflammatory complications including granuloma formation, late allergic reaction and chronic inflammation after dermatologic application of PCL-based fillers [36,37]. This reaction seemed to result from an immune overreaction of the host tissue to the product, which is related with underlying inflammatory status of the patient. Thus, the safety of the PCL/β-TCP scaffold in our cases should be proven in the long-term study

In our previous research, we reported on the three-dimensional internal structure of a scaffold using 3D printing [26,38–40]. In the case of our 3D printed scaffold, it has an internal structure in which pores with a size of several hundred micrometers are completely interconnected by a layer-by-layer fabrication method. When implanted into the body, these perfectly connected pores are advantageous for the penetration of surrounding cells, and also help the engraftment of regenerated tissue inside the artificial scaffold as blood vessels are connected.

The effect of the material composition and porosity of a scaffold on its properties, including cell proliferation and differentiation, stiffness, and degradation, has been discussed in the literature [11,41–47]. The addition of β-TCP in PCL was shown to improve the scaffold's mechanical performance and increase osteogenic cell proliferation and differentiation [41,42]. By increasing the β-TCP concentration in the scaffolds, significantly higher mineralization was achieved compared to the pure PCL [48]. In addition, the bioceramic composition in the PCL scaffold was shown to increase water absorption and induce hydrophilic properties, which can be useful to prevent nutrient loss during bone regeneration [45]. Other considerations are the porosity, pore size, and permeability of the scaffold, which plays a significant role in biological delivery and tissue regeneration [11,46,47]. Larger pore size and porosity could be beneficial for bone tissue growth but may affect the compressive strength and modulus of the scaffold. Bruyas et al. found that both an increasing amount of β-TCP and decreasing porosity augmented the modulus of the 3D printed scaffolds, while decreasing the elasticity [43].

In our experience, when the amount of β-TCP in PCL is increased, viscosity also increases, and as PCL/β-TCP blend viscosity affects scaffold printing speed, 3D printer feed rate reduces, and the polymer is exposed to more thermal energy. When the weight proportion of β-TCP in PCL was more than 20% and the pore size was set to larger than 500 μm, we observed that the printing accuracy and mechanical strengths decreased. Thus, we used the PCL/β-TCP scaffold with a ratio of 80:20 and pore size of 500 μm to achieve balance between β-TCP content and printing rate.

The degradation profile of the scaffold is another factor that should be considered. The PCL has extremely slow progress of degradation, ranging from 2 to 4 years, while the TCP has an unpredictable biodegradation profile, ranging from 6 to 24 months [44,49,50]. In general, it was reported that the PCL/β-TCP composites had a faster degradation rate than that of pure PCL. Yeo et al. reported the PCL–20% TCP scaffold gradually degraded within 6 months, while maintaining its pore interconnectivity for newly mature bone to form [24]. Initial degradation of β-TCP can produce calcium ions and enhance mineralization, thereby promoting osteogenic differentiation of adipose-derived stem cells. Bruyas et al. found that higher ceramic content of over 40% TCP might lead to structural integrity of the scaffold due to the extremely high rate of degradation [43]. We agreed on their opinion in that such a manipulation of the ceramic ratio to create an ideal bioresorbable plate to match the natural healing course of bone formation. From CT findings obtained during the six-month follow-up of clinical cases, we judged that the 80:20 proportion of PCL:β-TCP and 500 μm pore size of the implant were adequate to enhance earlier bone growth and maintain durability. Other animal studies also corroborated this view, reporting neovascularization, sufficient soft tissue ingrowth, and the absence of extensive inflammation with this pore size and porosity [49].

We concluded that bone regeneration was confirmed based on CT scan results six months after surgery. In particular, it was based on the bone mineral density value from the CT image. We thought that the bone mineral density value reflected not only the purely regenerated bone but also the density of the implanted scaffold as well. However, due to the radiolucent characteristic of the biodegradable polymer, the contribution to the bone mineral density value is insignificant. Nevertheless, histological analysis from the biopsy tissue might be required for confirming the obvious bone regeneration, but it has limitation due to ethical issues. On the other hand, according to a previous study conducted by our research team, an obvious bone regeneration result was confirmed eight weeks after transplantation in an animal experiment using the same PCL/TCP scaffold applied in this study [51].

We used various materials, including a mixture of demineralized bone matrix and blood controlled thermal responsive polymer. Demineralized bone matrix has been widely used as a mixture material to enhance bone union and new bone formation [52]. Various artificial materials, including oxidized-irradiated alginate hydrogel and hydroxyapatite were combined with the 3D scaffold. Some authors have reported the combined use of bone morphogenic proteins (rhBMP-2) to treat mandibular defects [50]. However, we should be reluctant to apply this material in patients who have undergone cancer ablation as it is unclear whether rhBMP-2 promotes or inhibits tumor generation [53].

The present study had the following limitations: (1) As we assessed the density in a region of interest containing both the implant and new bone, we did not obtain the actual bone density, which might be lower than the normal bony structure outside of the implant; (2) Although a degradation period from 2 to 4 years for PCL and 6 to 24 months for TCP are known, the speed of degradation will vary depending on the transplant site due to characteristic of hydrolysis. Therefore, a long-term follow up of more than 5 years is required for future studies; (3) The measured efficacy of PCL mesh in bone formation may have been confounded because we also applied osteoblastic agents. In the present study, we could not assess the efficacy of the combined mixture substances for bone formation, as we performed no comparative analysis. More structured investigation is necessary, with a prospective, comparative, controlled design.

## 5. Conclusions

The PCL/β-TCP scaffold can provide durable support and enhance bone formation in complex zygomatico-maxillary defects. The combined use of virtual surgical simulations, 3D printing techniques, and biodegradable implants may replace traditional non-absorbable implants because the method is more accurate and the materials more biocompatible.

**Author Contributions:** Conceptualization, J.-W.C.; methodology, Y.-C.K.; software, Y.-C.K.; validation, Y.-C.K.; formal analysis, Y.-C.K.; investigation, Y.-C.K., E.-J.L. and J.-H.S.; resources, J.-W.C.; data curation, Y.-C.K., J.-C.M. and H.-J.P.; writing—original draft preparation, W.-S.J.; writing—review and editing, Y.-C.K. and J.-H.S.; visualization, Y.-C.K.; supervision, J.-W.C.; project administration, J.-W.C. All authors have read and agreed to the published version of the manuscript.

**Funding:** This work was supported by the Demonstration Program of Industrial Technology for a 3D Printing Medical Device (P0008811) funded by the Korea Institute for Advancement of Technology (KIAT).

**Institutional Review Board Statement:** This study was conducted according to the Declaration of Helsinki and approved by the independent Ethics Committee/Institutional Review Board of the Asan Medical Center. (approval number: 2021-1292).

**Informed Consent Statement:** Written informed consent has been obtained from the patient(s) to publish this paper.

**Data Availability Statement:** Not applicable.

**Acknowledgments:** We specially thank to Eun-Ju Lee and Jin-Hyung Shim (Research Institute, T&R Biofab Co., Ltd., Seongnam-si, Korea, Department of mechanical engineering, Korea Polytechnic University, Sihueng-si, Korea) for adding the experimental evidence for the current study.

**Conflicts of Interest:** The funders had no role in the design of the study; in the collection, analyses, or interpretation of data; in the writing of the manuscript, or in the decision to publish the results.

## References

1. Chepeha, D.B.; Wang, S.J.; Marentette, L.J.; Bradford, C.R.; Boyd, C.M.; Prince, M.E.; Teknos, T.N. Restoration of the orbital aesthetic subunit in complex midface defects. *Laryngoscope* **2004**, *114*, 1706–1713. [CrossRef] [PubMed]
2. Fu, K.; Liu, Y.; Gao, N.; Cai, J.; He, W.; Qiu, W. Reconstruction of Maxillary and Orbital Floor Defect With Free Fibula Flap and Whole Individualized Titanium Mesh Assisted by Computer Techniques. *J. Oral Maxillofac. Surg.* **2017**, *75*, 1791.e1. [CrossRef] [PubMed]
3. Orringer, J.S.; Barcelona, V.; Buchman, S.R. Reasons for removal of rigid internal fixation devices in craniofacial surgery. *J. Craniofac. Surg.* **1998**, *9*, 40–44. [CrossRef] [PubMed]
4. Schmidt, B.L.; Perrott, D.H.; Mahan, D.; Kearns, G. The removal of plates and screws after Le Fort I osteotomy. *J. Oral Maxillofac. Surg.* **1998**, *56*, 184–188. [CrossRef]
5. Bell, R.B.; Kindsfater, C.S. The use of biodegradable plates and screws to stabilize facial fractures. *J. Oral Maxillofac. Surg.* **2006**, *64*, 31–39. [CrossRef]
6. Mackool, R.; Yim, J.; McCarthy, J.G. Delayed degradation in a resorbable plating system. *J. Craniofac. Surg.* **2006**, *17*, 194–197, discussion 197–198. [CrossRef]
7. Wiltfang, J.; Merten, H.A.; Schultze-Mosgau, S.; Schrell, U.; Wenzel, D.; Kessler, P. Biodegradable miniplates (LactoSorb): Long-term results in infant minipigs and clinical results. *J. Craniofac. Surg.* **2000**, *11*, 239–243, discussion 244–235. [CrossRef]
8. Tarsitano, A.; Battaglia, S.; Ciocca, L.; Scotti, R.; Cipriani, R.; Marchetti, C. Surgical reconstruction of maxillary defects using a computer-assisted design/computer-assisted manufacturing-produced titanium mesh supporting a free flap. *J. Craniomaxillofac. Surg.* **2016**, *44*, 1320–1326. [CrossRef]
9. Zhang, W.B.; Yu, Y.; Mao, C.; Wang, Y.; Guo, C.B.; Yu, G.Y.; Peng, X. Outcomes of Zygomatic Complex Reconstruction With Patient-Specific Titanium Mesh Using Computer-Assisted Techniques. *J. Oral Maxillofac. Surg.* **2019**, *77*, 1915–1927. [CrossRef]
10. Lethaus, B.; Kessler, P.; Boeckman, R.; Poort, L.J.; Tolba, R. Reconstruction of a maxillary defect with a fibula graft and titanium mesh using CAD/CAM techniques. *Head Face Med.* **2010**, *6*, 16. [CrossRef]
11. Kumar, A.; Mir, S.M.; Aldulijan, I.; Mahajan, A.; Anwar, A.; Leon, C.H.; Terracciano, A.; Zhao, X.; Su, T.L.; Kalyon, D.M.; et al. Load-bearing biodegradable PCL-PGA-beta TCP scaffolds for bone tissue regeneration. *J. Biomed. Mater. Res. Part B Appl. Biomater.* **2021**, *109*, 193–200. [CrossRef]
12. Mkhabela, V.J.; Ray, S.S. Poly(epsilon-caprolactone) nanocomposite scaffolds for tissue engineering: A brief overview. *J. Nanosci. Nanotechnol.* **2014**, *14*, 535–545. [CrossRef] [PubMed]
13. Cao, H.; Kuboyama, N. A biodegradable porous composite scaffold of PGA/beta-TCP for bone tissue engineering. *Bone* **2010**, *46*, 386–395. [CrossRef] [PubMed]

14. Walsh, W.R.; Vizesi, F.; Michael, D.; Auld, J.; Langdown, A.; Oliver, R.; Yu, Y.; Irie, H.; Bruce, W. Beta-TCP bone graft substitutes in a bilateral rabbit tibial defect model. *Biomaterials* **2008**, *29*, 266–271. [CrossRef] [PubMed]
15. Vaněček, V.; Klíma, K.; Kohout, A.; Foltán, R.; Jiroušek, O.; Šedý, J.; Štulík, J.; Syková, E.; Jendelová, P. The combination of mesenchymal stem cells and a bone scaffold in the treatment of vertebral body defects. *Eur. Spine J.* **2013**, *22*, 2777–2786. [CrossRef] [PubMed]
16. Shim, J.H.; Yoon, M.C.; Jeong, C.M.; Jang, J.; Jeong, S.I.; Cho, D.W.; Huh, J.B. Efficacy of rhBMP-2 loaded PCL/PLGA/β-TCP guided bone regeneration membrane fabricated by 3D printing technology for reconstruction of calvaria defects in rabbit. *Biomed. Mater.* **2014**, *9*, 065006. [CrossRef]
17. Shim, J.H.; Huh, J.B.; Park, J.Y.; Jeon, Y.C.; Kang, S.S.; Kim, J.Y.; Rhie, J.W.; Cho, D.W. Fabrication of blended polycaprolactone/poly (lactic-co-glycolic acid)/β-tricalcium phosphate thin membrane using solid freeform fabrication technology for guided bone regeneration. *Tissue Eng. Part A* **2013**, *19*, 317–328. [CrossRef]
18. Lee, S.; Choi, D.; Shim, J.H.; Nam, W. Efficacy of three-dimensionally printed polycaprolactone/beta tricalcium phosphate scaffold on mandibular reconstruction. *Sci. Rep.* **2020**, *10*, 4979. [CrossRef]
19. Mellor, L.F.; Nordberg, R.C.; Huebner, P.; Mohiti-Asli, M.; Taylor, M.A.; Efird, W.; Oxford, J.T.; Spang, J.T.; Shirwaiker, R.A.; Loboa, E.G. Investigation of multiphasic 3D-bioplotted scaffolds for site-specific chondrogenic and osteogenic differentiation of human adipose-derived stem cells for osteochondral tissue engineering applications. *J. Biomed. Mater. Res. B Appl. Biomater.* **2020**, *108*, 2017–2030. [CrossRef]
20. Khojasteh, A.; Behnia, H.; Hosseini, F.S.; Dehghan, M.M.; Abbasnia, P.; Abbas, F.M. The effect of PCL-TCP scaffold loaded with mesenchymal stem cells on vertical bone augmentation in dog mandible: A preliminary report. *J. Biomed. Mater. Res. B Appl. Biomater.* **2013**, *101*, 848–854. [CrossRef]
21. Boccaccini, A.R.; Roelher, J.A.; Hench, L.L.; Maquet, V.; Jérôme, R. A Composites Approach to Tissue Engineering. In Proceedings of the 26th Annual Conference on Composites, Advanced Ceramics, Materials, and Structures: B: Ceramic Engineering and Science Proceedings, Cocoa Beach, FL, USA, 13–18 January 2022; pp. 805–816. [CrossRef]
22. Brown, J.S.; Shaw, R.J. Reconstruction of the maxilla and midface: Introducing a new classification. *Lancet Oncol.* **2010**, *11*, 1001–1008. [CrossRef]
23. Sun, H.; Mei, L.; Song, C.; Cui, X.; Wang, P. The in vivo degradation, absorption and excretion of PCL-based implant. *Biomaterials* **2006**, *27*, 1735–1740. [CrossRef] [PubMed]
24. Yeo, A.; Rai, B.; Sju, E.; Cheong, J.J.; Teoh, S.H. The degradation profile of novel, bioresorbable PCL-TCP scaffolds: An in vitro and in vivo study. *J. Biomed. Mater. Res. A* **2008**, *84*, 208–218. [CrossRef] [PubMed]
25. Lakatos, É.; Magyar, L.; Bojtár, I. Material Properties of the Mandibular Trabecular Bone. *J. Med. Eng.* **2014**, *2014*, 470539. [CrossRef] [PubMed]
26. Kim, J.Y.; Ahn, G.; Kim, C.; Lee, J.S.; Lee, I.G.; An, S.H.; Yun, W.S.; Kim, S.Y.; Shim, J.H. Synergistic Effects of Beta Tri-Calcium Phosphate and Porcine-Derived Decellularized Bone Extracellular Matrix in 3D-Printed Polycaprolactone Scaffold on Bone Regeneration. *Macromol. Biosci.* **2018**, *18*, e1800025. [CrossRef]
27. Milne, N.; Fitton, L.; Kupczik, K.; Fagan, M.; O'Higgins, P. The role of the zygomaticomaxillary suture in modulating strain distribution within the skull of Macaca fascicularis. *HOMO J. Comp. Hum. Biol.* **2009**, 281.
28. Park, Y.J.; Cha, J.H.; Bang, S.I.; Kim, S.Y. Clinical Application of Three-Dimensionally Printed Biomaterial Polycaprolactone (PCL) in Augmentation Rhinoplasty. *Aesthetic Plast. Surg.* **2019**, *13*, 137–146. [CrossRef]
29. Han, H.H.; Shim, J.H.; Lee, H.; Kim, B.Y.; Lee, J.S.; Jung, J.W.; Yun, W.S.; Baek, C.H.; Rhie, J.W.; Cho, D.W. Reconstruction of Complex Maxillary Defects Using Patient-specific 3D-printed Biodegradable Scaffolds. *Plast. Reconstr. Surg. Glob. Open* **2018**, *6*, e1975. [CrossRef]
30. Kim, S.Y. Application of the three-dimensionally printed biodegradable polycaprolactone (PCL) mesh in repair of orbital wall fractures. *J. Craniomaxillofac. Surg.* **2019**, *47*, 1065–1071. [CrossRef]
31. Park, S.H.; Yun, B.G.; Won, J.Y.; Yun, W.S.; Shim, J.H.; Lim, M.H.; Kim, D.H.; Baek, S.A.; Alahmari, Y.D.; Jeun, J.H.; et al. New application of three-dimensional printing biomaterial in nasal reconstruction. *Laryngoscope* **2017**, *127*, 1036–1043. [CrossRef]
32. Kim, D.H.; Lee, I.H.; Yun, W.S.; Shim, J.H.; Choi, D.; Hwang, S.H.; Kim, S.W. Long-term efficacy and safety of 3D printed implant in patients with nasal septal deformities. *Eur. Arch. Oto-Rhino-Laryngol.* **2021**, 1–8. [CrossRef] [PubMed]
33. Sivaloganathan, S.; Amr, R.; Shrivastava, R.; Relwani, J. The Risotto sign - a severe inflammatory bursitis with rice body formation, complicating a rotator cuff repair with a bioabsorbable suture anchor. *JRSM Open* **2015**, *6*, 2054270414562986. [CrossRef] [PubMed]
34. Mastrokalos, D.S.; Paessler, H.H. Allergic reaction to biodegradable interference poly-L-lactic acid screws after anterior cruciate ligament reconstruction with bone-patellar tendon-bone graft. *Arthroscopy* **2008**, *24*, 732–733. [CrossRef] [PubMed]
35. Friedman, P.M.; Mafong, E.A.; Kauvar, A.N.; Geronemus, R.G. Safety data of injectable nonanimal stabilized hyaluronic acid gel for soft tissue augmentation. *Dermatol. Surg.* **2002**, *28*, 491–494. [CrossRef]
36. Skrzypek, E.; Górnicka, B.; Skrzypek, D.M.; Krzysztof, M.R. Granuloma as a complication of polycaprolactone-based dermal filler injection: Ultrasound and histopathology studies. *J. Cosmet. Laser Ther.* **2019**, *21*, 65–68. [CrossRef]
37. Chiang, C.H.; Peng, J.H.; Peng, H.P. Filler-induced granuloma from polycaprolactone-based collagen stimulator injection in the tear trough area: A case report. *J. Cosmet. Dermatol.* **2021**, *20*, 1529–1531. [CrossRef]

38. Yun, S.; Choi, D.; Choi, D.J.; Jin, S.; Yun, W.S.; Huh, J.B.; Shim, J.H. Bone Fracture-Treatment Method: Fixing 3D-Printed Polycaprolactone Scaffolds with Hydrogel Type Bone-Derived Extracellular Matrix and β-Tricalcium Phosphate as an Osteogenic Promoter. *Int. J. Mol. Sci.* **2021**, *22*, 9084. [CrossRef]
39. Bae, E.B.; Park, K.H.; Shim, J.H.; Chung, H.Y.; Choi, J.W.; Lee, J.J.; Kim, C.H.; Jeon, H.J.; Kang, S.S.; Huh, J.B. Efficacy of rhBMP-2 Loaded PCL/β-TCP/bdECM Scaffold Fabricated by 3D Printing Technology on Bone Regeneration. *Biomed. Res. Int.* **2018**, *2018*, 2876135. [CrossRef]
40. Park, H.; Kim, J.S.; Oh, E.J.; Kim, T.J.; Kim, H.M.; Shim, J.H.; Yoon, W.S.; Huh, J.B.; Moon, S.H.; Kang, S.S.; et al. Effects of three-dimensionally printed polycaprolactone/β-tricalcium phosphate scaffold on osteogenic differentiation of adipose tissue- and bone marrow-derived stem cells. *Arch. Craniofac. Surg.* **2018**, *19*, 181–189. [CrossRef]
41. Shin, Y.M.; Park, J.-S.; Jeong, S.I.; An, S.-J.; Gwon, H.-J.; Lim, Y.-M.; Nho, Y.-C.; Kim, C.-Y. Promotion of human mesenchymal stem cell differentiation on bioresorbable polycaprolactone/biphasic calcium phosphate composite scaffolds for bone tissue engineering. *Biotechnol. Bioprocess Eng.* **2014**, *19*, 341–349. [CrossRef]
42. Huang, B.; Caetano, G.; Vyas, C.; Blaker, J.J.; Diver, C.; Bártolo, P. Polymer-Ceramic Composite Scaffolds: The Effect of Hydroxyapatite and β-tri-Calcium Phosphate. *Materials* **2018**, *11*, 129. [CrossRef] [PubMed]
43. Bruyas, A.; Lou, F.; Stahl, A.M.; Gardner, M.; Maloney, W.; Goodman, S.; Yang, Y.P. Systematic characterization of 3D-printed PCL/β-TCP scaffolds for biomedical devices and bone tissue engineering: Influence of composition and porosity. *J. Mater. Res.* **2018**, *33*, 1948–1959. [CrossRef] [PubMed]
44. Feng, P.; Wu, P.; Gao, C.; Yang, Y.; Guo, W.; Yang, W.; Shuai, C. A Multimaterial Scaffold With Tunable Properties: Toward Bone Tissue Repair. *Adv. Sci.* **2018**, *5*, 1700817. [CrossRef] [PubMed]
45. Kim, Y.; Kim, G. Functionally graded PCL/β-TCP biocomposites in a multilayered structure for bone tissue regeneration. *Appl. Phys. A* **2012**, *108*, 949–959. [CrossRef]
46. Hollister, S.J. Porous scaffold design for tissue engineering. *Nat. Mater.* **2005**, *4*, 518–524. [CrossRef]
47. Polo-Corrales, L.; Latorre-Esteves, M.; Ramirez-Vick, J.E. Scaffold design for bone regeneration. *J. Nanosci. Nanotechnol.* **2014**, *14*, 15–56. [CrossRef]
48. Lu, L.; Zhang, Q.; Wootton, D.; Chiou, R.; Li, D.; Lu, B.; Lelkes, P.; Zhou, J. Biocompatibility and biodegradation studies of PCL/β-TCP bone tissue scaffold fabricated by structural porogen method. *J. Mater. Sci. Mater. Med.* **2012**, *23*, 2217–2226. [CrossRef]
49. Stal, S.; Hollier, L. The use of resorbable spacers for nasal spreader grafts. *Plast. Reconstr. Surg.* **2000**, *106*, 922–928, discussion 929–931. [CrossRef]
50. Schuckert, K.H.; Jopp, S.; Teoh, S.H. Mandibular defect reconstruction using three-dimensional polycaprolactone scaffold in combination with platelet-rich plasma and recombinant human bone morphogenetic protein-2: De novo synthesis of bone in a single case. *Tissue Eng. Part A* **2009**, *15*, 493–499. [CrossRef]
51. Shim, J.H.; Won, J.Y.; Park, J.H.; Bae, J.H.; Ahn, G.; Kim, C.H.; Lim, D.H.; Cho, D.W.; Yun, W.S.; Bae, E.B.; et al. Effects of 3D-Printed Polycaprolactone/β-Tricalcium Phosphate Membranes on Guided Bone Regeneration. *Int. J. Mol. Sci.* **2017**, *18*, 899. [CrossRef]
52. Zhang, M.; Matinlinna, J.P.; Tsoi, J.K.H.; Liu, W.; Cui, X.; Lu, W.W.; Pan, H. Recent developments in biomaterials for long-bone segmental defect reconstruction: A narrative overview. *J. Orthop. Translat.* **2020**, *22*, 26–33. [CrossRef] [PubMed]
53. Gao, Q.; Tong, W.; Luria, J.S.; Wang, Z.; Nussenbaum, B.; Krebsbach, P.H. Effects of bone morphogenetic protein-2 on proliferation and angiogenesis in oral squamous cell carcinoma. *Int. J. Oral Maxillofac. Surg.* **2010**, *39*, 266–271. [CrossRef] [PubMed]

Article

# Effect of Architected Structural Members on the Viscoelastic Response of 3D Printed Simple Cubic Lattice Structures

Ahmed Abusabir [1], Muhammad A. Khan [1,*], Muhammad Asif [2] and Kamran A. Khan [3,4,*]

[1] School of Aerospace, Transport and Manufacturing, Cranfield University, Cranfield MK43 0AL, UK; Ahmed.Abusabir@cranfield.ac.uk
[2] Department of Mechanical Engineering, National University of Sciences and Technology, Karachi 75350, Pakistan; muhammadasif@pnec.nust.edu.pk
[3] Advanced Digital & Additive Manufacturing Center, Khalifa University of Science and Technology, Abu Dhabi P.O. Box 127788, United Arab Emirates
[4] Department of Aerospace Engineering, Khalifa University, Abu Dhabi P.O. Box 127788, United Arab Emirates
* Correspondence: muhammad.a.khan@cranfield.ac.uk (M.A.K.); kamran.khan@ku.ac.ae (K.A.K.)

**Abstract:** Three-dimensional printed polymeric lattice structures have recently gained interests in several engineering applications owing to their excellent properties such as low-density, energy absorption, strength-to-weight ratio, and damping performance. Three-dimensional (3D) lattice structure properties are governed by the topology of the microstructure and the base material that can be tailored to meet the application requirement. In this study, the effect of architected structural member geometry and base material on the viscoelastic response of 3D printed lattice structure has been investigated. The simple cubic lattice structures based on plate-, truss-, and shell-type structural members were used to describe the topology of the cellular solid. The proposed lattice structures were fabricated with two materials, i.e., PLA and ABS using the material extrusion (MEX) process. The quasi-static compression response of lattice structures was investigated, and mechanical properties were obtained. Then, the creep, relaxation and cyclic viscoelastic response of the lattice structure were characterized. Both material and topologies were observed to affect the mechanical properties and time-dependent behavior of lattice structure. Plate-based lattices were found to possess highest stiffness, while the highest viscoelastic behavior belongs to shell-based lattices. Among the studied lattice structures, we found that the plate-lattice is the best candidate to use as a creep-resistant LS and shell-based lattice is ideal for damping applications under quasi-static loading conditions. The proposed analysis approach is a step forward toward understanding the viscoelastic tolerance design of lattice structures.

**Keywords:** 3D lattice structure; simple cubic lattice structures; plate-based lattice; shell-based lattice; truss-based lattice; ABS; PLA; 3D printing; FFF; viscoelastic behavior; relaxation; creep; cyclic loading

## 1. Introduction

A new generation of engineering materials, known as lattice structures (LSs), has recently found applications in biomedical [1], aerospace [2] and automotive [3]. Notable properties of LSs include their low density and high specific thermal, electrical and mechanical properties, energy absorption, and ability to reduce noise/vibration [4–6]. The overall response of LSs depend on the relative density, solid base material, and topology of the microstructure. For damping and energy absorption applications, a better understanding of the relationship between microstructure of the LS and their effective viscoelastic properties is required to obtain desired performance [7,8].

LSs consist of a solid skeleton and air pores. The architecture of microstructure influences their mechanical behaviors. Numerous architectures were proposed in the literature to describe the microstructure of LS. The architected LSs are classified into two categories: open-cell and closed-cell foams, with either a random or periodic arrangement [9]. Earlier design of three dimensional (3D) networks of LSs are usually designed using discrete structural members such as struts or truss members. The microstructure, such as, rhombic dodecahedron [10], tetrakaidecahedron [11,12], cubic [6,13,14], Kelvin [15], Gibson-Ashby [16] and gyroids [17] have been studied. Analytical solutions for the effective response of these LSs were obtained through beam theory for elastic behavior [12,18–20] and viscoelastic behavior [21–23]. For more complicated architected LS, finite element homogenization method has been used to predict the elastic [24–29] and viscoelastic [22,30] responses.

Recently, three-dimensional network structures have been developed with interesting geometries derived from atomic crystal structures system [31]. The network of these lattice structures can be constructed with different structural members such as truss-, plate-, or shell-based (triply periodic minimal-surface (TPMS)) [32,33]. Out of these structural members, the plate-based lattice structures [34,35], offer superior stiffness which makes them excellent candidates for load-bearing applications. However, the shell based LSs such as TPMS demonstrated good energy absorption characteristics. Tancogne-Dejean et al. [36] showed that the specific energy absorption of plate-based LSs is around 45% greater than that of truss-based LSs. The elastic and viscoelastic properties of these lattice structures have been studied and investigated using the finite element method (FEM). Khan et al. used micromechanical homogenization approach to compute the apparent viscoelastic behavior such as creep, relaxation under quasi-static loading and dynamic behavior under cyclic excitation [37,38], and [39]. Previous studies highlighted the excellent viscoelastic response of the architected LSs [40]. Comprehensive studies have been conducted using theoretical and simulation approaches to investigate the properties of cellular solids; however, very limited experimental investigations have been undertaken to determine the viscoelastic response of polymeric LSs [15,40,41]. Moreover, the effect of architected structural member and base material on the viscoelastic response of 3D printed lattice structure has not been investigated

The revolution and growth in additive manufacturing have allowed the fabrication of complex and precise geometries of LSs. Additive manufacturing (AM) offers high flexibility of design and rapid prototyping. In the recent review article, it has been discussed that AM can reduce the production cost of complex components and can be implemented not only for prototyping but also production using different approaches in design [41]. Additive Manufacturing technology has enabled the porosity and architecture of cellular solids to be controlled; therefore, the density and mechanical properties can be tailored [42] for several applications [43]. Additive manufacturing includes several processes; however, the 3D printing technology using material extrusion (MEX) process [44] has been widely used to fabricate complex geometries such as cellular solids. Moreover, the base materials have significant influenced on the design of LSs. The LSs should be able to contribute to the functional purpose of structure with excellent damping performance, strength-to-weight ratio, and others. Thermoplastic polymers have been widely utilized in the fabrication of cellular solids due to their adaptability for 3D printing and their unique properties. The most utilized polymers are acrylonitrile butadiene styrene (ABS), and polylactic acid (PLA) [7,8]. The comparison of the flexural properties of ABS, PLA and a PLA–wood composite manufactured through MEX process has been presented [45]. Several authors have extensively studied the manufacturing of PLA using MEX process such an in-process monitoring of temperature evolution, multiscale damage and fatigue modeling of PLA [46–48]. The influence of process parameters has also been investigated on the mechanical properties [49], impact resistance properties [50] and interlayer adhesion on the tensile strength of 3D printed PLA [51].

In this study, the effect of architected structural member geometry and base material on the viscoelastic response of 3D printed lattice structure has been experimentally investigated. The LSs possessing simple cubic symmetry based on plate-, truss-, and shell-type structural members were considered to describe the microstructure of the LSs. The proposed LSs were fabricated with two materials, i.e., PLA and ABS using the material extrusion (MEX) process. The quasi-static compression response of lattice structures was investigated, and mechanical properties were obtained. Then, the creep, relaxation and cyclic viscoelastic response of the lattice structure were characterized and some interesting conclusions were presented.

## 2. Methodology
### 2.1. Design of Lattice Architecture and Manufacturing

In this study, the three lattice microstructures of simple cubic family were considered. The three designs are named as simple cubic truss-based lattice (SCTL), simple cubic plate-based lattice (SCPL), and simple cubic shell-based lattice (SCSL). The SCTL, SCPL, and SCSL unit cells consist of struts, plate and shell, respectively. The arrangement of these structural members yield simple cubic LSs. Solidworks software was used to model the considered designs. The 3D designs were made with overall dimensions of $25 \times 25 \times 25$ mm$^3$. The investigation was conducted using two polymeric materials: Polylactic acid (PLA), and acrylonitrile butadiene styrene (ABS). Raw materials of ABS and PLA were procured in the form of filament with 1.75 mm thickness. The specifications of the utilized materials are shown below in Table 1 as provided by the manufacturing company.

Table 1. Specifications of PLA and ABS filaments.

| Materials | Thickness | Density | Young's Modulus | Strain at Break | Melting Temperature | Printing Temperature | Brand |
|---|---|---|---|---|---|---|---|
| ABS | 1.75 mm | 1.03 g/cm$^3$ | 2 GPa | 9% | 245 °C | 220–270 °C | RS Pro |
| PLA | 1.75 mm | 1.25 g/cm$^3$ | 2.7 GPa | 2% | 150 °C | 190–220 °C | Raise3D |

Additive manufacturing based on material extrusion (MEX) process, was adopted to fabricate all specimens. In this study, we employed the Raised3D Pro2 printer, which is equipped with a 0.4 nozzle. Several attempts were made to attain the best designs in terms of lightweight, manufacturability, and flexibility. The printing parameters that were given using software Idea Maker are shown below in Table 2.

Table 2. Parameters of 3D printing.

| Materials | Printing Temperature | Heated Bed Temperature | Printing Speed | Extrusion Width | Infill Topology |
|---|---|---|---|---|---|
| ABS | 250 °C | 100 °C | 50 mm/s | 0.4 mm | Lines |
| PLA | 205 °C | 60 °C | 50 mm/s | 0.4 mm | Lines |

For all specimens, the faces of the infill were perpendicular to the direction of the build (out-of-plane). All samples were printed with a raft platform to ensure the flatness of the base and stability throughout the printing process. Concerning solid infill density, all candidates were designed with 27% solid infill density. Table 4 shows the unit cell CAD design, the LS with array of $5 \times 5 \times 5$ unit cells, the design and printing parameter, and the fabricated LSs made of PLA and ABS. Throughout this study, the investigated samples will be referenced by the assigned ID codes shown in Table 3.

Table 3. ID codes of the 3D printed specimens.

| Material | Geometry | Code |
|---|---|---|
| ABS | Simple cubic Plate-based lattice | ABS/Plate-based |
| ABS | Simple cubic Truss-based lattice | ABS/Truss-based |
| ABS | Simple cubic Shell-based lattice | ABS/Shell-based |
| PLA | Simple cubic Plate-based lattice | PLA/Plate-based |
| PLA | Simple cubic Truss-based lattice | PLA/Truss-based |
| PLA | Simple cubic Shell-based lattice | PLA/Shell-based |

Table 4. Details of considered designs.

| Type | Unit Cell | Thickness | Infill Density | Lattice Structure | PLA Sample | ABS Sample |
|---|---|---|---|---|---|---|
| Shell-based lattice | | 0.5 mm | 27% | | | |
| Truss-based lattice | | 1.1 mm | 27% | | | |
| Plate-based lattice | | 0.5 mm | 27% | | | |

## 2.2. Experiments

Four experiments were performed to understand the mechanical properties and time-dependent behavior of the 3D printed polymeric LSs, i.e., quasi-static compression test, stress relaxation test, creep test, and compressive cyclic loading test, as shown in Figure 1. The experiments were conducted using an Instron universal testing machine with 5KN and 30KN load cells. The crosshead speed was 2.5 mm/min in all tests, chosen based on ASTM D1621-16 [52]. A pre-load was applied to guarantee a full initial contact between plates and specimen; all tests were conducted at room temperature. The experimental setup is shown in Figure 1. Pre-experimenting, the relative density of considered specimens were measured using a weight scale and Equation (1)

$$\bar{\rho} = \frac{\rho_c}{\rho} \quad (1)$$

where $\bar{\rho}$: relative density, $\rho_c$: density of cellular solid, $\rho$: density of solid material.

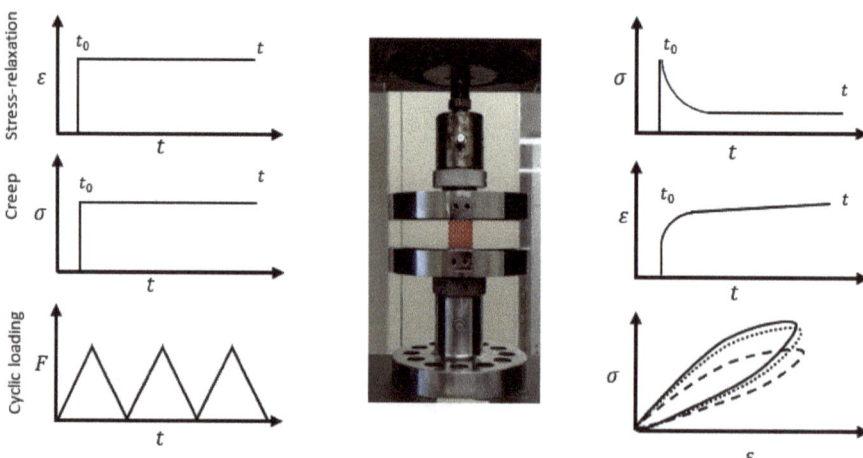

Figure 1. Experiment set-up and loading program.

### 2.2.1. Quasi-Static Compression Test

First, the quasi-static compression test was performed until fracture. The quasi-static compression test was performed according to ASTM D1621-16 "Standard Test Method for Compressive Properties of Rigid Cellular Plastics". The specimens were placed between the compression plates ensuring that the specimen centerline was aligned with the load cell centerline. Pre-loading was applied to ensure the stability of the samples and full initial contact between plates and specimens. The LSs were compressed at a constant crosshead speed of 2.5 mm/min and the effective stress–strain behavior was recorded. Many interesting characteristics of LS such as elastic modulus ($E$) and specific stiffness ($C$) were calculated using Equations (2) and (3).

$$E = \frac{\sigma}{\varepsilon} \quad (2)$$

$$C = \frac{E}{\rho} \quad (3)$$

### 2.2.2. Stress Relaxation Test

A stress relaxation test is necessary to understand the viscoelasticity behavior (time-dependent response), in which the specimen is compressed and held at a certain displacement; accordingly, the stress relaxation response is recorded as a function of time. The relaxation response can be measured by calculation stress-relaxation modulus using Equation (4).

$$E_{sr} = \frac{\sigma_t}{\varepsilon_0} \quad (4)$$

The samples made from the different materials were compressed to the same strain level called effective strain. The effective strain should be on or below the yield point, which was determined using the data obtained from the quasi-static compression test. It was considered to be a value below the least yield limit among the three samples. Table 5 shows the effective strain levels used during stress relaxation test. The stress relaxation test were performed according to ASTM E328 − 21: Standard Test Methods for Stress Relaxation for Materials and Structures [53]. In this study, the displacement was applied on the specimen at the strain rate of 2.5 mm/min until reaching the desired displacement. The position (displacement) was held constant for 30 min and the stress relaxation response was recorded as a function of time.

**Table 5.** Parameters of stress relaxation test.

| Sample | Hold at (Displacement) | Time for Holding |
|---|---|---|
| ABS/Truss-based lattice | 0.375 mm | 30 min |
| ABS/Plate-based lattice | 0.375 mm | 30 min |
| ABS/Shell-based lattice | 0.375 mm | 30 min |
| PLA/Truss-based lattice | 0.625 mm | 30 min |
| PLA/Plate-based lattice | 0.625 mm | 30 min |
| PLA/Shell-based lattice | 0.625 mm | 30 min |

### 2.2.3. Creep Test

Viscoelastic behavior can also be measured by creep testing, in which constant stress is applied for a period of time and changes in strain are observed as a function of time. The viscoelastic behavior can be measured by finding creep compliance (J) using Equation (5). The creep test was performed according to ASTM D2990 − 17: Standard Test Methods for Tensile, Compressive, and Flexural Creep and Creep-Rupture of Plastics [54]. Table 6 shows the forces levels used during creep test. Here, the sample was compressed with a strain rate of 2.5 mm/min to the predetermined load limit and held constant for 30 min.

While constant stress was applied, the strain will continue to increase with time and therefore recorded.

$$J_t = \frac{\varepsilon_t}{\sigma_0} \tag{5}$$

**Table 6.** Parameters of creep test.

| Sample | Hold at (Load) | Time for Holding |
|---|---|---|
| ABS/Truss-based lattice | 600 N | 30 min |
| ABS/Plate-based lattice | 600 N | 30 min |
| ABS/Shell-based lattice | 600 N | 30 min |
| PLA/Truss-based lattice | 1500 N | 30 min |
| PLA/Plate-based lattice | 1500 N | 30 min |
| PLA/Shell-based lattice | 1500 N | 30 min |

### 2.2.4. Compressive Cyclic Loading Test

The viscoelastic phenomenon and energy dissipation behavior of cellular materials can be observed by loading and unloading a specimen at a constant strain rate. The compressive cyclic loading test involves an appropriate repeating pattern of loading-unloading. The test may be conducted with a peak strain-controlled, or peak stress-controlled technique. In this study, the experiments were carried out with a peak stress-controlled method and the specimens were compressed with a strain rate of 2.5 mm/min to the predetermined load limit. The testing parameters are illustrated in Table 7. In total, three loading-unloading cycles were applied, and the load-displacement hysteresis loop were recorded. OriginLab software was used to calculate the area under the hysteresis curve, which represents the amount of energy absorption.

**Table 7.** Parameters of compressive cyclic loading test.

| Sample | Maximum Load | Number of Cycles |
|---|---|---|
| ABS/Truss-based lattice | 600 N | 3 Cycles |
| ABS/Plate-based lattice | 600 N | 3 Cycles |
| ABS/Shell-based lattice | 600 N | 3 Cycles |
| PLA/Truss-based lattice | 1500 N | 3 Cycles |
| PLA/Plate-based lattice | 1500 N | 3 Cycles |
| PLA/Shell-based lattice | 1500 N | 3 Cycles |

## 3. Results and Discussion

In this section, the data obtained from the experiments described above will be shown, analyzed, and discussed. A weight scale was used to measure the weight of the 3D printed specimens, then the relative density was calculated using Equation 1 as shown in Table 8.

**Table 8.** Weight, density, and relative density of the 3D printed specimen.

| Specimen | ABS/Shell | ABS/Plate | ABS/Truss | PLA/Shell | PLA/Plate | PLA/Truss |
|---|---|---|---|---|---|---|
| Weight (g) | 3.95 | 4.01 | 3.98 | 4.94 | 5.1 | 4.99 |
| Density (g/cm$^3$) | 0.253 | 0.257 | 0.255 | 0.316 | 0.326 | 0.319 |
| Relative density (g/cm$^3$) | 0.246 | 0.250 | 0.248 | 0.253 | 0.261 | 0.255 |

The measured values show that all ABS samples having almost the same weight with a variation of ±0.06 (1.5%), similarly shown in all PLA specimens with a variation of ±0.16 (3%). The equality in weights verifies that the initial designs have the same solids infill density and the excellent accuracy of the manufacturing process. Several factors may have contributed to the slight variations, such as the uncertainty of the scaling device, the surrounding conditions in the lab, or minor uncertainties in the design or fabrication process.

## 3.1. Quasi-Static Compression Test

Figure 2 shows the compression stress–strain curves for the investigated LSs. The stress–strain curve provides the mechanical behavior of LSs and could help to find the Young's modulus and yield strength. The main purpose of this test is to obtain the linear stress–strain limit so that the effective load, and strain levels can be identified for creep, stress relaxation, and cyclic loading-unloading tests. It can be observed that the overall compressive behavior of LSs depends mainly on its microstructural design and relative density, and the mechanical properties of the base material. Generally, the higher the density, the higher the collapse stress. As defined early, PLA has a higher density than ABS, 1.25 and 1.03 g/cm$^3$, respectively. Therefore, the fracture stress of the PLA samples is higher than that of the ABS specimens, as illustrated in Figure 2. With regards to the effect of the architected structural member geometry, it is evident that plate-based lattices are stiffer than others, followed by truss-based lattices then shell-based lattices made of the same material and relative density.

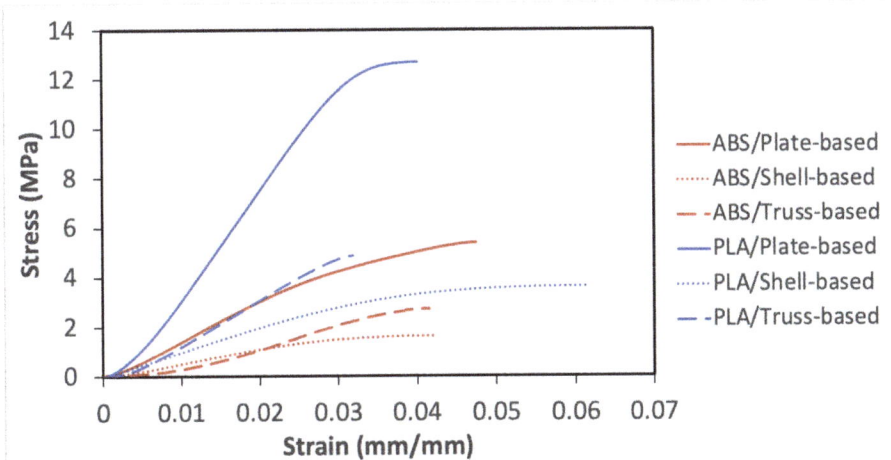

**Figure 2.** Compression stress–strain curves for the investigated samples.

The Young's modulus values were determined through the tangent value of the initial slope of the stress–strain curves, by using Equation 2 and the values Young's Modulus are shown in Table 9. The plate-based lattice in both materials has the highest Young's modulus values, and the least value of Young's modulus belongs to the shell-based lattice. Moreover, another interesting property that can be obtained from the stress–strain curve is the specific stiffness, whereby the stiffness-to-density ratio can be measured using Equation 3; specific stiffness values are shown in Table 9.

Another important point to be noticed that the yield limit was not clearly defined as the LSs demonstrated nonlinear stress–strain response. The method of offset point was used to compute the yield point, that indicates the limit of elastic behavior and the beginning of plastic deformation. Table 9 shows the yield stress for the considered samples, which was important to be identified for subsequent experiments.

Table 9. Obtained properties from quasi-static compression test.

| Specimen | Fracture Stress (MPa) | Young's Modulus (MPa) | Specific Stiffness (MPa/(g/cm$^3$)) | Yield Limit Load (N) |
|---|---|---|---|---|
| ABS/Plate-based lattice | 5.38 | 168 | 672 | 2563 |
| ABS/Truss-based lattice | 2.7 | 70 | 275 | 1481 |
| ABS/Shell-based lattice | 1.64 | 59.5 | 242 | 781 |
| PLA/Plate-based lattice | 12.7 | 443 | 1697 | 7250 |
| PLA/Truss-based lattice | 4.9 | 177.8 | 697 | 2750 |
| PLA/Shell-based lattice | 3.64 | 93.75 | 370 | 1812 |

We investigated the architected structural member geometry on the deformation mechanism. All the three structures were deformed under uni-axial compression and representative pictures were taken during the tests at different strain levels as presented in Figure 3. Noticeably, there is no physical failure in the identified yield point as shown in the 1st row in Figure 3. Moreover, it was observed that buckling occurred when compressive strain reached to some critical value and consequently led to rapid and dramatic changes of the material microstructure, as illustrated in the 2nd row in Figure 3 (in which all three structures demonstrated clear buckling). Subsequently, the middle region of structural members reached to a completely collapsed and then the deformation progressed to the neighboring cells. The plate-based lattice deformation occurred by compressing layers over each other, while truss-based lattice deformed due to buckling of its struts, whereas shell-based lattice deformed by squeezing its unit cells.

Generally, it was observed that all samples have deformed in a stretching-dominated manner; however, each specimen has its characteristics. For examples, the high stiffness in plate-based lattice is due to its plates involvement to carry load capacity and the integration or configuration of the plate-based structure. On the other hand, when a truss-based lattice experiences a compression load, and most of the load is carried by struts located in the longitudinal direction of the force, which means more stress concentration in thin struts. Therefore, vertical struts are the first to fail via buckling. Moreover, shell-based lattice has a novel geometry that doesn't contain struts or walls, the advantages of its architecture were observed during the experiment, whereby it exhibited great extension, resulting from the uniform distribution of the stresses.

### 3.2. Stress Relaxation Test

The stress relaxation experiment was undertaken according to the procedure explained in above methodology section. Equal effective strain was applied in each sample made of the same material, based on the outcomes of quasi-static compression test, the elastic limit of PLA samples is higher than ABS samples. Therefore, PLA samples experienced higher initial stress than ABS, as shown in Figure 4.

As can be seen in Figure 4, the stress relaxation curves can be divided into three stages. The first stage is the effective elastic stage, in which the specimens were compressed to the predetermined displacement and then held for 30 min. This initial displacement determined the starting point of stress relaxation. Then, the stress relaxation started after the first stage and can also be divided into two stages: transient stage and stable stage, representing the regions of decreasing stress relaxation rate and near-constant stress relaxation rate, respectively.

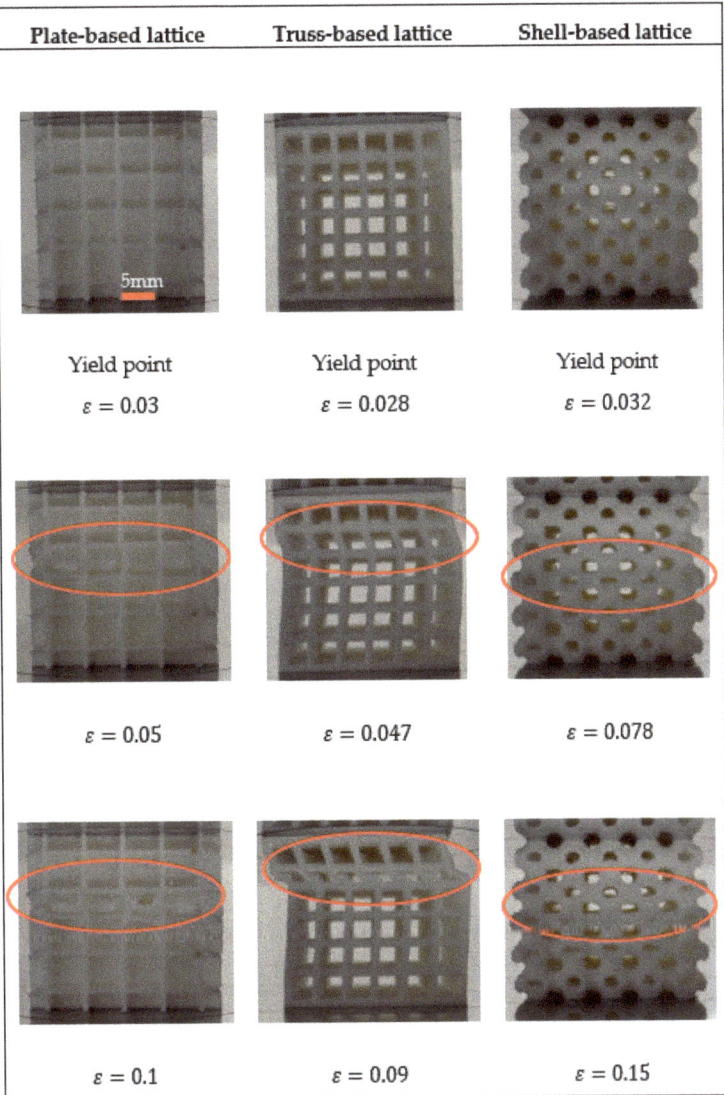

**Figure 3.** Deformation mechanism of the samples under investigation.

Figure 4 shows the plate-based lattice experienced the greatest stress to deform to the predetermined strain level, followed by the truss-based lattice. In contrast, the shell-based lattice demonstrated the least load bearing capacity. These results are due to the stiff plate-based structure, which is aligned with the conclusions drawn from the quasi static test. As shown in Figure 4, all considered samples exhibited different stress relaxation behavior over time, which demonstrates that different viscoelastic mechanism exists in each specimen.

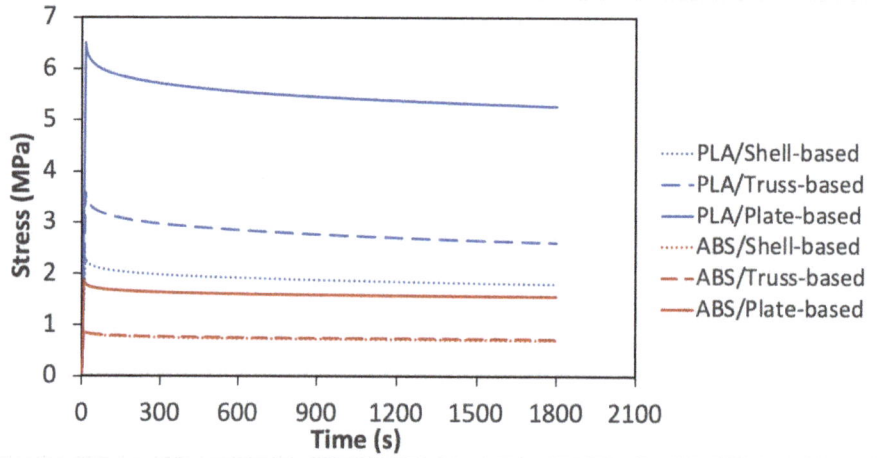

**Figure 4.** Stress relaxation response of all considered specimens.

For further analysis, the percentage of the normalized stress was calculated and shown in Table 10. It was found that the shell-based lattice outperformed the truss-based lattice and the plate-based lattice in terms of normalized stress over time. In addition, to determine the viscoelastic response from the stress relaxation test, the stress relaxation moduli were calculated using Equation 4. Then, the stress-relaxation moduli were converted to the relative moduli to compare based on the two considered materials as listed in Table 10. From the calculated values, it can be seen that the shell-based lattices have the greatest viscoelastic behavior, followed by the truss-based lattice, then the plate-based lattice. The outperformance of the shell-based lattice is due to its smooth geometry and curvature interconnections, by which the stress concentration is reduced, and the applied stress distributed uniformly. However, the stiffness of plate-based lattice has an adverse effect on the viscoelastic response. From the relative modulus values, it can be concluded that ABS samples have better viscoelasticity than that of PLA, resulting from the less stiffness and better elongation of ABS.

**Table 10.** Obtained properties from the stress relaxation test.

| Specimen | Normalized Stress (%) | Stress Relaxation Modulus (MPa) | Relative Modulus (MPa) |
|---|---|---|---|
| ABS/Plate-based lattice | 17% | 104 | 0.62 |
| ABS/Truss-based lattice | 19% | 48.67 | 0.69 |
| ABS/Shell-based lattice | 21% | 47.3 | 0.79 |
| PLA/Plate-based lattice | 19% | 210.8 | 0.48 |
| PLA/Truss-based lattice | 21% | 104.4 | 0.59 |
| PLA/Shell-based lattice | 23% | 72 | 0.77 |

*3.3. Creep Test*

The creep experiment was conducted following the procedure discussed earlier in the methodology section. The data obtained from the creep test are plotted in Figure 5. The shell-based lattice experienced the highest initial strain level, while the least value of applied strain belongs to the plate-based lattice. Those results are because all samples made of the same material have compressed to the same effective stress level and conform to the conclusions of previous experiments. The shell-based lattice was the compliant, while plate-based lattice was the stiffest.

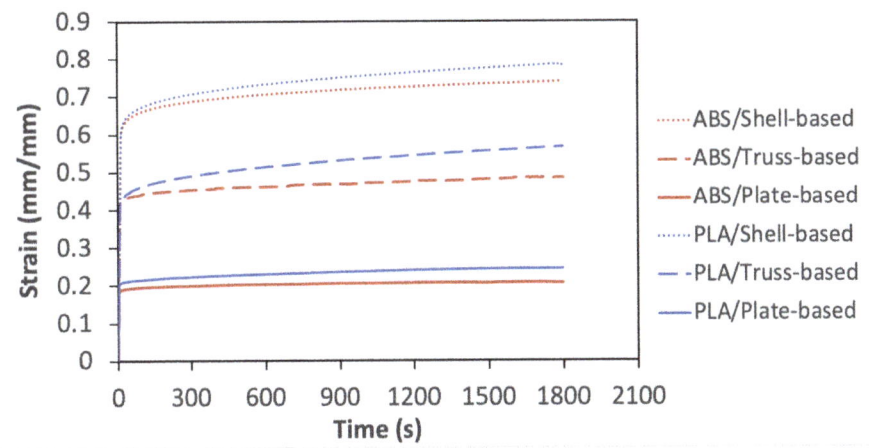

**Figure 5.** Strain–time plots of creep test for all considered specimens.

Additionally, the creep curves can be divided into three stages: the first stage is the elastic deformation stage, in which a uniaxial compression load was applied at a constant rate to the specimen until it reached the predetermined stress level and then be held. In this stage, the slope of PLA specimens is higher than that of ABS specimens due to the higher stiffness of PLA, which required more strain energy. The creep started after the first stage and can be divided into two stages: the transient stage and near-stable stage. All samples demonstrated creep deformation over time, which verifies the nature of viscoelastic behavior. However, only plate-based LSs demonstrated steady state creep strain for the considered testing time. The percentage of the creep strain increase was calculated and shown in Table 11. All shell-based lattice outperformed the truss-based lattice and plate-based lattice in terms of creep response. Moreover, the creep compliance was calculated using Equation 5. whereby the greatest compliance behavior belongs to the shell-based lattices, followed by the truss-based lattice, then the plate-based lattice. The is again because of the smooth interconnection of the shell-based lattice and uniform stress distribution and transfer from one cell layer to another. It is concluded that the viscoelastic behavior of ABS is better than that of PLA due to the softness and elongation of ABS.

**Table 11.** Obtained properties from the creep test.

| Specimen | Strain Increased (%) | Strain Compliance (1/MPa) |
|---|---|---|
| ABS/Plate-based lattice | 10% | 0.0086 |
| ABS/Truss-based lattice | 15% | 0.0202 |
| ABS/Shell-based lattice | 19% | 0.0308 |
| PLA/Plate-based lattice | 17% | 0.0041 |
| PLA/Truss-based lattice | 24% | 0.0095 |
| PLA/Shell-based lattice | 26% | 0.0130 |

*3.4. Compressive Cyclic Loading Test*

The compressive cyclic loading experiment was conducted following the procedure described in the methodology section. Figures 6 and 7 show plots of the load vs. displacement values for ABS and PLA samples. All tested specimens demonstrated a viscoelastic behavior and formed a hysteresis loop. The shape of the hysteresis curves dictates the energy dissipation capacity of LSs. The samples can be ranked by estimating the area inside the hysteresis loop; the wider loop means the better damping performance, energy dissipation capacity, or viscoelastic behavior. Figure 8 shows the estimation of the area of the hysteresis loop for all samples, which was calculated using OriginLab software. The results show that the shell-based lattice has a wider hysteresis loop, then the truss-based lattice and the plate-based lattice, respectively. Thus, the shell-based LS exhibits the greatest energy dissipation performance. This phenomenon shows that the energy dissipation of a hysteresis loop increases with the growth of the displacement as the PLA samples were compressed to a displacement level higher than that of the ABS samples, as illustrated in Figure 8.

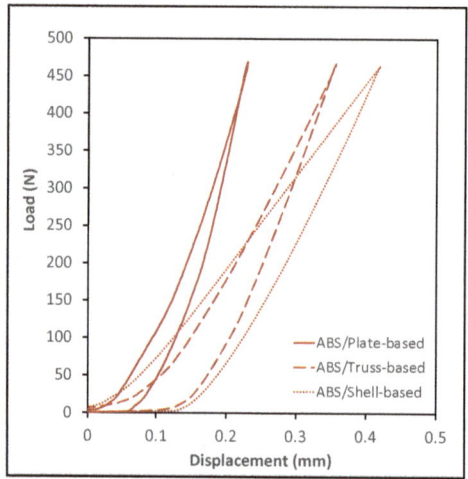

**Figure 6.** Cyclic loading of ABS specimens.

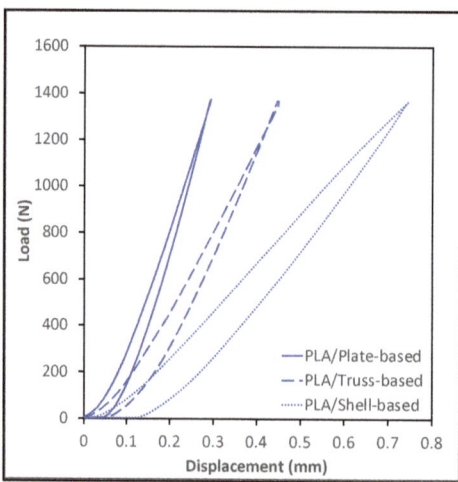

**Figure 7.** Cyclic loading of PLA specimens.

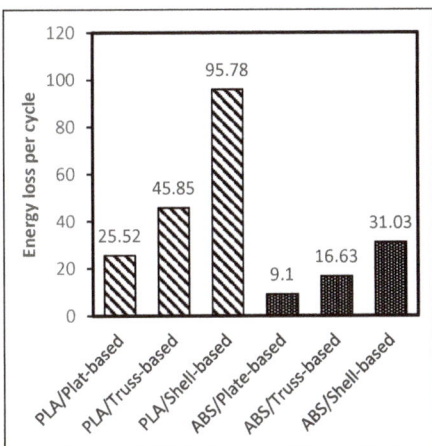

**Figure 8.** Area of hysteresis curves of the considered specimens.

In the end, a table is formulated comparing the specific elastic properties of the proposed architecture with those available in the literature, as shown in Table 12. There is abundant of studies available but here we mainly selected few architectures having cubic symmetry and made from polymeric materials such as ABS, PA and PLA using material extruding process (MEX). Table 12 shows that the specific Young's modulus of the PLA/Plate-based lattice have properties like the ones obtained from PA2200/Sheet-based IWP TPMS structures. However, as per the considered cellular materials shown in table below the sheet based Neovius TPMS structures has the highest specific Young's modulus. There is no experimentl data available in the literature that investigate the viscoelastic behavior of cellular materials with cubic symmetry, though few studies are available that characterize the time dependent response of bulk material made of PA2200 using Selective Laser Sintering technology (SLS) [55]. The authors are actively working in this area and more studies are ongoing related to the time dependent response of cellular materials.

**Table 12.** Elastic properties comparison of cubic symmetry cellular materials.

| Polymer | Architecture | E/Es | Reference: |
|---|---|---|---|
| ABS | Plate-based lattice | 0.084 | Current Work |
|  | Truss-based lattice | 0.035 |  |
|  | Shell-based lattice | 0.030 |  |
| PLA | Plate-based lattice | 0.164 |  |
|  | Truss-based lattice | 0.066 |  |
|  | Shell-based lattice | 0.035 |  |
| PLA | Honeycomb-Hexagonal | 0.067 | Leon et al. [56] |
|  | Honeycomb-Triangular | 0.122 |  |
| ABS | Honeycomb-Trianglular | 0.048 | Monkova [57] |
| PA2200 | TPMS sheet Primitive | 0.082 | Abueidda [58] |
|  | TPMS sheet IWP | 0.163 |  |
|  | TPMS sheet Neovius | 0.184 |  |
| PA1102 | TPMS ligament Diamond | 0.039 | Abou-Ali [59] |
|  | TPMS ligament Gyroid | 0.048 |  |
|  | TPMS ligament IWP | 0.030 |  |

## 4. Conclusions

In this study, the effect of the architected structural member's geometry on the viscoelastic behavior of lattice structures with simple cubic crystal symmetry was investigated. The structural members of simple cubic LS were designed with three architectures: plate-based LS, truss-based LS, and shell-based LS. Three-dimensional (3D) printing based on material extrusion (MEX) process technology was utilized to fabricate the considered designs. The behavior of LSs were investigated for two different materials, namely, PLA and ABS. The LSs mechanical response was obtained under quasi-static compression, stress-relaxation, creep, and compressive cyclic loading tests. The obtained data was analyzed and the following conclusions are summarized:

1. From the quasi-static compression test, it was found that the plate-based LS has the greatest stiffness and strength. The shell-based LS has excellent extension but least strength. Moderate properties are observed in the truss-based lattice with a rapid fracture mechanism. In terms of materials, PLA showed greater stiffness and strength than ABS, which is due to its higher density. However, ABS showed better viscoelastic behavior at the same infill density.
2. The shell-based has the greatest normalized stress and strain over time, which indicates its remarkable viscoelastic behavior, followed by truss-based lattice then plate-based lattice. In addition, the results of compressive cyclic loading testing showed that the shell-based lattice had formed a wide load-displacement hysteresis curves, meaning it has the greatest damping performance, and energy dissipation capacity. Whereas truss-based ranked in the second, followed by the plate-based LS. By comparing the ABS and PLA materials, the better viscoelastic behavior belongs to ABS, due to its elongation and flexibility.
3. A wide variety of material properties can be achieved by controlling the design of cellular solids. A material with maximum stiffness, as demonstrated in the plate-based lattice, is valuable as an engineering material for stiffness-dominated applications and lightweight structures. Whereas a material with excellent energy dissipation response, as observed in the shell-based lattice, is a great choice to be utilized where an application requires to be designed with bending-dominated behavior.
4. This study provides the comparison of viscoelastic behavior of simple cubic LSs made of different structural members. This research methodology will open up new research paths where the researchers can explore the effect of different types of symmetries on the isotropic and anisotropic viscoelastic properties of LSs.

**Author Contributions:** Conceptualization, K.A.K. and M.A.K.; methodology, K.A.K. and M.A.K.; software, A.A.; validation, A.A. and K.A.K.; formal analysis, A.A.; investigation, K.A.K.; resources, M.A.K.; data curation, A.A.; writing—original draft preparation, A.A.; writing—review and editing, K.A.K., M.A.K. and M.A.; visualization, A.A.; supervision, M.A.K.; project administration, M.A.K.; funding acquisition, M.A.K. and K.A.K. All authors have read and agreed to the published version of the manuscript.

**Funding:** This research was funded by Abu Dhabi Education Council grant number 8434000349/AARE19232.

**Institutional Review Board Statement:** Not applicable.

**Informed Consent Statement:** Not applicable.

**Data Availability Statement:** Not applicable.

**Acknowledgments:** The author Kamran A Khan would like to acknowledge the funding support by ASPIRE/ADEK through Abu Dhabi Award for Research Excellence (AARE-2019) under project number 8434000349/AARE19-232 and the APC was funded by Cranfield University.

**Conflicts of Interest:** The authors declare no conflict of interest.

## References

1. Bauer, J.; Hengsbach, S.; Tesari, I.; Schwaiger, R.; Kraft, O. High-Strength Cellular Ceramic Composites with 3D Microarchitecture. *Proc. Natl. Acad. Sci. USA* **2014**, *111*, 2453–2458. [CrossRef] [PubMed]
2. Wadley, H.N. Multifunctional Periodic Cellular Metals. *Philos. Trans. R. Soc. Lond. Math. Phys. Eng. Sci.* **2006**, *364*, 31–68. [CrossRef] [PubMed]
3. Vigliotti, A.; Pasini, D. Mechanical Properties of Hierarchical Lattices. *Mech. Mater.* **2013**, *62*, 32–43. [CrossRef]
4. Abueidda, D.W.; Abu Al-Rub, R.K.; Dalaq, A.S.; Lee, D.-W.; Khan, K.A.; Jasiuk, I. Effective Conductivities and Elastic Moduli of Novel Foams with Triply Periodic Minimal Surfaces. *Mech. Mater.* **2016**, *95*, 102–115. [CrossRef]
5. Lee, D.-W.; Khan, K.A.; Abu Al-Rub, R.K. Stiffness and Yield Strength of Architectured Foams Based on the Schwarz Primitive Triply Periodic Minimal Surface. *Int. J. Plast.* **2017**, *95*, 1–20. [CrossRef]
6. Gibson, L.J.; Ashby, M.F. *Cellular Solids: Structure and Properties*; Cambridge University Press: London, UK, 1997.
7. Bates, S.R.G.; Farrow, I.R.; Trask, R.S. 3D Printed Polyurethane Honeycombs for Repeated Tailored Energy Absorption. *Mater. Des.* **2016**, *112*, 172–183. [CrossRef]
8. Del Rosso, S.; Iannucci, L. On the Compressive Response of Polymeric Cellular Materials. *Materials* **2020**, *13*, 457. [CrossRef]
9. Pan, C.; Han, Y.; Lu, J. Design and Optimization of Lattice Structures: A Review. *Appl. Sci.* **2020**, *10*, 6374. [CrossRef]
10. Plateau, J. *Statique Expérimentale et Théorique Des Liquides Soumis Aux Seules Forces Moléculaires*; Gauthier-Villars: Paris, France, 1873; Volume 2.
11. Zhu, H.X.; Knott, J.F.; Mills, N.J. Analysis of the Elastic Properties of Open-Cell Foams with Tetrakaidecahedral Cells. *J. Mech. Phys. Solids* **1997**, *45*, 319–343. [CrossRef]
12. Li, K.; Gao, X.-L.; Roy, A.K. Micromechanics Model for Three-Dimensional Open-Cell Foams Using a Tetrakaidecahedral Unit Cell and Castigliano's Second Theorem. *Compos. Sci. Technol.* **2003**, *63*, 1769–1781. [CrossRef]
13. Gibson, L.J. Biomechanics of Cellular Solids. *J. Biomech.* **2005**, *38*, 377–399. [CrossRef] [PubMed]
14. Roberts, A.P.; Garboczi, E.J. Elastic Properties of Model Random Three-Dimensional Open-Cell Solids. *J. Mech. Phys. Solids* **2002**, *50*, 33–55. [CrossRef]
15. Khan, K.A.; Abu Al-Rub, R.K. Viscoelastic Properties of Architected Foams Based on the Schoen IWP Triply Periodic Minimal Surface. *Mech. Adv. Mater. Struct.* **2018**, *27*, 775–788. [CrossRef]
16. Pettermann, H.E.; Hüsing, J. Modeling and Simulation of Relaxation in Viscoelastic Open Cell Materials and Structures. *Int. J. Solids Struct.* **2012**, *49*, 2848–2853. [CrossRef]
17. Khaderi, S.N.; Deshpande, V.S.; Fleck, N.A. The Stiffness and Strength of the Gyroid Lattice. *Int. J. Solids Struct.* **2014**, *51*, 3866–3877. [CrossRef]
18. Warren, W.E.; Kraynik, A.M. Linear Elastic Behavior of a Low-Density Kelvin Foam With Open Cells. *J. Appl. Mech.* **1997**, *64*, 787–794. [CrossRef]
19. Kim, H.S.; Al-Hassani, S.T.S. A Morphological Elastic Model of General Hexagonal Columnar Structures. *Int. J. Mech. Sci.* **2001**, *43*, 1027–1060. [CrossRef]
20. Daphalapurkar, N.P.; Hanan, J.C.; Phelps, N.B.; Bale, H.; Lu, H. Tomography and Simulation of Microstructure Evolution of a Closed-Cell Polymer Foam in Compression. *Mech. Adv. Mater. Struct.* **2008**, *15*, 594–611. [CrossRef]
21. Zhu, H.X.; Mills, N.J. Modelling the Creep of Open-Cell Polymer Foams. *J. Mech. Phys. Solids* **1999**, *47*, 1437–1457. [CrossRef]
22. Mills, N.J. Finite Element Models for the Viscoelasticity of Open-Cell Polyurethane Foam. *Cell. Polym.* **2006**, *25*, 293–316. [CrossRef]
23. Huang, J.S.; Gibson, L.J. Creep of Polymer Foams. *J. Mater. Sci.* **1991**, *26*, 637–647. [CrossRef]
24. Gong, L.; Kyriakides, S. Compressive Response of Open Cell Foams Part II: Initiation and Evolution of Crushing. *Int. J. Solids Struct.* **2005**, *42*, 1381–1399. [CrossRef]
25. Hohe, J.; Becker, W. Geometrically Nonlinear Stress–Strain Behavior of Hyperelastic Solid Foams. *Comput. Mater. Sci.* **2003**, *28*, 443–453. [CrossRef]
26. Luxner, M.H.; Stampfl, J.; Pettermann, H.E. Numerical Simulations of 3D Open Cell Structures—Influence of Structural Irregularities on Elasto-Plasticity and Deformation Localization. *Int. J. Solids Struct.* **2007**, *44*, 2990–3003. [CrossRef]
27. Luxner, M.H.; Woesz, A.; Stampfl, J.; Fratzl, P.; Pettermann, H.E. A Finite Element Study on the Effects of Disorder in Cellular Structures. *Acta Biomater.* **2009**, *5*, 381–390. [CrossRef] [PubMed]
28. Degischer, H.-P.; Kriszt, B. *Handbook of Cellular Metals*; Wiley-VCH: Weinheim, Germany, 2002.
29. Ai, L.; Gao, X.-L. Evaluation of Effective Elastic Properties of 3D Printable Interpenetrating Phase Composites Using the Meshfree Radial Point Interpolation Method. *Mech. Adv. Mater. Struct.* **2016**, *25*, 1241–1251. [CrossRef]
30. Markert, B. A Biphasic Continuum Approach for Viscoelastic High-Porosity Foams: Comprehensive Theory, Numerics, and Application. *Arch. Comput. Methods Eng.* **2008**, *15*, 371–446. [CrossRef]
31. Torquato, S.; Donev, A. Minimal Surfaces and Multifunctionality. *Proc. Math. Phys. Eng. Sci.* **2004**, *460*, 1849–1856. [CrossRef]
32. Al Hassanieh, S.; Alhantoobi, A.; Khan, K.A.; Khan, M.A. Mechanical Properties and Energy Absorption Characteristics of Additively Manufactured Lightweight Novel Re-Entrant Plate-Based Lattice Structures. *Polymers* **2021**, *13*, 3882. [CrossRef]
33. Andrew, J.J.; Schneider, J.; Ubaid, J.; Velmurugan, R.; Gupta, N.K.; Kumar, S. Energy Absorption Characteristics of Additively Manufactured Plate-Lattices under Low-Velocity Impact Loading. *Int. J. Impact Eng.* **2021**, *149*, 103768. [CrossRef]
34. Liu, Y. Mechanical Properties of a New Type of Plate–Lattice Structures. *Int. J. Mech. Sci.* **2021**, *192*, 106141. [CrossRef]

35. Tancogne-Dejean, T.; Diamantopoulou, M.; Gorji, M.B.; Bonatti, C.; Mohr, D. 3D Plate-Lattices: An Emerging Class of Low-Density Metamaterial Exhibiting Optimal Isotropic Stiffness. *Adv. Mater.* **2018**, *30*, e1803334. [CrossRef] [PubMed]
36. Tancogne-Dejean, T.; Li, X.; Diamantopoulou, M.; Roth, C.C.; Mohr, D. High Strain Rate Response of Additively-Manufactured Plate-Lattices: Experiments and Modeling. *J. Dyn. Behav. Mater.* **2019**, *5*, 361–375. [CrossRef]
37. Wineman, A.S.; Rajagopal, K.R. *Mechanical Response of Polymers: An Introduction*; Cambridge University Press: London, UK, 2000.
38. Christensen, R. *Theory of Viscoelasticity: An Introduction*; Elsevier: Amsterdam, The Netherlands, 2012.
39. Findley, W.N.; Lai, J.S.; Onaran, K. *Creep and Relaxation of Nonlinear Viscoelastic Materials*, New ed.; Dover Publications: New York, NY, USA, 2011; ISBN 978-0-486-66016-5.
40. Khan, K.A.; Al-Rub, R.K.A. Time Dependent Response of Architectured Neovius Foams. *Int. J. Mech. Sci.* **2017**, *126*, 106–119. [CrossRef]
41. Khorasani, M.; Ghasemi, A.; Rolfe, B.; Gibson, I. Additive Manufacturing a Powerful Tool for the Aerospace Industry. *Rapid Prototyp. J.* **2021**, *28*, 87–100. [CrossRef]
42. Henriques, I.R.; Rouleau, L.; Castello, D.A.; Borges, L.A.; Deü, J.-F. Viscoelastic Behavior of Polymeric Foams: Experiments and Modeling. *Mech. Mater.* **2020**, *148*, 103506. [CrossRef]
43. Gonzalez Alvarez, A.; Evans, P.L.; Dovgalski, L.; Goldsmith, I. Design, Additive Manufacture and Clinical Application of a Patient-Specific Titanium Implant to Anatomically Reconstruct a Large Chest Wall Defect. *Rapid Prototyp. J.* **2021**, *27*, 304–310. [CrossRef]
44. ISO—ISO/ASTM 52900:2015—Additive Manufacturing—General Principles—Terminology. Available online: https://www.iso.org/standard/69669.html (accessed on 29 November 2021).
45. Travieso-Rodriguez, J.A.; Jerez-Mesa, R.; Llumà, J.; Gomez-Gras, G.; Casadesus, O. Comparative Study of the Flexural Properties of ABS, PLA and a PLA–Wood Composite Manufactured through Fused Filament Fabrication. *Rapid Prototyp. J.* **2020**, *27*, 81–92. [CrossRef]
46. Vanaei, H.R.; Deligant, M.; Shirinbayan, M.; Raissi, K.; Fitoussi, J.; Khelladi, S.; Tcharkhtchi, A. A Comparative in-process Monitoring of Temperature Profile in Fused Filament Fabrication. *Polym. Eng. Sci.* **2021**, *61*, 68–76. [CrossRef]
47. Vanaei, H.R.; Shirinbayan, M.; Deligant, M.; Khelladi, S.; Tcharkhtchi, A. In-Process Monitoring of Temperature Evolution during Fused Filament Fabrication: A Journey from Numerical to Experimental Approaches. *Thermo* **2021**, *1*, 332–360. [CrossRef]
48. Vanaei, H.R.; Shirinbayan, M.; Vanaei, S.; Fitoussi, J.; Khelladi, S.; Tcharkhtchi, A. Multi-Scale Damage Analysis and Fatigue Behavior of PLA Manufactured by Fused Deposition Modeling (FDM). *Rapid Prototyp. J.* **2020**, *27*, 371–378. [CrossRef]
49. Afonso, J.A.; Alves, J.L.; Caldas, G.; Gouveia, B.P.; Santana, L.; Belinha, J. Influence of 3D Printing Process Parameters on the Mechanical Properties and Mass of PLA Parts and Predictive Models. *Rapid Prototyp. J.* **2021**, *27*, 487–495. [CrossRef]
50. Kumar Mishra, P.; Ponnusamy, S.; Reddy Nallamilli, M.S. The Influence of Process Parameters on the Impact Resistance of 3D Printed PLA Specimens under Water-Absorption and Heat-Treated Conditions. *Rapid Prototyp. J.* **2021**, *27*, 1108–1123. [CrossRef]
51. von Windheim, N.; Collinson, D.W.; Lau, T.; Brinson, L.C.; Gall, K. The Influence of Porosity, Crystallinity and Interlayer Adhesion on the Tensile Strength of 3D Printed Polylactic Acid (PLA). *Rapid Prototyp. J.* **2021**, *27*, 1327–1336. [CrossRef]
52. D20 Committee. *Test Method for Compressive Properties of Rigid Cellular Plastics*; ASTM International: West Conshohocken, PA, USA, 2017.
53. E28 Committee. *Test Methods for Stress Relaxation for Materials and Structures*; ASTM International: West Conshohocken, PA, USA, 2017.
54. D20 Committee. *Test Methods for Tensile, Compressive, and Flexural Creep and Creep-Rupture of Plastics*; ASTM International: West Conshohocken, PA, USA, 2017.
55. Kozior, T. Rheological Properties of Polyamide PA 2200 in SLS Technology. *Teh. Vjesn.* **2020**, *27*, 1092–1100. [CrossRef]
56. León-Becerra, J.; González-Estrada, O.A.; Quiroga, J. Effect of Relative Density in In-Plane Mechanical Properties of Common 3D-Printed Polylactic Acid Lattice Structures. *ACS Omega* **2021**, *6*, 29830–29838. [CrossRef] [PubMed]
57. Monkova, K.; Monka, P.; Tkac, J.; Torok, J.; Monkova, K.; Suba, O.; Zaludek, M. Research of Young's Modulus of the Simple Lattice Structures Made from Plastics. In Proceedings of the 2019 IEEE 10th International Conference on Mechanical and Aerospace Engineering (ICMAE), July 2019; pp. 555–558.
58. Abueidda, D.W.; Bakir, M.; Abu Al-Rub, R.K.; Bergström, J.S.; Sobh, N.A.; Jasiuk, I. Mechanical Properties of 3D Printed Polymeric Cellular Materials with Triply Periodic Minimal Surface Architectures. *Mater. Des.* **2017**, *122*, 255–267. [CrossRef]
59. Abou-Ali, A.M.; Al-Ketan, O.; Rowshan, R.; Abu Al-Rub, R. Mechanical Response of 3D Printed Bending-Dominated Ligament-Based Triply Periodic Cellular Polymeric Solids. *J. Mater. Eng. Perform.* **2019**, *28*, 2316–2326. [CrossRef]

Article

# Effect of Printing Process Parameters on the Shape Transformation Capability of 3D Printed Structures

Matej Pivar, Diana Gregor-Svetec and Deja Muck *

Chair of Information and Graphic Arts Technology, Faculty of Natural Sciences and Engineering, University of Ljubljana, Snežniška 5, 1000 Ljubljana, Slovenia; matej.pivar@ntf.uni-lj.si (M.P.); diana.gregor@ntf.uni-lj.si (D.G.-S.)
* Correspondence: deja.muck@ntf.uni-lj.si

**Abstract:** The aim of our research was to investigate and optimise the main 3D printing process parameters that directly or indirectly affect the shape transformation capability and to determine the optimal transformation conditions to achieve predicted extent, and accurate and reproducible transformations of 3D printed, shape-changing two-material structures based on PLA and TPU. The shape-changing structures were printed using the FDM technology. The influence of each printing parameter that affects the final printability of shape-changing structures is presented and studied. After optimising the 3D printing process parameters, the extent, accuracy and reproducibility of the shape transformation performance for four-layer structures were analysed. The shape transformation was performed in hot water at different activation temperatures. Through a careful selection of 3D printing process parameters and transformation conditions, the predicted extent, accuracy and good reproducibility of shape transformation for 3D printed structures were achieved. The accurate deposition of filaments in the layers was achieved by adjusting the printing speed, flow rate and cooling conditions of extruded filaments. The shape transformation capability of 3D printed structures with a defined shape and defined active segment dimensions was influenced by the relaxation of compressive and tensile residual stresses in deposited filaments in the printed layers of the active material and different activation temperatures of the transformation.

**Keywords:** additive manufacturing; 4D printing; PLA; TPU; printability; shape transformation

Citation: Pivar, M.; Gregor-Svetec, D.; Muck, D. Effect of Printing Process Parameters on the Shape Transformation Capability of 3D Printed Structures. *Polymers* 2022, 14, 117. https://doi.org/10.3390/polym14010117

Academic Editor: Hamid Reza Vanaei

Received: 19 November 2021
Accepted: 25 December 2021
Published: 29 December 2021

**Publisher's Note:** MDPI stays neutral with regard to jurisdictional claims in published maps and institutional affiliations.

**Copyright:** © 2021 by the authors. Licensee MDPI, Basel, Switzerland. This article is an open access article distributed under the terms and conditions of the Creative Commons Attribution (CC BY) license (https://creativecommons.org/licenses/by/4.0/).

## 1. Introduction

Four-dimensional printing evolved from the 3D printing concept by incorporating the fourth dimension, i.e., the ability of the 3D printed object to change over time, to transform its geometry after being produced [1]. Today, 4D printing is becoming an increasingly interesting and widespread field of research. Some of the research is focused on the printing process parameters. Other research is more focused on the development of new, programmable printing materials that can change shape as a response to external stimuli. Based on the stimuli that can trigger a response, materials can be classified into thermo-responsive, magneto-responsive, chemo-responsive, photo-responsive or mechano-responsive materials [2].

One of the simplest options for 4D printing applications is the use of materials that are thermo-responsive, e.g., shape memory polymers (SMPs). SMPs have been used in 3D printing since 2013 [3]. These are dual-shape materials belonging to a group of actively moving polymers [4]. The shape memory effect is not an intrinsic property; it results from the combination of polymer morphology and specific processing. SMPs can be programmed into temporary shapes and return to their original shapes [5]. The mechanism of the shape memory effect (SME) can be described by two systems within the polymer, these being the net points and switching segments [6]. The net points, consisting of the more ordered, entangled or crystalline structure, act as the memory component of the polymer network that wants to return the SMP to its original shape [6]. The switching segments, consisting of less entangled structures, act to keep the SMP in its programmed shape [7]. Similarly to SMPs, the stimuli that induce time-dependent behaviour are temperature changes or water exposure [8].

The shape transformation depends on the residual stresses created during the 3D printing process in the thermoplastic materials. During the extrusion process, the thermoplastic materials are in a viscoelastic state—the high temperature enables the stretching and alignment of polymer chains in the direction of the material flow through the extrusion nozzle. After the material leaves the extrusion nozzle, it begins to cool and solidify. If the cooling is rapid, the polymer chains are forced to keep their extended state, which causes the development of internal stresses. When the 3D printed object is reheated above its glass transition temperature ($T_g$), the polymer chains start to rearrange, during which stress can be released, which causes shrinkage or changes in the shape of the 3D printed object [9–12].

Thermoplastic SMPs can be printed with the most widespread and cost-effective 3D printing technology, i.e., fused deposition modelling (FDM). The structure printed in the FDM process that will be capable of shape transformation is printed using one or a combination of two thermoplastic materials. In the first case, the shape transformation is controlled by a different orientation of deposited filament layers of the polymer [10–14], the so-called active material. In the second case, it is controlled by the multi-material structure, consisting of active and passive segments. For the active segment, a combination of polymers that differ in thermal transition temperatures and have different physical and mechanical properties is used [9,15]. One of the polymers, the active one, shrinks when heated above its $T_g$, and the second polymer, called the passive material, remains unchanged and serves only as a support for the active material to twist in a certain direction and plane. However, only one, i.e., the passive material, is used for the passive, inactive segment of the 3D printed structure. One article reported that single layer mono-material structures produce highly varying, unpredictable bending-twisting motions which are not desirable [15].

Polylactide (PLA) has shown shape memory properties based on the physical entanglements of polymer chains that maintain the structure of the 3D printed object [4,6,16,17]. After stretching into a temporary shape, the entanglements can recover to the unstretched state at $T_g$ when the material becomes highly elastic [7]. Another thermoplastic material studied in 4D printing is thermoplastic polyurethane (TPU). The process parameters for filament extrusion and printing influencing shape memory behaviour were studied [1,18–20] using FDM to print a shape memory TPU for a thermoactivated self-folding part. Five parameters that are important for shape memory properties were studied, i.e., material/surface of the platform, material/surface temperature, printing speed, liquefier temperature and delay time for printing each layer. In our research, two different thermoplastics materials, i.e., PLA and TPU, were used to fabricate 3D printed shape-changing structures.

To be able to control the shape transformation, high printability and controlled transformation conditions must be achieved. The shape transformation capability is defined by extent, accuracy and reproducibility. The extent of transformation is defined by changing the angle or radius of the active segment of the 3D printed structure. It depends on the type and properties of the thermoplastic material; the shape of the active segment; 3D printing parameters that affect the generation of residual stresses; and transformation conditions defined by the type of the thermal stimulus, activation temperature and activation time. The shape transformation accuracy is defined by the transformation of the 3D printed structure in the predicted plane, depending on the direction of filament deposition. The reproducibility of the shape transformation is defined by the accuracy of the transformations of several active segments printed and transformed under the same conditions. The reproducibility and accuracy of the shape transformation depend on the properties of 3D printing filaments; printing process parameters which affect the accuracy of extruded filament deposition in layers; and transformation conditions. Among the research dealing with 3D printing filaments and printing process parameters, there is little about the accuracy and reproducibility of shape transformation, which is a topic of our research. We wanted to provide some insight into the optimisation of 3D printing process parameters that directly or indirectly affect the print quality, and consequently the quality of the shape transformation.

## 2. Materials and Methods

*2.1. Materials*

The thermoplastic materials used in the research were PLA and TPU, obtained from Plastika Trček (Ljubljana, Slovenia) in the form of monofilaments with a 1.75 mm diameter. These two commercial filaments were selected due to their different physical and mechanical properties, glass transition (Tg) and melting (Tm) temperature. White coloured PLA with Tg at around 60 °C and Tm between 150–160 °C was used for the fabrication of a part of an active segment, whereas black coloured TPU with the shore hardness A89, Tg bellow 0 °C and Tm at around 180 °C was used for the fabrication of the passive segment of 3D printed structures.

Drying of Thermoplastic Materials

Thermoplastic materials are mostly hygroscopic and must be dried before 3D printing to achieve good printability. Evaporation of moisture during extrusion causes the formation of pores in the printed object, which affects their geometry and mechanical performance [21–23]. To remove the initial moisture in filaments, drying in an oven was performed—PLA at 45 °C and TPU at 50 °C for 24 h. After the drying, the filaments were stored in an airtight container filled with a desiccant and fed to the 3D printer through a PTFE tube to protect them from environmental humidity.

*2.2. 3D Printing*

All 3D printed structures, hereinafter referred to as 3D test samples, were fabricated using a 3D printer ZMorph VX (ZMorph S.A., Wroclaw, Poland). Prior to the printing, 3D modelling of 3D test samples was performed with the software Blender. A G-code file to produce the 3D printed samples was generated from STL files using the software Slic3R.

Several printing process parameters were kept constant throughout the experiment to avoid their influence on shape transformation, among them the extrusion nozzle diameter and layer height. It was reported that with a higher layer height, lower residual stress and thus lower shrinking ratio are achieved [10,11,24]. In our study, an extrusion nozzle with the diameter of 300 µm was used and each layer was 200 µm in height.

To obtain high printability, it is necessary to calibrate the printer, to level the print bed and the extrusion nozzles height above it. If the printer is not calibrated, the nozzle cannot extrude the material properly, the first few layers can be compressed or may not stick to the platform [25]. To level the print bed, a ZMorph probe for a semi-automatic calibration was used in our case. To set the extrusion nozzle height, a single layer 3D test sample with the layer height of 200 µm was printed and its height was measured using a Holex digital caliper.

*2.3. Optimisation of 3D Printing Process Parameters*

The optimisation of printing process parameters is important to ensure the quality and dimensional accuracy of the printed object [26,27]. In our study, the flow rate, printing speed and cooling conditions were optimised to ensure high printability and reproducible residual stress formation in printed layers. The extrusion temperature was set to the lowest possible recommended by the producer. It was reported that with a lower extrusion temperature, a larger shrinkage in the filament length can be achieved [10]. The extrusion temperature for PLA and TPU was set to 195 and 230 °C, respectively.

2.3.1. Printing Speed

The printing speed influences the shape transformation. Higher printing speed produces higher residual stress and as a result, higher shrinkage ratio [9,15,28–30]. Higher printing speed also leads to a less accurate extrusion process and lower print quality [31]. We assume that an increase in printing speed influences the accuracy of the shape transformation performance.

To achieve the largest possible and most accurate shape transformation, the highest printing speed at which a satisfactory filament deposition accuracy was achieved was set as optimal speed. It was determined separately for both materials, based on the image analysis of 3D printed test samples (Figure 1). The recommended printing speeds for PLA and TPU are according to the producer 30–120 mm/s and 10–30 mm/s, respectively. To eliminate as many factors as possible which influence the shape transformation, the printing speed was set to the same value for both materials. Single wall 3D test samples, consisting of sequences of curved lines, were printed at 10, 20, 30 and 60 mm/s. The shape of curved lines, the width of filaments and precision of filament deposition at junctions were analysed with image analyses, as shown in Figure 1.

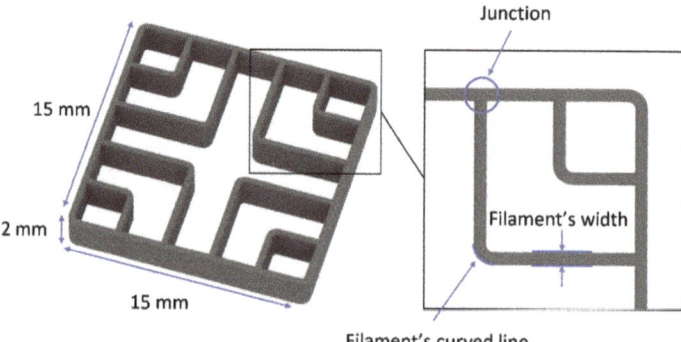

**Figure 1.** Scheme of 3D printed test sample.

2.3.2. Flow Rate

By varying the ratio of extrusion speed to printing speed, more or less material can be extruded, and if they are not appropriately synchronised, problems with the flow rate can occur [32]. In the case of over-extrusion, the defined shape of the printed object results in lower dimensional accuracy [33]. Deposited filaments are deformed and overlap, which affect the stress formation. Over-extrusion also increases the contact between the deposited filaments and the bonding between the polymer chains and provides high strength of the printed objects [34]. Another problem that occurs due to an insufficient material flow and orientation of deposited filaments are the voids between the deposited filaments, which reduce the strength of printed objects [34,35].

For the shape-changing structures, the extruded filaments must be deposited as evenly and precisely as possible, with constant shape without overlapping and deformation, and the size of voids between them must be as small as possible. In Figure 2, a schematic representation of the theoretically ideal filament deposition is presented, which could be achieved by changing the extrusion width and flow rate at the same layer height. The extrusion width determines the position of each filament deposited, and the flow rate determines the amount of the extruded material. In our research, the extrusion width settings were left as default, only the flow rate was changed. The flow rate was controlled with an extrusion multiplier. The extrusion multiplier influences the amount of the material extruded in the unit of length travelled by the printhead with a given speed [33]. The extrusion multiplier value 1 means 100%, whereas 1.1 means 110% material flow. In our research, different values were studied to optimise filament deposition.

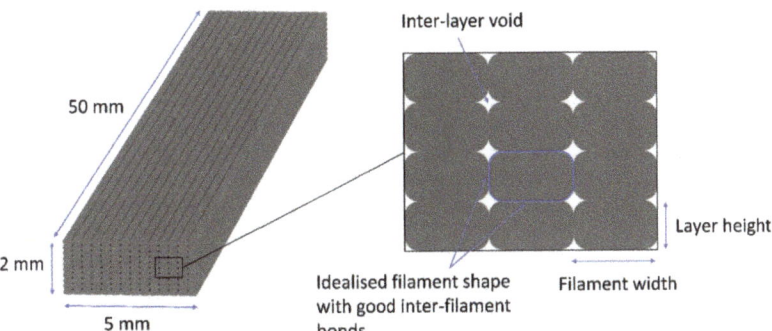

**Figure 2.** Scheme of theoretically ideal filament deposition.

The flow rate was determined separately for PLA and TPU at a predetermined extrusion temperature and printing speed. The average value of the filament diameter measured with a digital caliper at several locations was entered into the slicing software to reduce the impact of the filament diameter deviation on flow rate. It was reported that the inconsistency across filament length will change the rate of material extrusion, resulting in dimensional imprecision [36]. For the determination of the flow rate, a 3D test sample of size 5 (x) × 50 (y) × 2 (z) mm was fabricated (Figure 2). A linear pattern with the infill density of 100%, oriented longitudinally in all layers was used.

The filament deposition was determined by determining the distribution and coverage of voids at the cross-sectional area of the 3D printed test specimens using the ImageJ software. The images, taken with a stereo microscope Nikon SMZ800 with a built-in high-resolution camera Nikon D850 (FX) (Nikon Europe, Amsterdam, Netherlands), with the lightning adapted to each material, were cropped to 2400 × 1400 pixels and converted to grayscale (8 bit). The images were processed using the Auto Local Threshold algorithm in ImageJ [37]. Based on a comparative analysis of different methods for deciding the threshold level (Bernsen, Contrast, Mean, Media, Midgrey, Niblack, Otsu, Phansalkar and Sauvola) and determining the radius of the pixel conversion area, the Midgrey method with the radius of 100 pixels for PLA and 70 pixels for TPU was chosen. With this method, the most accurate binary image of voids of all different sizes was achieved. The size distribution of voids was determined and is shown as a histogram

### 2.3.3. Cooling Conditions

The cooling conditions of the extruded filament are a very important printing process parameter, which influences the mechanical properties, visual quality and formation of residual stress. Immediately after the extruded filament leaves the extrusion nozzle, it begins to cool and solidify. The process of cooling is affected by the cooling fan speed parameter and the temperature of the extrusion nozzle, print bed and consequently ambient air. An article reported that the time interval between printing two adjacent layers also has a significant effect on the cooling and strength of the printed object [38]. Rapid cooling and fast solidification lead to limited chain diffusion and weak bonding between the deposited filaments and layers. In this case, lower strength is achieved [38]. In the case of insufficient cooling, the 3D printed object can deform [39]. During the printing, the surrounding air can heat up, which affects the solidification and formation of residual stress. It was reported that different environmental temperatures lead to different thermal gradients in the printed specimen. The specimen temperature decreases with the distance from the build plate and leads to different thermal expansions in different layers, causing warping defects. Cooler and higher layers shrink more due to a larger temperature difference between the specimen temperature and glass transition temperature of the material than lower and warmer layers [23].

In our research, the cooling fan speed was set to the average fan speed recommended by the producer, for PLA to 35% and for TPU to 20%, respectively. To control the cooling as much as possible and to eliminate the influence of the room air temperature, the printer enclosure was used.

In the works by Byoungkwon et al. [8] and Kačergis et al. [15], it was shown that the temperature of the print bed affects the shape transformation ability. Printing on a cooler print bed reduces the chain mobility, increases residual stress, leading to higher shape transformation performance. However, in the case of too low bed temperature, warping, bending of edges and even detachment of the first layer from the print bed can occur [25]. It was reported, that the optimal adhesion of printed objects to the print bed is achieved when the print bed is heated slightly above the Tg of the polymer material [40]. In our study, the temperature of the print bed was set to 60 °C (Tg of PLA) to achieve the dimensional accuracy of printed structures without warping and bending and good polymer chain diffusion between the deposited filaments. It was also reported that better adhesion is achieved by applying suitable glue [41]. To ensure adequate adhesion, the Dimafix spray adhesive (DIMA 3D, Valladolid, Spain) was used. To analyse the heating of the print bed, heat maps were taken with a Seek Thermal Reveal PRO camera (Seek Thermal Inc., Santa Barbara, CA, USA).

To achieve good accuracy and reproducibility of transformation, it is important that the cooling of the extruded filament stays constant throughout the printing process to create the reproducible residual stress in deposited filaments in all layers. To analyse the cooling of extruded filaments, a 3D test sample of size 200 (x) × 220 (y) × 0.8 (z) mm was prepared. A linear pattern with the infill density of 100%, oriented longitudinally in all layers, was used. The air temperature was measured in the vicinity of the 3D printed test sample at the distance of 1 to 2 mm. The temperature was measured with a FLUKE 287 instrument (Fluke Corporation, Everett, WA, USA), separately for PLA and TPU prints, and at printing both materials simultaneously. The measurements were performed within three hours with every minute reading.

*2.4. Shape Transformation Capability*

2.4.1. Extent of Shape Transformation

The extent of the shape transformation was determined on the samples printed with optimised printing process parameters in water at different activation temperatures. It was reported that with a higher activation temperature, a larger shrinkage in the filament length can be achieved [10]. A four-layered 3D test sample with the size of 10 (x) × 60 (y) × 0.8 (z) mm and an active segment length of 15 mm (Figure 3a) was printed. In the study by Byoungkwon et al. [9] and by Shunsuke et al. [42], it was shown that the length of the active segment influences the shape transformation performance. With a larger active segment, a larger angle of transformation can be achieved. Our related preliminary research revealed that the optimal length of the active segment was 15 mm. The length of the passive segment on both sides of the active segment was set to 22.5 mm to be able to determine the transformation angle precisely (Figure 3b). The active segment was built from 3 layers of PLA and 1 layer of TPU, as also reported in the research by Byoungkwon et al. [9], since by increasing the number of layers, i.e., thickness of the active segment, the shape transformation performance deteriorates [28,43]. In our research, the passive segment of the 3D printed test sample consisted of 4 layers of TPU. A linear pattern with the infill density of 100%, oriented longitudinally in all layers, was used.

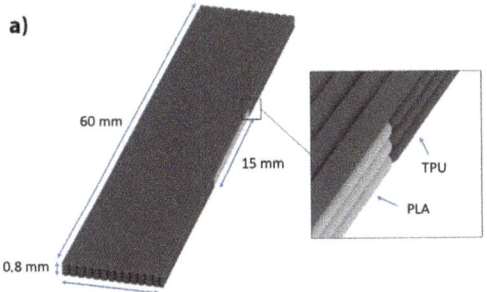 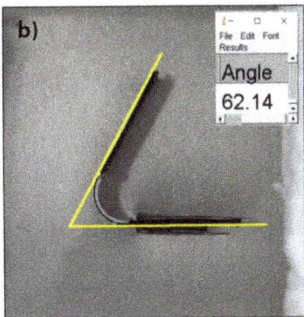

**Figure 3.** (**a**) Scheme of 3D printed test sample, (**b**) determination of extent (angle) of transformation.

The shape transformation in water was performed in a bath with controlled heating, with ±1 °C accuracy. It was reported that hot water as a trigger provides uniform heating and high controllability with little gravitational effects, leading to better accuracy and repeatability; however, the triggering conditions must remain consistent [14]. The temperature during the transformation was monitored in the immediate vicinity of the 3D printed test sample. The activation temperatures used were 60, 70, 80 and 90 °C. The extent of the shape transformation, determined by measuring the transformation angle, was tested at different time intervals, i.e., 60 min at 60 °C, 15 min at 70 °C, 10 min at 80 °C and 5 min at 90 °C. The image analysis was applied to determine the change in the transformation angle. The images were captured with a Sony rx100V camera in a certain time interval/every 30 s (Figure 3b). From the measured angles, supplementary angles were determined and then applied to show the change in the angle as a function of the transformation time.

2.4.2. Accuracy of Shape Transformation

In the study by Bona et al. [13], it was shown that the newly printed layer solidifies and shrinks on the previously deposited already hardened layer. Thus, the first printed layer undergoes compressive residual stress, and the upper layer undergoes tensile residual stress. When the stresses are released, the thermal deformation occurs in the opposite direction. The tensile deformation occurs in the first printed layers and the compressive deformation in the upper layers, resulting in sample bending. We assume that the differences in residual stresses affect the accuracy of the transformation and cause an unpredictable deformation during the transformation.

The influence of the optimisation of the 3D printing process and transformation conditions on the accuracy of the shape transformation capability was analysed. Our previous research has shown that a higher undesired deformation occurs at a thinner, longer and wider 3D printed structure. To eliminate as many factors as possible, 5-layered 3D test sample made entirely from PLA, with the size of 10 (x) × 50 (y) × 1 (z) mm was fabricated. A linear pattern with the infill density of 100%, oriented longitudinally in all layers, was used.

The accuracy of the shape transformation performance or the unpredictable deformation was determined on two sets of samples. The first set was composed from the test samples printed at optimised 3D printing process parameters, as previously discussed (hereafter called optimised printing or printing at optimised printing conditions). For comparison, the second set of samples was fabricated, using the same printing process parameter settings, though without 3D printer calibration (hereafter called non-optimised printing conditions). Afterwards, 3D printed test samples from both sets of samples were exposed to hot water at two different temperatures, i.e., 70 °C for 15 min and 90 °C for 5 min. For each series of testing, 5 specimens were prepared.

The dimensions of 3D printed test samples, before and after the exposure to hot water, were measured to 0.01 mm accuracy using a digital caliper to determine the directional strain ($\varepsilon$) and effective thermal expansion coefficient ($\alpha$) in each printing direction (x, y and z) by using Equation (1).

$$\alpha = \frac{\varepsilon}{\Delta T} = \frac{1}{\Delta T}\frac{(L_f - L_0)}{L_0} \qquad (1)$$

where $L_0$ and $L_f$ are the directional dimensions of printed samples before and after the exposure to hot water with the temperature change of $\Delta T$.

In few cases, the dimensions could not be measured accurately with a caliper due to an excessive deformation of 3D printed structures. Therefore, test samples were captured with a Shining 3D OptimScan-5M inspection 3D scanner (SHINING 3D Technology, Hangzhou, China), with the scanning accuracy of 0.015 mm and their dimensions determined using the Blender software. The deflection in the vertical direction of printing (z direction) was determined using the 3D Scan-Optim software by fitting the 3D printed test specimen exposed to hot water to the reference sample (unexposed 3D printed test sample). 3D printed test samples were cut and the cross-sectional area was captured with a stereomicroscope Nikon SMZ800 with a built-in high-resolution camera Nikon D850 (FX). The surface coverage and circularity of voids and the dimensions of the deposited filaments in layers were determined with image analysis using the ImageJ software. The measurements were performed on ten randomly selected deposited filaments in each layer.

2.4.3. Reproducibility of Shape Transformation

The reproducibility of the shape transformation was determined on the 3D printed test sample containing four identical active segments of 11.7 mm in length and passive segments of different lengths (10 and 20 mm). The previous testing of the length of the active segment and model prediction namely showed that with the length of 11.7 mm, the transformation angle of 90° could be reached. Five specimens were analysed. The transformation was performed in water at the temperature of 70 °C and a time of 15 min. After the shape transformation, the 3D printed test sample was captured with a high-resolution camera Nikon D850 (FX). The transformation angle was determined for each individual active segment with image analysis with the ImageJ software.

## 3. Results
### 3.1. Optimisation of Printing Process Parameters
#### 3.1.1. Printing Speed

Our research confirmed that the printing speed affects the deposition of filaments, and thus print quality. Irregularities occurred in the curved parts of the filaments, in the width of them and at junctions of two adjacent filaments due to the poor synchronisation of the printing speed and extrusion speed. The results of image analysis showed that satisfactory deposition of PLA filaments was achieved at printing speeds of 20–30 mm/s (Figure 4). At higher speeds, the extruded filaments deformed at the junctions and curved parts when deposited, and the width of the filaments in a layer was not the same and constant, as seen from Figure 4d. For TPU, however, irregularities occurred at the junction at the printing speed of just 20 mm/s (Figure 5b). The cause for this problem was oozing that could not be fixed due to their viscoelastic state at a certain extrusion temperature. Finally, the printing speed of 22 mm/s was determined as the maximum possible for TPU. At higher printing speeds, the feeding of the filament into the extrusion nozzle presented a problem due to the insufficient rigidity of the filament, interrupting the extrusion process. Elastomers, e.g., TPU, are prone to buckling during printing, which limits the printing speed. To have as few variables as possible that affect the deposition of extruded filaments when printing both materials at the same time, and to achieve the largest extent and most accurate transformation, both materials were printed at 22 mm/s.

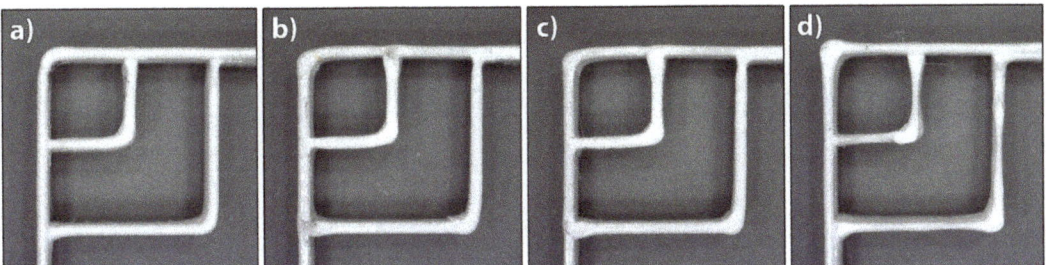

**Figure 4.** Images of 3D printed PLA test samples printed at different printing speeds: (**a**) 10 mm/s, (**b**) 20 mm/s, (**c**) 30 mm/s and (**d**) 60 mm/s.

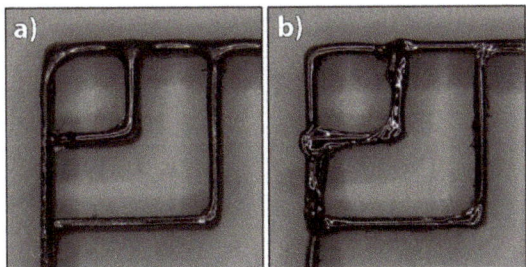

**Figure 5.** Images of 3D printed TPU test samples printed at different printing speeds: (**a**) 10 mm/s and (**b**) 20 mm/s.

3.1.2. Flow Rate

The influences of flow rate on the deposition of extruded filaments and size of voids in 3D printed structures were accessed at three different extrusion multiplier settings for both materials: 1.0 (default setting), 1.05 and 1.1 for PLA; and 1.1, 1.15 and 1.2 for TPU. The cross-sectional area of 3D printed test samples printed at different extrusion multipliers and their binary images are shown in Figure 6 for PLA and Figure 7 for TPU.

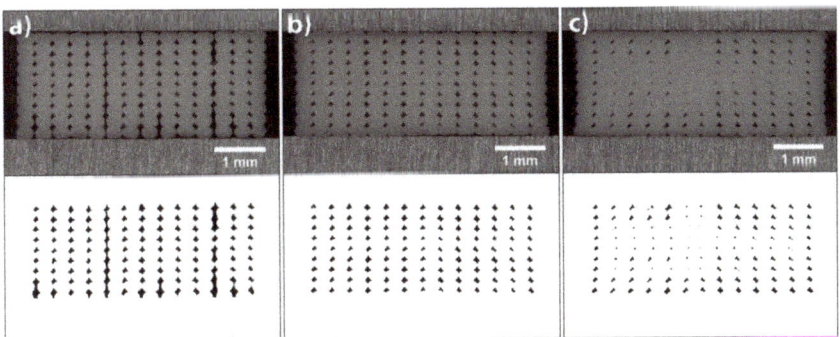

**Figure 6.** Images of cross-sectional area of 3D printed test samples from PLA and binary images belonging to them at extrusion multipliers: (**a**) 1, (**b**) 1.05 and (**c**) 1.1.

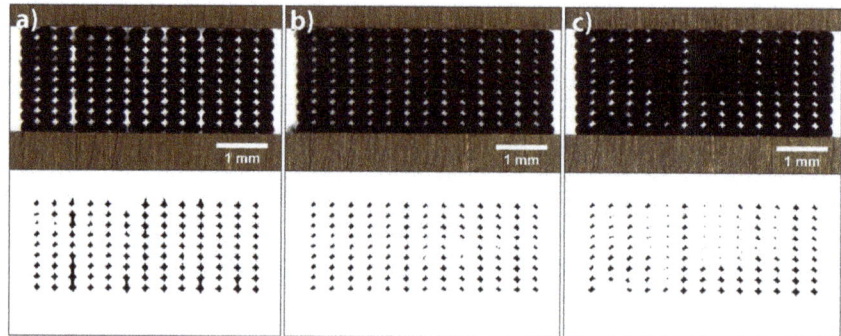

**Figure 7.** Images of cross-sectional area of 3D printed test samples from TPU and binary images belonging to them at extrusion multipliers: (**a**) 1.1, (**b**) 1.15 and (**c**) 1.2.

When printing PLA with default settings, the flow rate was too low. From the image of the cross-sectional area of the 3D printed test sample and its binary image, it can be seen that only limited contact and bonding between the filaments was present (Figure 6a). The total coverage of voids was determined to be 5.37%, and the average size was 8753.99 (±11,267.85) µm². The size distribution histogram shown in Figure 8a is symmetrical, with a wide range and one outlier. This outlier was a void larger than 15,000 µm², caused by an insufficient flow rate. Increasing the extrusion multiplier by 5% improved the bonding of extruded filaments; all filaments were joined and the voids were reduced in size (Figure 6b). The size distribution histogram remained symmetrical, without outliers. The total coverage of voids was determined to be 3.66%, and the average size was 5343.63 (±1373.46) µm². In the case of an additional increase by 5%, however, the extruded filaments began to deform; the voids disappeared (Figure 6c). This phenomenon occurred in the central part of each 3D printed test sample due to the filament deposition within each layer, which was set from outside towards inside. The total coverage of voids was determined to be 2.03%, and the average size was 2707.81 (±2069.229) µm². The size distribution histogram is asymmetric, and the outliers, below 1000 µm² in size, stand out (Figure 8a). These outliers represent the voids in the central part; they started to close due to too much material being extruded.

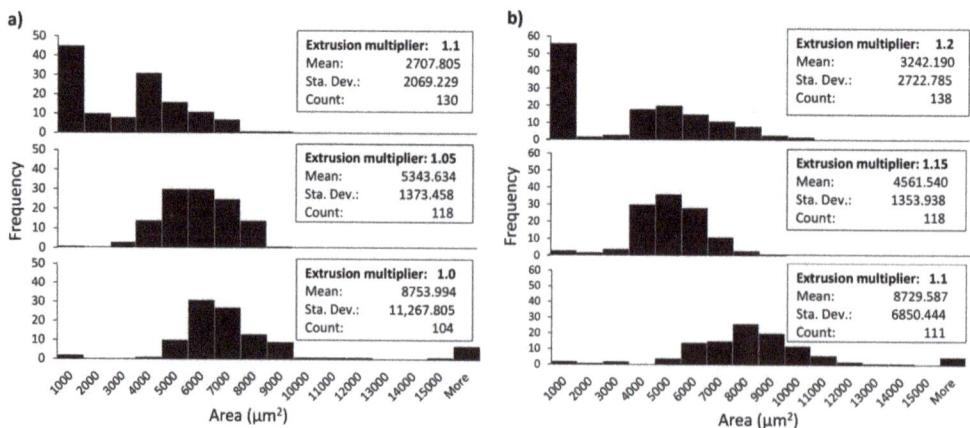

**Figure 8.** Size distribution of voids at different extrusion multipliers: (**a**) PLA and (**b**) TPU.

The analysis of the size and distribution of voids showed that the optimal flow rate in the case of PLA was achieved with the extrusion multiplier of 1.05. The voids were mostly even in size and shape, which indicates even and precise deposition of the filaments in all layers. This is very important for the transformation accuracy and reproducibility of 3D printed structures. The deformation of the shape of individual filaments during the printing and their deposition when printing the active PLA segment influences the residual stress relaxation and leads to unwanted deformations of 3D printed structures.

When printing TPU at default settings and at the extrusion multiplier of 1.05, under-extrusion occurred, resulting in filaments being too far away to merge. Also, at the extrusion multiplier 1.1, the flow rate was too low, resulting in a limited contact and bonding between the filaments, as it is seen in Figure 7a. The total coverage of voids was determined to be 5.79% and the average size 8729.59 (±6850.44) µm$^2$. The size distribution histogram shows a broad range and has a similar shape to the histogram for the PLA printed with default settings (Figure 8). In the case of a 5% increase in the flow rate, the total coverage of voids was reduced to 3.24% and the average size to 4561.54 (±1353.94) µm$^2$. From the binary images and histogram, some smaller voids were clearly present, caused by the impurities in the voids and the deformation of the filaments (Figure 7b). By further increasing the flow rate, the size of voids reduced and the bonding of extruded filaments increased, whereas the filaments began to deform and deposited unevenly (Figure 7c). The total coverage of voids was determined to be 2.66%, and average size was 3242.19 (±2722.79) µm$^2$. The size of voids was lower than 1000 µm$^2$ in 40% of cases. As with PLA, the voids began to double due to the deformation of the filaments. The optimal flow rate in the case of TPU was achieved with the extrusion multiplier of 1.15. Though TPU is not an active material, the quantity of the extruded material is important, since it influences the transformation performance of the active material. A larger amount of the extruded material gives higher resistance to transformation and reduces the extent of transformation.

It is evident that the voids were quite large and could be reduced; however, in our case, this was not the priority. It was more important to obtain evenly deposited and interconnected filaments, with as little deformation as possible. A higher flow rate, besides reducing the size of voids and deforming the shape of the filaments, could lead to the dimensional inaccuracy of 3D printed structures. For the shape transformation ability, the ratio between the amounts of TPU and PLA materials is important. If the amount of TPU is higher, this could result in higher resistance and hinder the shape transformation performance. Therefore, the amount of the extruded material is even more important.

3.1.3. Cooling Conditions

Measuring the ambient air temperature in the immediate vicinity of the 3D printed test specimen confirmed that the air temperature was below the Tg of the active PLA material, which prevented rapid residual stress relaxation during the printing process. Figure 9 shows the temperature of ambient air as a function of time for printing PLA and TPU separately, and PLA and TPU together.

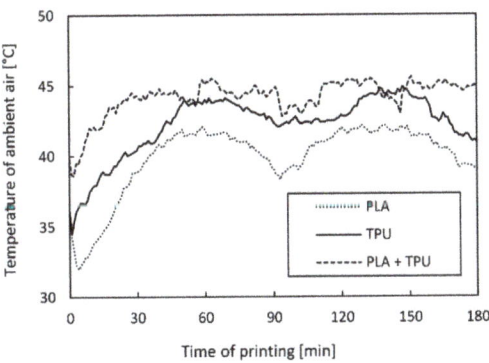

Figure 9. Temperature of ambient air in vicinity of 3D printed test sample as a function of time.

As seen from Figure 9, the air temperature is slightly higher in the initial phase due to the heating of the extrusion nozzle before the printing when the cooling fan is switched off. As soon as the printing process begins, the fan is turned on and the temperature begins to drop to a certain point; then it starts to rise again before levelling out. A quite large fluctuation in the air temperature was observed, with a CV of 6.1% when printing PLA, slightly lower at that of TPU (5.1%), and below 3% during the simultaneous printing of both materials. The air temperature fluctuations depend on the position of the printhead and temperature of the working extrusion nozzle. The highest temperature was measured when the printhead was in the middle of the print bed. This is shown at around 60 and 135 min in the graph where individual PLA and TPU curves have the highest peaks. Along the sides of the print bed, the air temperature dropped significantly. This can be seen from the curves for the initial part of the printing at about 90 min, and for the final part of the printing, at about 180 min. The findings can also be confirmed from the series of images in Figure 10, which show the heat maps of the print bed, measured with a thermal camera. The heat maps show that the print bed was unevenly heated. The temperature was 59 °C in the warmest parts and 48 °C in the coldest parts (edges), although the print bed was heated to 60 °C. Such changes in temperature influence the formation of different residual stress in the active PLA material and consequently affect the shape transformation ability.

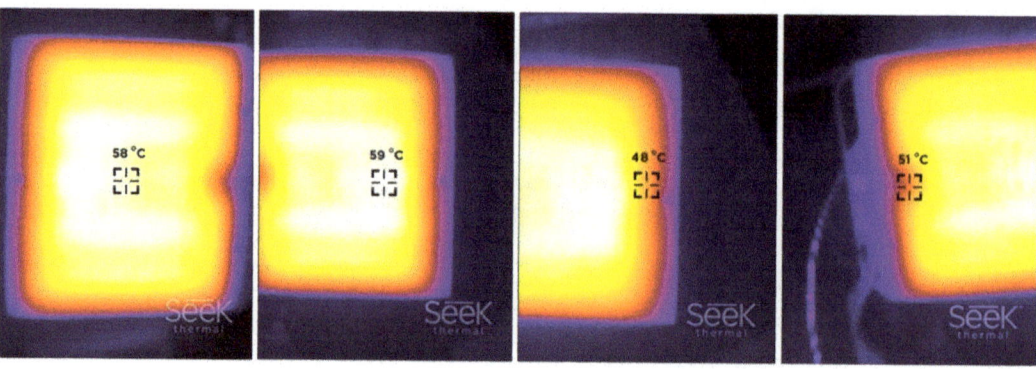

**Figure 10.** Heat maps of print bed.

When printing a single material, a higher ambient air temperature (up to 44.9 °C) was measured for TPU, as the extrusion nozzle was heated to a higher temperature. Moreover, in this case, the air temperature did not exceed the Tg of the active PLA. When PLA and TPU were printed simultaneously, the air temperature was slightly higher, i.e., up to 45.6 °C.

*3.2. Shape Transformation Capability*

3.2.1. Extent of Shape Transformation

The research showed that the activation temperature influences the extent of shape transformation. In Figure 11, the changes in the transformation angle of the 3D printed test sample exposed to water as a function of the transformation time at different activation temperatures are presented. From the slopes and shapes of the curves, it can be deduced that the transformation rate and the achieved final angle were higher at higher activation temperatures. The shape transformation was quicker initially; then, depending on the activation temperature, it slowed down and stopped at a certain point, when the relaxation of residual stresses in all layers of the active part of the 3D printed test sample was reached.

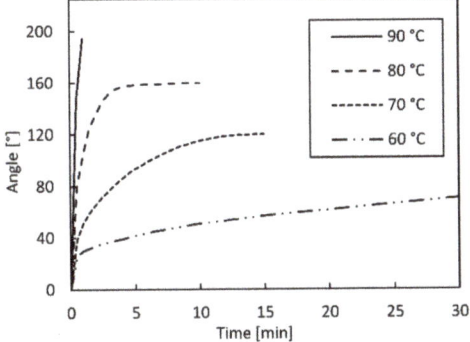

**Figure 11.** Transformation angle as a function of transformation time.

The lower the temperature of the water, the more time needed for the shape transformation. At 60 °C, the relaxation of residual stresses took about 60 min, although after 50 min, the change in the transformation angle was barely evident (about 1°). As the activation temperature rose, the relaxation of residual stresses accelerated and was completed earlier. At the activation temperature of 70 °C, it ended in 12 to 15 min; at 80 °C, after 8 min. When the temperature of 90 °C was used, the transformation was very fast, although it was difficult to determine the final transformational angle due to the length of the passive segment, which hindered the transformation (Figure 12d). From Table 1 and Figure 12, it is evident that a higher activation temperature causes a higher final transformation angle. For the water temperature of 60 °C, the transformation angle was around 83°, whereas it was around 195° for 90 °C water. The determination of the final transformation angles represent the basis for the determination of model prediction, which is the topic of another publication.

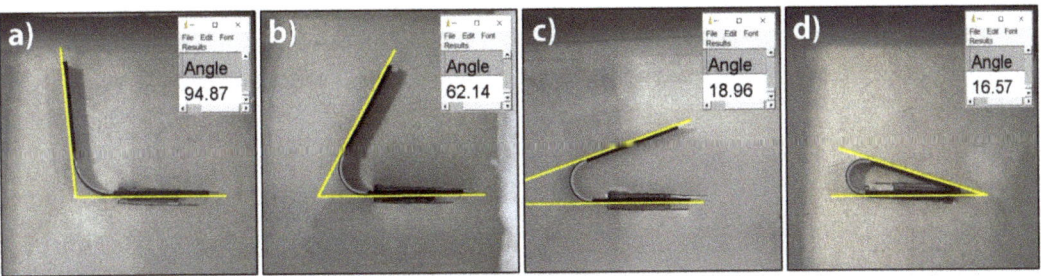

**Figure 12.** Determination of final transformation angles in water: (a) 60 °C, (b) 70 °C, (c) 80 °C and (d) 90 °C.

**Table 1.** Transformation angles at different activation temperatures in hot water.

|  | Water Temperature | | | |
| --- | --- | --- | --- | --- |
|  | 60 °C | 70 °C | 80 °C | 90 °C |
| Angle min. [°] | 80.6 | 117.9 | 157.4 | 194.0 |
| Angle max. [°] | 85.1 | 120.9 | 161.1 | 196.6 |
| Average [°] | 83.1 | 119.8 | 159.6 | 195.2 |
| CV [%] | 2.28 | 1.35 | 1.22 | 0.67 |

The results of the research showed that any fluctuations of the activation temperature have a strong effect on the relaxation of residual stresses and the final transformation angle. Therefore, it is very important that the activation temperature remains as constant as possible.

An example of the shape transformation of a 3D printed test sample exposed to water at 90 °C can be seen in the Video S1 (Supplementary Materials).

3.2.2. Accuracy of Shape Transformation

In Figure 13 and Table 2, the changes in the dimensions of the 3D printed test samples after the shape transformation in water are shown. It is evident that the thermal expansion coefficient ($\alpha$) is negative in the longitudinal direction of 3D printing (y direction) and positive in the vertical direction (z direction) for all printing and shape transformation modes. This suggests that the residual stress relaxation is an anisotropic dimensional change of the 3D printed test samples. In the transverse direction (x direction), the thermal expansion coefficient is positive for optimised printing and negative for non-optimised printing. This means that different printing conditions affect the accuracy of the transformation. The coefficient of thermal expansion is higher at a higher activation temperature, meaning that the active PLA would shrink more, resulting in higher shape transformation. This was already found when determining the extent of the shape transformation of shape-changing structures.

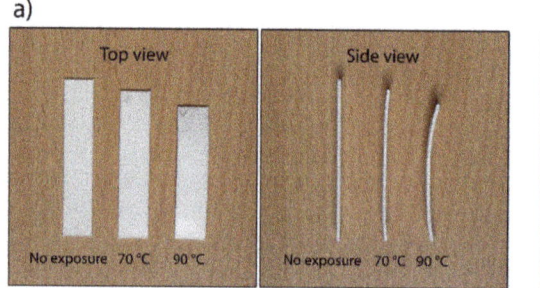 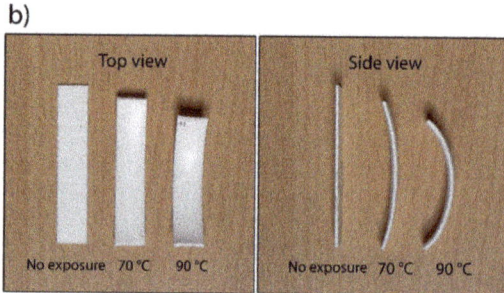

**Figure 13.** Test samples printed in (**a**) optimised printing conditions and (**b**) non-optimised printing conditions—no exposure (printed and not transformed), 70 and 90 °C (printed and transformed in water at 70 and 90 °C).

**Table 2.** Linear elongation and thermal expansion coefficient of 3D printed test samples in x, y and z directions.

| Printing Conditions | Activation T [°C] | Direction | ε | $\alpha$ ($\times 10^{-3}$ °C$^{-1}$) |
|---|---|---|---|---|
| Optimised | 70 | Transverse (x) | 0.013 | 0.285 |
| | | Longitudinal (y) | −0.064 | −1.432 |
| | | Vertical (z) | 0.064 | 1.417 |
| | 90 | Transverse (x) | 0.058 | 0.887 |
| | | Longitudinal (y) | −0.152 | −2.342 |
| | | Vertical (z) | 0.160 | 2.460 |
| Non-optimised | 70 | Transverse (x) | −0.024 | −0.539 |
| | | Longitudinal (y) | −0.072 | −1.596 |
| | | Vertical (z) | 0.084 | 1.858 |
| | 90 | Transverse (x) | −0.021 | −0.327 |
| | | Longitudinal (y) | −0.121 | −1.869 |
| | | Vertical (z) | 0.181 | 2.791 |

The binary images of the cross-sectional area photos are 2400 × 700 pixels in size and are shown in Figure 14. The total coverage of voids determined for the sample printed with optimised printing settings was 1.58%, and via non-optimised printing it was only

0.15%. 3D printing at optimised printing settings resulted in even and accurate deposition of filaments, with an even distribution of voids across the layers (Figure 14a). In non-optimised printing, the filaments were deposited more unevenly and inaccurately, resulting in smaller voids due to the deformation of the filaments. Some voids could not be captured even when adjusting contrast and brightness with global thresholding (Figure 14b).

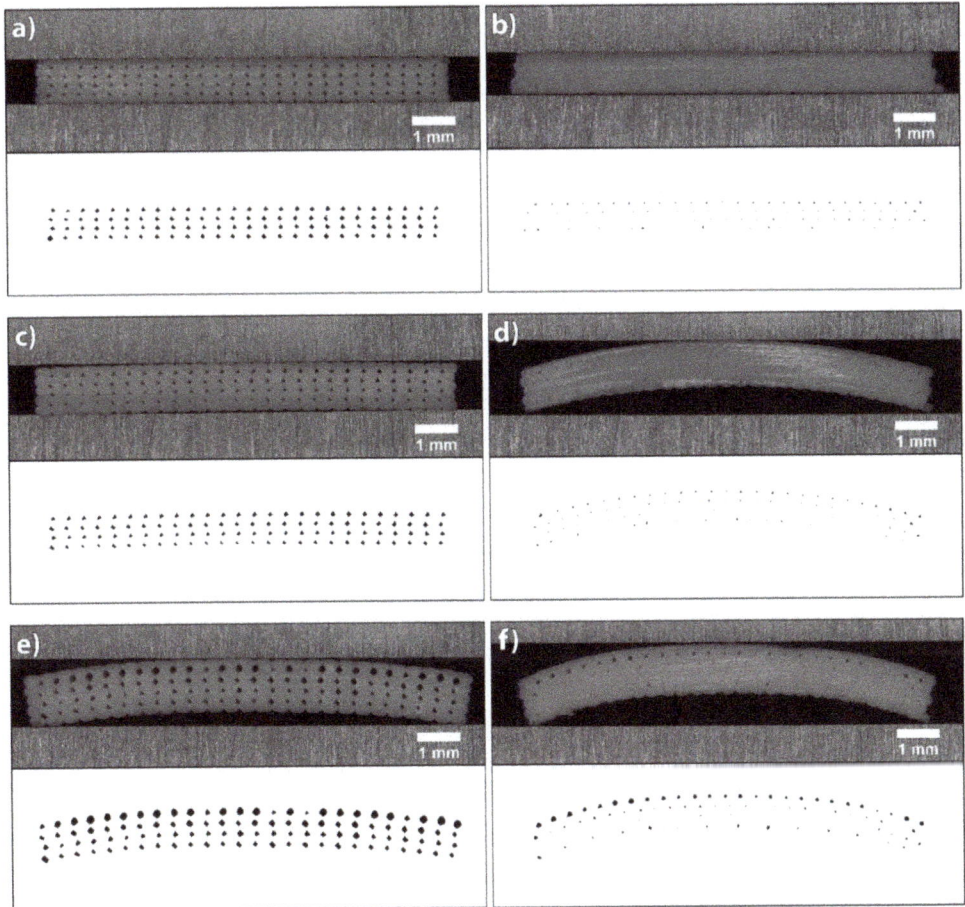

**Figure 14.** Cross-sectional areas and associated binary images of 3D printed test samples: (**a**) optimised printing conditions, (**b**) non-optimised printing conditions, (**c**) optimised printing and transformed at 70 °C, (**d**) non-optimised printing and transformed at 70 °C, (**e**) optimised printing and transformed at 90 °C, (**f**) non-optimised printing and transformed at 90 °C.

After the shape transformation in water at the temperature of 70 °C (Figure 14c), the total coverage of voids in samples printed at optimised printing settings was 1.41%, which is only 0.17% less than the total coverage determined immediately after the printing. The difference is very small and is difficult to explain. The size of the voids may have been reduced due to the thermal expansion of the filaments in x and z directions, or the cause may have been the inhomogeneity of the sample or measurement uncertainty. Slight curvature or deformation of the 3D printed test sample in z direction was seen. The maximum measured deviation was 35.25 µm. With the deviation being so small, it had no impact on the transformation accuracy. The transformation in water at the temperature of

90 °C resulted in an increase of the total coverage of voids (3.38%). The voids were larger and of different shapes, and the filaments in the upper layers were deformed (Figure 14e). Furthermore, the curvature of the 3D printed test sample increased to 307.21 µm.

For non-optimised printing, the total levels of coverage of voids determined for samples before and after the transformation in water at the temperature of 70 °C were the same, i.e., 0.15%. Additionally, in this case, the smallest voids could not be captured after thresholding. However, a very large twist/bend occurred in the z direction of printing, with a maximum measured deviation of 624.57 µm (Figure 14d). The total coverage of voids of the sample exposed to water at the temperature of 90 °C increased to 0.58% due to the larger size of the voids. The curvature of the sample increased as well, the maximum measured deviation being 749.24 µm (Figure 14f).

The analysis of the circularity of voids determined between the deposited layers in the samples printed and transformed under different conditions is shown in Figure 15. The shape of the voids after the transformation in water at 70 °C remained mostly the same as after the printing for both printing conditions, whereas at 90 °C, the voids took the form of a regular circle. For the samples transformed in water at 90 °C, the circularity of the voids increased after each deposited layer and peaked in the 4th row, i.e., between the 4th and 5th layer. Moreover, the size of voids increased from the 1st to the 4th row, as is clearly shown in Figure 14e.

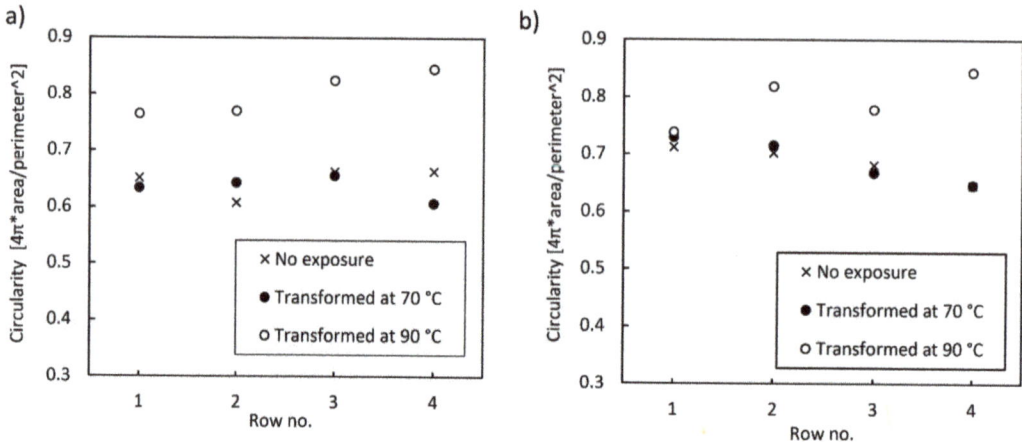

**Figure 15.** Circularity of voids in rows between layers determined on 3D printed test samples: (**a**) optimised printing conditions, (**b**) non-optimised printing conditions.

The determination and results of the analysis of height (h) and width (w) of deposited filaments in the samples printed under optimised printing conditions and exposed to water are shown in Figures 16 and 17. The analysis of the samples printed under non-optimised printing conditions could not be performed, as the filaments were too deformed for an accurate analysis. Nevertheless, we could visually determine that the filaments in the last deposited layer had the most regular shape, whereas the filaments in the first layer were the most deformed and compressed, most likely due to the incorrectly set offset of the extrusion nozzles from the print bed.

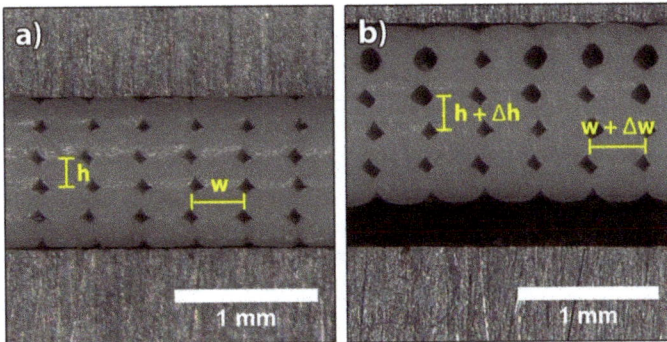

**Figure 16.** Images of the cross-section of a 3D printed sample for determination of height (h) and width (w) of deposited filaments: (**a**) no exposure, (**b**) after transformation at 90 °C.

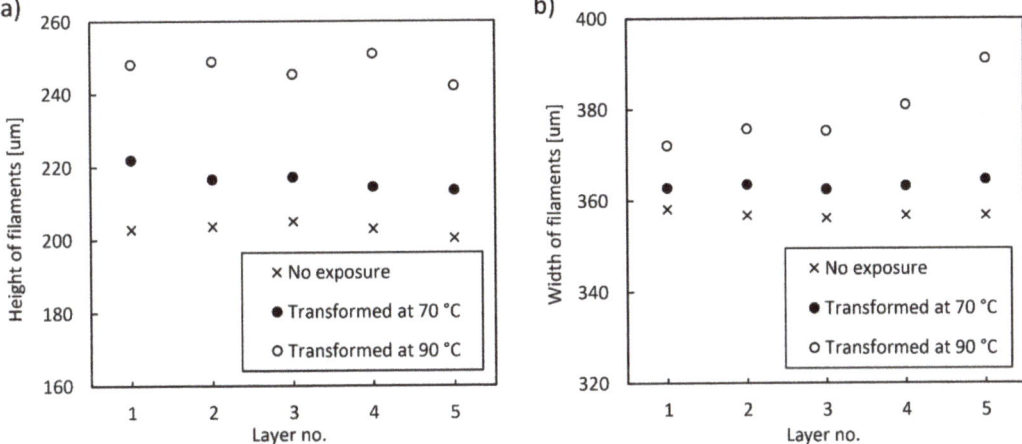

**Figure 17.** Dimensions of deposited filaments in layers determined in samples without exposure and after exposure to water at 70 and 90 °C: (**a**) heights of deposited filaments, (**b**) widths of deposited filaments.

The average measured height of filaments determined in all five layers in the sample after the printing was 202.9 ± 5.24 μm, which indicates that the printing process parameters were optimised well, as the predicted height was 200 μm and the deviation was very small. After the transformation in water at 70 °C, the height of the filaments increased on average to 216.8 ± 6.45 μm, which coincides with a small thermal expansion coefficient (Table 2). After the transformation in water at 90 °C, the height increased to 247.1 ± 7.26 μm, which coincides with a higher thermal expansion coefficient. The cause for the thermal expansion is the compression of the extruded filament to the height of the layer. After leaving the extrusion nozzle with the diameter of 300 μm, the extruded filament must be compressed to the predicted layer height of 200 μm, whereby internal stresses are created inside the filament. During the shape transformation process, the residual stresses are released, the rearrangement of the polymer chains leads to changes in dimensions and the height of the filaments increases.

The widths of the filaments show similar values for the 3D printed test samples before and after the exposure to water at 70 °C. Differences, however, occurred in the 4th and 5th layer after the exposure to water at 90 °C. After the printing, the average width of the filaments of every layer was 356.9 ± 6.83 μm, which after the transformation at 70 °C increased slightly to 363.3 ± 6.84 µm, coinciding with the thermal expansion coefficient (Table 2). After the transformation at 90 °C, the width increased to 374.4 ± 8.54 µm in the first three layers, whereas in the 4th and 5th layers, the average widths were slightly higher, i.e., 381 ± 8.76 µm and 391.1 ± 6.66 µm, respectively. As a result, the upper two layers of the sample expanded more, causing the bending/twisting or deformation of the sample. The cause of the thermal expansion in the width of the filaments is related to the contraction of the filaments in the longitudinal direction. During the transformation/relaxation process, the filaments shrink in the longitudinal direction, and at the same time, the height and width of the filaments increase.

The graph in Figure 18a shows the deflection of the 3D printed test samples in the vertical direction in regard to the xy-plane. The starting point of the coordinate system (z = 0, y = 0) represents the origin, and the upper surface of the undeformed specimen represents the reference. The curves on the graph, however, represent the deviations of the 3D printed test samples after the shape transformation in hot water from the xy-plane. The samples printed with the optimised 3D printing process parameters showed a negative deviation from the xy-plane, as they bent/twisted downward during the transformation. This can be explained by the difference in the temperature between filaments in adjacent layers. In the first layer, they were deposited on a relatively cold print bed (measured temperature 48–59 °C), so higher internal stresses were created in the material than in the remaining layers, where the temperature was higher and the cooling of the deposited filaments slower, enabling the residual stress relaxation. As a result, greater shrinkage of the first layer and twisting of the 3D printed test sample in the negative direction occurred.

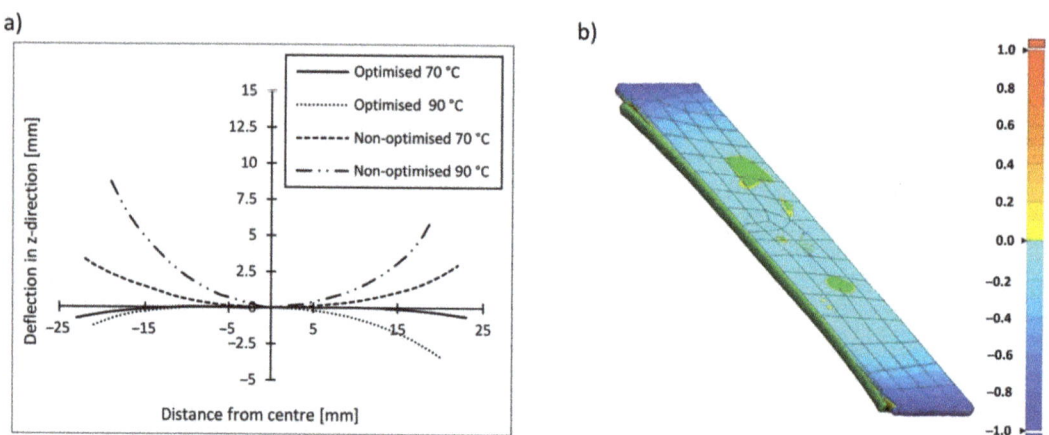

**Figure 18.** (a) Deflection in vertical direction of 3D printed test sample, (b) colour matching 3D plot of the sample transformed 70 °C with respect to reference.

The test samples printed with non-optimised printing, however, twisted in a positive direction. Bending in the positive direction is attributed to the inaccurate deposition of extruded filaments and their deformation. The deposition of the extruded filaments in non-optimised printing was not accurate enough; the filaments were differently stretched and deformed, which resulted in the differences in the temperature distribution, as is seen in the colour matching 3D plot (Figure 18b). The deflection from the xy-plane was higher when the printing process is not optimised and when it was performed at a higher activation temperature, as already determined with the extent of the shape transformation. At the temperature of 90 °C, the bending/twisting motion was not the same at both ends of specimen, suggesting a difference in the structure. The smallest deformation was detected in the optimised 3D printing process and at the activation temperature of 70 °C.

### 3.2.3. Reproducibility of Shape Transformation

Figure 19 shows five 3D printed test samples transformed into the predicted shape of a square with rounded edges. The average transformation angle of the 20 active segments from specimens was 90.4° with the deviation of 5.9°. Two active segments deviated by more than 10°, and two between 6 and 8°. Different times for the transformation of the active segments in the same specimen were noticed, although the same final transformation angle was reached at the end of the shape transformation process. It is very important that the transformation reaches the final angle and stops there. If we could determine intermediate angles, due to different transformation speeds, we would also get different angles in the end. The reason for the transformations in some cases still ending with incorrect angles is still being investigated. We assume that the main causes were the unevenly heated heat bed, which affected the residual stresses, as explained in Section 3.1.3, and the time interval between the printing of two adjacent layers of a different active segment. The cause may also have been related to the geometry of the transformation of the 3D printed test sample and to water resistance during the transformation. Figure 19e shows that one active segment achieved a larger transformation angle than 90° if it was connected to a shorter passive element (10 mm), since it had less water resistance. Figure 19d shows the opposite. Additionally, some parts of the sample touched the bottom wall of the bath and inhibited the transformation.

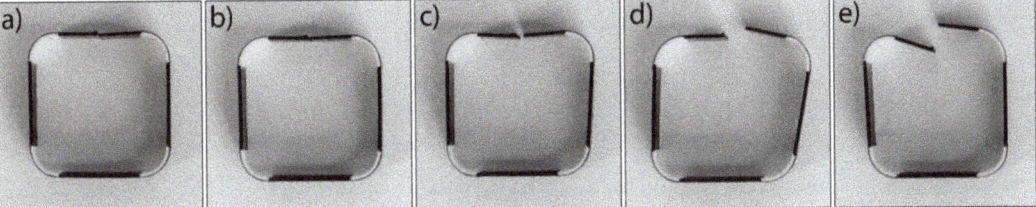

**Figure 19.** 3D printed test samples after shape transformation in water at 70 °C: (**a**) and (**b**) ideal case; (**c**) small deflection of angle; (**d**) and (**e**) larger deflection of angle from 90°.

## 4. Discussion

Printability is a very important factor influencing the shape transformation of 3D printed structures. Prior to the optimisation of the 3D printing process parameters, the calibration of the 3D printer was an important step to obtain an even and accurate distribution of filaments, with an even distribution of voids across the layers. By carefully selecting 3D printing process parameters, accuracy and good reproducibility of shape transformation for multi-material (PLA/TPU) 3D printed structures was obtained. High precision of filament deposition in layers was achieved by adjusting the printing speed, flow rate and cooling conditions of extruded filaments for both polymers printed simultaneously. The optimal printing speed which gave satisfactory deposition was 22 mm/s, which was the highest achievable for TPU and lowest feasible for PLA. For the optimal flow rate, the extrusion multiplier was selected with which the deposited filaments were as even in size and shape as possible, evenly deposited and spaced apart. The voids between them had to be as small as possible, and the connections between the deposited filaments as optimal as possible. The optimal flow rate was achieved with the extrusion multiplier of 1.05 for PLA and 1.15 for TPU. 3D printing was performed with a closed printer; the cooling speed was set to the average fan speed recommended by the producer; and the temperature of the print bed was set to 60 °C to ensure as constant cooling conditions as possible, to create the same residual stresses in the deposited filaments in all layers.

The shape transformation performance in hot water of a multi-material (PLA/TPU) 3D printed structure with a defined shape and defined active segment dimensions was substantially influenced by activation temperature. The extent and transformation rate increased with activation temperature and the uncontrolled deformation of the 3D printed structure. The activation temperature of 70 °C resulted in achieving the highest shape transformation accuracy.

To obtain the high transformation accuracy for multi-material (PLA/TPU) 3D printed shape-changing structures, the optimisation of printing process parameters is necessary, and exposure to water at the temperature of 70 °C is recommended. To achieve reproducibility in the case of printing PLA and TPU with a FDM 3D printer, we recommend using a print bed with evenly heating, or an unheated print bed and installing a local heating element to control and maintain constant ambient air temperature throughout the printing process to eliminate the formation of various residual stresses in the active PLA. The shape transformation must be performed in still water with the temperature remaining as constant as possible, and without the touching of other specimens or bath walls. In the case of touching the bottom of the bath, additives could be added to the water to increase the density of the water to reduce the effect of gravity.

Based on a pre-calibrated 3D printer and careful consideration of all the above parameters of printing flat four-layer structures with a combination of PLA and TPU materials, high-quality shape transformation was achieved. Based on repeatable empirical measurements under water exposure at 70 °C, the mathematical or analytical model of a bilayer beam composed of two materials with different thermal expansion coefficients can be established that predicts the transformation of printed structures that will change at different angles or radii of curvature by only changing the lengths of the active segments.

**Supplementary Materials:** The following supporting information can be downloaded at: https://www.mdpi.com/article/10.3390/polym14010117/s1, Video S1: An example of the shape transformation of a 3D printed test sample exposed to water at 90 °C.

**Author Contributions:** M.P.: Methodology, investigation, validation, writing—original draft preparation, D.G.-S.: data curation, writing—reviewing and editing, D.M.: conceptualisation, supervision, writing—reviewing and editing. All authors have read and agreed to the published version of the manuscript.

**Funding:** The research was co-funded by the Slovenian Research Agency (Programme P2-0213), Infrastructural Centre RIC UL-NTF).

**Institutional Review Board Statement:** Not applicable.

**Informed Consent Statement:** Not applicable.

**Data Availability Statement:** Data are contained within the article.

**Conflicts of Interest:** The authors declare no conflict of interest.

## References

1. Monzón, M.D.; Paz, R.; Pei, E.; Ortega, F.; Suárez, L.A.; Ortega, Z.; Alemán, M.E.; Plucinski, T.; Clow, N. 4D printing: Processability and measurement of recovery force in shape memory polymers. *Int. J. Adv. Manuf. Technol.* **2017**, *89*, 1827–1836. [CrossRef]
2. Rosales, C.A.G.; Kim, H.; Duarte, M.F.G.; Chavez, L.; Castañeda, M.; Tseng, T.-L.B.; Lin, Y. Characterization of shape memory polymer parts fabricated using material extrusion 3D printing technique. *Rapid Prototyp. J.* **2019**, *25*, 322–331. [CrossRef]
3. Momeni, F.; Hassani, N.S.M.M.; Liu, X.; Ni, J. A review of 4D printing. *Mater. Des.* **2017**, *122*, 42–79. [CrossRef]
4. Behl, M.; Andersen, L. Shape-memory polymers. *Mater. Today* **2007**, *10*, 20–28. [CrossRef]
5. Xiao, R.; Nguyen, T.D. Thermo-mechanics of Amorphous Shape-memory Polymers. *Proc. IUTAM* **2015**, *12*, 154–161. [CrossRef]
6. Andersen, L.; Steffen, K. Shape-Memory polymers. *Angew. Chem. Int. Ed.* **2002**, *41*, 2034–2057.
7. Carrell, J.; Gruss, G.; Gomez, E. Four-dimensional printing using fused-deposition modeling: A review. *Rapid Prototyp. J.* **2020**, *26*, 855–869. [CrossRef]
8. Sun, Y.-C.; Wan, Y.; Nam, R.; Chu, M.; Naguib, H.E. 4D-printed hybrids with localized shape memory behaviour: Implementation in a functionally graded structure. *Sci. Rep.* **2019**, *9*, 18574. [CrossRef]
9. An, B.; Tao, Y.; Gu, J.; Cheng, T.; Chen, X.A.; Zhang, X.; Zhao, W.; Do, Y.; Takahashi, S.; Wu, H.-Y.; et al. Thermorph: Democratizing 4D Printing of Self-Folding Materials and Interfaces. In Proceedings of the 2018 CHI Conference on Human Factors in Computing Systems, Montreal, QC, Canada, 21–26 April 2018; Paper no. 260. pp. 1–12.
10. Van Manen, T.; Janbaz, S.; Zadpoor, A.A. Programming 2D/3D shape-shifting with hobbyist 3D printers. *Mater. Horiz.* **2017**, *4*, 1064–1069. [CrossRef]

11. Wang, G.; Yang, H.; Yan, Z.; Ulu, N.G.; Tao, Y.; Gu, J.; Kara, L.B.; Yao, L. 4DMesh: 4D Printing Morphing Non-Developable Mesh Surfaces. In Proceedings of the ACM Symposium on User Interface Software and Technology, Berlin, Germany, 14–17 October 2018. [CrossRef]
12. Song, J.; Feng, Y.; Wang, Y.; Zeng, S.; Hong, Z.; Qiu, H.; Tan, J. Complicated deformation simulating on temperature-driven 4D printed bilayer structures based on reduced bilayer plate model. *Appl. Math. Mech.* **2021**, *42*, 1619–1632. [CrossRef]
13. Goo, B.; Hong, C.-H.; Park, K. 4D printing using anisotropic thermal deformation of 3D-printed thermoplastic parts. *Mater. Des.* **2020**, *188*, 108485. [CrossRef]
14. Wang, G.; Tao, Y.; Capunaman, O.B.; Yang, H.; Yao, L. A-line: 4D Printing Morphing Linear Composite Structures. In Proceedings of the 37th Annual ACM Conference on Human Factors in Computing Systems, Glasgow, UK, 4–9 May 2019. [CrossRef]
15. Kačergis, L.; Mitkus, R.; Sinapius, M. Influence of fused deposition modeling process parameters on the transformation of 4D printed morphing structures. *Smart Mater. Struct.* **2019**, *28*, 105042. [CrossRef]
16. Senatov, F.S.; Niaza, N.K.; Zadorozhnyy, M.Y.; Maksimkin, A.V.; Kaloshkin, S.D.; Estrin, Y.Z. Mechanical properties and shape memory effect of 3D-printed PLA-based porous scaffolds. *J. Mech. Behav. Biomed. Mater.* **2016**, *57*, 139–148. [CrossRef] [PubMed]
17. Mehrpouya, M.; Vahabi, H.; Janbaz, S.; Darafsheh, A.; Mazur, T.R.; Ramakrishna, S. 4D printing of shape memory polylactic acid (PLA). *Polymer* **2021**, *230*, 124080. [CrossRef]
18. Yang, Y.; Chen, Y.; Wei, Y.; Li, Y. 3D printing of shape memory polymer for functional part fabrication. *Int. J. Adv. Manuf. Technol.* **2016**, *84*, 2079–2095. [CrossRef]
19. Raasch, J.; Ivey, M.; Aldrich, D.; Nobes, D.; Ayranci, C. Characterization of polyurethane shape memory polymer processed by material extrusion additive manufacturing. *Addit. Manuf.* **2015**, *8*, 132–141. [CrossRef]
20. Bodaghi, M.; Damanpack, A.; Liao, W. Adaptive metamaterials by functionally graded 4D printing. *Mater. Des.* **2017**, *135*, 26–36. [CrossRef]
21. Zaldivar, R.; Mclouth, T.; Ferrelli, G.; Patel, D.; Hopkins, A.; Witkin, D. Effect of initial filament moisture content on the microstructure and mechanical performance of ULTEM® 9085 3D printed parts. *Addit. Manuf.* **2018**, *24*, 457–466. [CrossRef]
22. Valerga, A.P.; Batista, M.; Salguero, J.; Girot, F. Influence of PLA Filament Conditions on Characteristics of FDM Parts. *Materials* **2018**, *11*, 1322. [CrossRef] [PubMed]
23. Fang, L.; Yan, Y.; Agarwal, O.; Yao, S.; Seppala, J.E.; Kang, S.H. Effects of Environmental Temperature and Humidity on the Geometry and Strength of Polycarbonate Specimens Prepared by Fused Filament Fabrication. *Materials* **2020**, *13*, 4414. [CrossRef] [PubMed]
24. Yu, Y.; Liu, H.; Qian, K.; Yang, H.; McGehee, M.; Gu, J.; Luo, D.; Yao, L.; Zhang, Y.J. Material characterization and precise finite element analysis of fiber reinforced thermoplastic composites for 4D printing. *Comput. Des.* **2020**, *122*, 102817. [CrossRef]
25. Song, R.; Telenko, C. Material Waste of Commercial FDM Printers under Realstic Conditions. In Proceedings of the 27th Annual International Solid Freeform Fabrication Symposium, Austin, TX, USA, 8–10 August 2016; p. 1217.
26. Mohamed, O.A.; Masood, S.; Bhowmik, J.L. Optimization of fused deposition modeling process parameters: A review of current research and future prospects. *Adv. Manuf.* **2015**, *3*, 42–53. [CrossRef]
27. Dey, A.; Yodo, N. A Systematic Survey of FDM Process Parameter Optimization and Their Influence on Part Characteristics. *J. Manuf. Mater. Process.* **2019**, *3*, 64. [CrossRef]
28. Rajkumar, A.R.; Shanmugam, K. Additive manufacturing-enabled shape transformations via FFF 4D printing. *J. Mater. Res.* **2018**, *33*, 4362–4376. [CrossRef]
29. Bodaghi, M.; Noroozi, R.; Zolfagharian, A.; Fotouhi, M.; Norouzi, S. 4D Printing Self-Morphing Structures. *Materials* **2019**, *12*, 1353. [CrossRef] [PubMed]
30. Noroozi, R.; Bodaghi, M.; Jafari, H.; Zolfagharian, A.; Fotouhi, M. Shape-Adaptive Metastructures with Variable Bandgap Regions by 4D Printing. *Polymers* **2020**, *12*, 519. [CrossRef]
31. Goulas, A.; Zhang, S.; Cadman, D.A.; Järveläinen, J.; Mylläri, V.; Whittow, W.G.; Vardaxoglou, J.C.; Engström, D.S. The Impact of 3D Printing Process Parameters on the Dielectric Properties of High Permittivity Composites. *Designs* **2019**, *3*, 50. [CrossRef]
32. Geng, P.; Zhao, J.; Wu, W.; Ye, W.; Wang, Y.; Wang, S.; Zhang, S. Effects of extrusion speed and printing speed on the 3D printing stability of extruded PEEK filament. *J. Manuf. Process.* **2019**, *37*, 266–273. [CrossRef]
33. Ćwikła, G.; Grabowik, C.; Kalinowski, K.; Paprocka, I.; Ociepka, P. The influence of printing parameters on selected mechanical properties of FDM/FFF 3D-printed parts. *IOP Conf. Ser. Mater. Sci. Eng.* **2017**, *227*, 012033. [CrossRef]
34. Blok, L.G.; Longana, M.L.; Yu, H.; Woods, B.K.S. An investigation into 3D printing of fibre reinforced thermoplastic composites. *Addit. Manuf.* **2018**, *22*, 176–186. [CrossRef]
35. Sharma, M.; Ziemi, S. Anisotropic Mechanical Properties of ABS Parts Fabricated by Fused Deposition Modelling. In *Mechanical Engineering*; InTech: Rijeka, Croatia, 2012.
36. Hernandez, D. Factors Affecting Dimensional Precision of Consumer 3D Printing. *Int. J. Aviat. Aeronaut. Aerosp.* **2015**, *2*, 2. [CrossRef]
37. Tronvoll, S.A.; Welo, T.; Elverum, C.W. The effects of voids on structural properties of fused deposition modelled parts: A probabilistic approach. *Int. J. Adv. Manuf. Technol.* **2018**, *97*, 3607–3618. [CrossRef]
38. Kuznetsov, V.E.; Solonin, A.N.; Tavitov, A.; Urzhumtsev, O.; Vakulik, A. Increasing strength of FFF three-dimensional printed parts by influencing on temperature-related parameters of the process. *Rapid Prototyp. J.* **2020**, *26*, 107–121. [CrossRef]

39. Kogut, P.; Kalinowski, K.; Grabowik, C.W.; Ćwikła, G.; Paprocka, I. Algorithms of control parameters selection for automation of FDM 3D printing process. In Proceedings of the IManE&E 2017 MATEC Web of Conferences, EDP Sciences, Iasi, Romania, 24–27 May 2017; Volume 112, p. 05011. [CrossRef]
40. Spoerk, M.; Gonzalez-Gutierrez, J.; Sapkota, J.; Schuschnigg, S.; Holzer, C. Effect of the printing bed temperature on the adhesion of parts produced by fused filament fabrication. *Plast. Rubber Compos.* **2018**, *47*, 17–24. [CrossRef]
41. Krotkỳ, J.; Honzíková, J.; Moc, P. Deformation of print PLA material depending on the temperature of reheating printing pad. *Manuf. Technol.* **2016**, *16*, 1213–2489. [CrossRef]
42. Yamamura, S.; Iwase, E. Hybrid hinge structure with elastic hinge on self-folding of 4D printing using a fused deposition modeling 3D printer. *Mater. Des.* **2021**, *203*, 109605. [CrossRef]
43. Wang, Y.; Li, X. An accurate finite element approach for programming 4D-printed self-morphing structures produced by fused deposition modeling. *Mech. Mater.* **2020**, *151*, 103628. [CrossRef]

Article

# 3D Bioprinting of Polycaprolactone-Based Scaffolds for Pulp-Dentin Regeneration: Investigation of Physicochemical and Biological Behavior

Zohre Mousavi Nejad [1], Ali Zamanian [1,*], Maryam Saeidifar [1], Hamid Reza Vanaei [2] and Mehdi Salar Amoli [3,4]

[1] Biomaterials Research Group, Department of Nanotechnology and Advance Materials, Materials and Energy Research Center, Karaj 31787-316, Iran; z.mousavinejad@merc.ac.ir (Z.M.N.); saeidifar@merc.ac.ir (M.S.)
[2] Arts et Metiers Institute of Technology, CNAM, LIFSE, HESAM University, 75013 Paris, France; hamidreza.vanaei@ensam.eu
[3] OMFS-IMPATH Research Group, Department of Imaging & Pathology, Campus Sint-Rafaël, KU Leuven, Kapucijnenvoer 33, 3000 Leuven, Belgium; mehdi.salaramoli@kuleuven.be
[4] Campus Group T, Materials Technology TC, KU Leuven, Andreas Vesaliusstraat 13-Box 2600, 3000 Leuven, Belgium
* Correspondence: a-zamanian@merc.ac.ir

**Abstract:** In this study, two structurally different scaffolds, a polycaprolactone (PCL)/45S5 Bioglass (BG) composite and PCL/hyaluronic acid (HyA) were fabricated by 3D printing technology and were evaluated for the regeneration of dentin and pulp tissues, respectively. Their physicochemical characterization was performed by field emission scanning electron microscopy (FESEM) equipped with energy dispersive spectroscopy (EDS), Fourier-transform infrared spectroscopy (FTIR), X-ray diffraction (XRD), atomic force microscopy (AFM), contact angle, and compressive strength tests. The results indicated that the presence of BG in the PCL/BG scaffolds promoted the mechanical properties, surface roughness, and bioactivity. Besides, a surface treatment of the PCL scaffold with HyA considerably increased the hydrophilicity of the scaffolds which led to an enhancement in cell adhesion. Furthermore, the gene expression results showed a significant increase in expression of odontogenic markers, e.g., dentin sialophosphoprotein (DSPP), osteocalcin (OCN), and dentin matrix protein 1 (DMP-1) in the presence of both PCL/BG and PCL/HyA scaffolds. Moreover, to examine the feasibility of the idea for pulp-dentin complex regeneration, a bilayer PCL/BG-PCL/HyA scaffold was successfully fabricated and characterized by FESEM. Based on these results, it can be concluded that PCL/BG and PCL/HyA scaffolds have great potential for promoting hDPSC adhesion and odontogenic differentiation.

**Keywords:** 3D bioprinting; tissue engineering; pulp-dentin; polycaprolactone; 45S5 Bioglass; hyaluronic acid

## 1. Introduction

Tooth loss can be caused by a range of incidents and complications, including trauma, periodontal disease, or tooth decay [1,2]. Several approaches are currently used to address the problem of missing teeth, such as dentures, dental bridges, or dental implants, all of which are nonbiological methods and entail further complications. As tooth decay is one of the most common causes of tooth loss, the effective treatment of pulp necrosis has been the focus of various treatment strategies. With root canal therapy being the most widely used treatment option, the following consequences, such as brittleness of the teeth, have given rise to research on the regeneration of dental pulp as a promising alternative [3]. Tissue engineering is an approach that combines support materials and cells aimed at the regeneration of different tissues [4–6]. In this strategy, scaffolds provide mechanical support along with biological cues required for cells to form the new tissues, hence, playing a central part in the regeneration strategy [6–8]. Due to the limitations of current regenerative endodontic treatments, such as variability in the outcome, various studies have focused on the development of tooth tissue engineering scaffolds, supporting the viability and growth of cells in dental pulp and dentin, encouraging them to regenerate damaged tissue [9]. A range of techniques has been used traditionally to manufacture scaffolds for tissue engineering, including salt leaching [10], solvent casting [11], gas foaming [12], freeze casting [13], freeze drying [14], and electrospinning [15,16]. However, there are limitations associated with these methods, mainly, (1) restricted control over the microstructure (size, shape, spatial distribution, and interconnectivity of the pores), (2) difficulty with removing residual solvent from the final structure and, (3) inability to replicate complex structures [17]. A layer-by-layer deposition of materials, known as additive manufacturing or 3D printing, enables the production of three dimensional constructs with complex shapes in a significantly more facile manner compared to other techniques [18–22]. Moreover, the use of 3D printing techniques for regeneration applications of various tissues including pulp [23] and dentin [24] has been attracting increasing attention due to its promising results.

Presenting a host of opportunities, 3D-printed scaffolds made out of polymers, ceramics, or composites are being widely investigated as candidates for dental tissue engineering [24–26]. Polycaprolactone (PCL) is a biocompatible, biodegradable, and printable polymer with a reasonably high mechanical strength, which has been approved by the FDA to be used in medical devices [27]. Because of these favorable properties, PCL has been the most widely used material among the candidate materials [24]. However, PCL suffers from disadvantages such as being hydrophobic or lack of support for cell adhesion [7]. To address these challenges, various methods and techniques have been proposed, such as surface treatment with hyaluronic acid (HyA) [28] or supplementing with additional active materials. Based on the literature, HyA is a promising biomaterial for use in pulp-dentin regeneration due to its ability to enhance cellular metabolism leading to increased deposition of the mineralized matrix deposition by human dental pulp stem cells (hDPSCs) [29]. In addition, 45S5 Bioglass (BG) which was developed by Hench et al. for first the time in 1969 [30], is a silicate glass containing 45% $SiO_2$, 24.5% CaO, 24.5% $Na_2O$, and 6% $P_2O_5$, in wt% [31]. This material has a great ability to bond with host tissue which makes it a potential candidate for use in both soft and hard tissue regeneration applications [32]. In the last years, a lot of studies have dealt with the incorporation of BG particles as a reinforcement for polymeric scaffolds to improve the mechanical properties, bioactivity, and biocompatibility of the scaffolds [33,34]. For these reasons and while most of the studies focused on 3D printing of dental tissues, reconstruction of either dental pulp [35,36], or dentin [24,25], it was demonstrated that due to the intertwined nature of these tissues, a successful tissue engineering approach requires development of hybrid scaffolds supporting regeneration of both tissues simultaneously.

In this present study, two different PCL-based scaffolds were fabricated by 3D printing technique and were evaluated physicochemically and biologically by field emission scanning electron microscopy (FESEM), atomic force microscopy (AFM), water contact angle, cell viability, cell adhesion and gene expression, e.g., PCL/BG and PCL/HyA scaffolds with the aim of supporting dentin and pulp regeneration, respectively. Since pulp and dentin have a close relationship during the life of the tooth, an ideal scaffold for successful tooth tissue engineering is a bilayer scaffold where each layer differs in the geometry and material. To examine the feasibility of this idea, a novel biphasic 3D-printed scaffold was designed and successfully fabricated.

## 2. Materials and Methods
### 2.1. Preparation of BG Powder

To produce 25 g of BG powder, briefly, 41.9 mL tetraethyl orthosilicate (TEOS, $Si(OC_2H_5)_4$; Merck, Darmstadt, Germany) was added to 62.4 mL nitric acid (1M) in a glass beaker. The mixture was stirred for 1 h to complete the hydrolysis process. Then, 3.6 mL triethyl phosphate (TEP, $(C_2H_5)_3PO_4$; Merck, Darmstadt, Germany), 25.2 g calcium nitrate tetrahydrate (Ca(NO$_3$)$_2$·4H$_2$O; Merck, Germany) and 16.9 g sodium nitrate (NaNO$_3$; Merck, Darmstadt, Germany) were sequentially added to the stirring mixture at 45 min intervals. The prepared sol was stored in a sealed container at room temperature for three days, followed by aging at 70 °C for one day and drying at 120 °C for one day. Eventually, the dried gel was stabilized in a furnace in a planetary ball mill at 300 rpm using a ball-to-powder mass ratio of 5 and a milling time of 1 h. Yttria-stabilized zirconia vial and 3 mm balls were utilized. The 45S5 powder was prepared for use in the PCL/BG scaffold fabrication process by sieving on sieve No. 270 (53 µm).

### 2.2. Fabrication of 3D-Printed PCL, PCL/HyA, and PCL/BG Scaffolds

Figure 1 delineates the method used to fabricate 3D-printed PCL, PCL/HyA, and PCL/BG scaffolds. Initially, the PCL/BG composite film was fabricated by making 5% ($w/v$) PCL solution (Merck, Darmstadt, Germany) in DCM (Dichloromethane; Merck, Darmstadt, Germany) on a magnetic stirrer for 3 h at 40 °C. Next, the required amount of BG to make a PCL:BG ratio of 70:30 was gradually added to the PCL solution, and the mixture was left under stirring for 1 h until a milky-white-colored suspension was obtained. Subsequently, to obtain dry films, the suspension was cast into glass petri dishes and placed in a clean environment at room temperature. To print the scaffolds, dried films were cut into 5 mm slices and loaded into 3D printer (3D BIOPRINTER N2, 3DPL Co. Ltd., Tehran, Iran) cartridges at a temperature of 90 °C, pressure 6 bar, and speed of 2 mm/s. The pure PCL film was prepared through the same protocol by casting PCL solution. The printing process was the same as PCL/BG scaffolds. To fabricate PCL/HyA scaffolds, a two-stage technology was used, consisting of plasma treatment of pure PCL scaffolds (LFG 40, Diener Electronic, Ebhausen, Germany) and subsequent immobilization of HyA on its surface. PCL scaffolds were placed in the chamber (frequency of 40 kHz, power of 100 W, and pressures of 0.6 mbar), and both top and bottom sides were exposed to plasma for 5 min (total exposure time = 10 min). The aim of plasma treatment in this study was to activate the PCL scaffold surface before immersion in HyA solution. To coat the activated scaffolds with HyA, first 4 mg/mL HyA (1.2 MDa, bloomage Freda Biopharm Co., Ltd., Jinan, China) solution in distilled water was prepared and stored at 4 °C for 24 h. Then, plasma-treated scaffolds were immersed in HyA solution on a stirrer for 12 h. Finally, the scaffolds were freeze-dried (FD-10, Pishtaz Engineering Co., Tehran, Iran) at a temperature of −58 °C and pressure 0.5 Torr for 24 h.

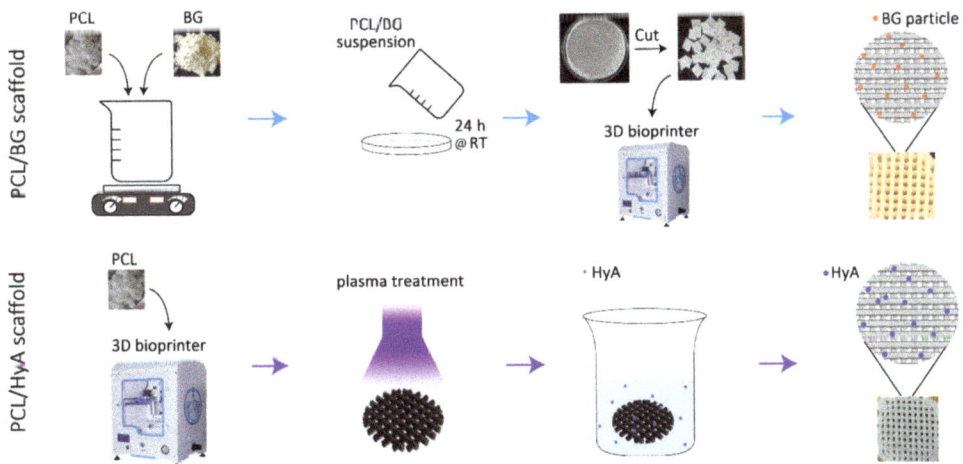

**Figure 1.** Schematic diagram of 3D-printed PCL/BG and PCL/HyA scaffold fabrication as artificial matrices for dentin and pulp tissue engineering, respectively. Abbreviations used in this Figure are as follows: PCL: polycaprolactone, BG: 45S5 bioactive glass, HyA: hyaluronic acid, h: hour, RT: room temperature.

### 2.3. Characterization of 45S5 Bioglass Powder

The microstructure and apatite formation ability of the BG powder was characterized using a TESCAN MIRA3 Field emission scanning electron microscopy (FESEM) equipped with energy dispersive spectroscopy (EDS). Additionally, Fourier-transform infrared spectroscopy (FTIR) was performed before and after immersion in SBF. According to the in vitro standard described by Kokubo et al., 1 g of BG powder was immersed in 20 mL SBF (in a 50 mL falcon tube) and kept in a humidified 37 °C/5%$CO_2$ incubator for 14 days [37]. SBF solution was refreshed twice a week, simulating the circulation of biological fluids inside a human body. After 14 days, the sample was transferred to a glass plate and allowed to be dried at 40 °C for 12 h. Samples were coated with gold and analyzed at an accelerating voltage of 15 kV. The heat treated 45S5 powder was characterized by X-ray diffraction (XRD), using a Philips PW 3710 X-rays diffractometer equipped with $CuK_\alpha$ radiation, ($\lambda$ = 1.5405 Å) operating at 40 kV and 30 mA. The chemical composition of the synthesized BG powder was analyzed by XRF using Philips PW 1480 XRF Spectrometer. Thermogravimetric analysis (TGA)-differential thermal analysis (DTA) was undertaken from 50 °C to 900 °C using TGA instrument (STA 504, TA Instruments) at a heating rate of 10 °C $min^{-1}$.

### 2.4. Characterization of Scaffolds

FESEM/EDS were used to analyze the surface morphology and chemical composition of the samples. All the samples were sputter coated with gold for 150 s. Hydrophilicity was evaluated using a contact angle measuring (CAM) device. Images were taken using DFK 23U618 USB 3.0 Color Industrial Camera using a 2X lens. Briefly, a 4 µL droplet of distilled water was deposited at the center of the PCL, PCL/HyA, and PCL/BG scaffolds. The contact angle was measured 1 min after deposition through image processing. Three samples were analyzed for each group.

To study the in vitro bioactivity of the samples, they were immersed in simulated body fluid (SBF) solution and then placed inside an incubator at 37 °C for 14 days. The formation of apatite crystals on the surface of the 3D-printed scaffolds was examined by FESEM/EDS. To evaluate the mechanical properties of the scaffolds, universal compressive strength test system (STM20, Santam Engineering Design Co., Tehran, Iran). Electromechanical compression testing machine equipped with a 5 kN load cell at a compression rate of 1 mm/min was used. Three replicates were tested for each sample.

To further evaluate the impact of HyA grafting on the surface structure of PCL scaffolds, the surface topography and roughness of PCL and PCL/HyA scaffolds were examined and compared using atomic force microscopy (AFM, AP 010, Park Scientific Instruments, Suwon, South Korea). AFM images were obtained by scanning the surface in contact mode (scan rate 0.1 Hz). To compare topologies of each surface, the arithmetical mean roughness, Ra values, were determined in three random areas per sample.

### 2.5. In Vitro Cell Viability Assay

The cytotoxicity and cell viability analyses were performed on PCL/BG and PCL/HyA scaffolds using MTT (3-(4,5-dimethylthiazol-2-yl)-2,5-diphenyltetrazolium bromide) assay. Initially, 3D-printed scaffolds were cut in dimensions of 5 mm × 5 mm. The sterilized scaffolds were placed in 96-well culture plates and incubated in a 5% $CO_2$ incubator at 37 °C in cell culture medium in triplicate. After 1 day, a cell suspensions of human gingival fibroblast (HGF) cells (Pasteur Institute of Iran, Tehran, Iran) containing the cell density of $5 \times 10^4$ cells/well were added in each well and left undisturbed for 24 h. Afterwards, 100 µL of MTT solution (5 mg/mL) was added to the wells to be incubated. After 4 h of incubation, the supernatant was removed carefully, 100 µL DMSO was added to each well, and the optical density was measured using an automatic microplate reader (BIO-TEK, VT,

USA) at a wavelength of 570 nm. The wells without 3D-printed scaffolds were applied as control. The cell viability was calculated using the formula below [38,39]:

Cell viability (%) = [mean OD of test group/mean OD of control group] × 100

### 2.6. Cell Adhesion Assay

To assess the cell adhesion on 3D-printed PCL/BG and PCL/HyA scaffolds (disc with diameter 5 mm and thickness of 2 mm), hDPSCs (Pasteur Institute of Iran, Tehran, Iran) were seeded on the scaffolds with a density of $5 \times 10^4$ cells/well and incubated in a $CO_2$ incubator at 37 °C for 2 days. Following incubation, the cell-cultured samples were extracted and rinsed with PBS solution to remove non-attached cells. The cells were fixed using 4% paraformaldehyde followed by dehydration with ethanol solutions of ascending concentrations.

### 2.7. Gene Expression Analysis

After 21 days of hDPSCs culture on 3D-printed PCL/BG and PCL/HyA scaffolds (disc with diameter 5 mm and thickness of 2 mm with pore size of 200 ± 5 μm and 300 ± 5 μm for PCL/HyA and PCL/BG, respectively) with a density of $5 \times 10^5$ cells/well, the total RNA was extracted from samples using a Qiagen RNeasy Mini kit (Qiagen, Seoul, South Korea), and then, converted to complementary DNA (cDNA) with a first-strand cDNA using the TaKaRa RNA PCR Kit (AMV) Ver.3.0 (Takara Bio., San Jose, CA, USA). The differentiation of hDPSCs was monitored by measuring mRNA expression levels of differentiation markers, including osteocalcin (OCN), dentin matrix protein 1 (DMP1), and dentin sialophosphoprotein (DSPP). The selected housekeeping gene was β-actin for all real-time polymerase chain reaction (PCR) runs.

Primer sequences for OCN, DMP 1, DSPP, and β-actin were designed based on published cDNA sequences (Table 1). The cells were cultured for a total of 3 weeks, with the differentiation medium being changed every 3–4 days. Each measurement was assessed in triplicate.

**Table 1.** Real-time PCR primer sequences of the genes coding osteocalcin (OCN), dentin sialophosphoprotein (DSPP), and dentin matrix protein 1 (DMP1) and β-actin.

| Gene | Primer Sequence | |
|---|---|---|
| | Forward | Reverse |
| OCN | 5′-GCAAAGGTGCAGCCTTTGTG-3′ | 5′-GGCTCCCAGCCATTGATACAG-3′ |
| DSPP | 5′-CCATTCCAGTTCCTCAAAGC-3′ | 5′-TGGCATTTAACTCCTGTA C-3′ |
| DMP1 | 5′-TTCTTTGTGAACTACGGAGG-3′ | 5′-TTGATACCTGGTTACTGGGA-3′ |
| β-actin | 5′-CTTCCTTCCTGGGCATG-3′ | 5′-GTCTTTGCGGATGTCCAC-3′ |

### 2.8. Statistical Analysis

All data were expressed as means ± standard deviation and represented at least three independent experiments. All data were analyzed using a two-way ANOVA test. $p$-values < 0.05 were considered significant. All analyses were carried out using GraphPad Prism version 9.0 for Windows (GraphPad Software, San Diego, CA, USA).

## 3. Results and Discussion

### 3.1. Characterization of 45S5 Bioglass Powder

The TGA-DTA of the BG powder was carried out to obtain the right sintering temperature. As shown in Figure 2A, the mass loss occurred in three stages. The first mass loss happened between 85 °C and 165 °C, demonstrated by an endothermic peak at 114 °C in the DTA curve assigned to the elimination of physically absorbed water, which was not removed in the drying process. Another mass loss in 250–310 °C range could be attributed to the removal of chemically absorbed water. The third mass loss took place at the

530–620 °C interval was attributed to eliminating the residual nitrates and condensation of silanol groups. The TGA trace exhibited a mass stability after 630 °C, reflected by an endothermic peak caused by glass transition, followed by an exothermic peak emerging at 650 °C. These curves confirmed that the residuals could be removed before 650 °C, which is also shown in other reports [40–43]. The result from the TGA-DTA allowed us to set the temperature of 650 °C for stabilization of the sample.

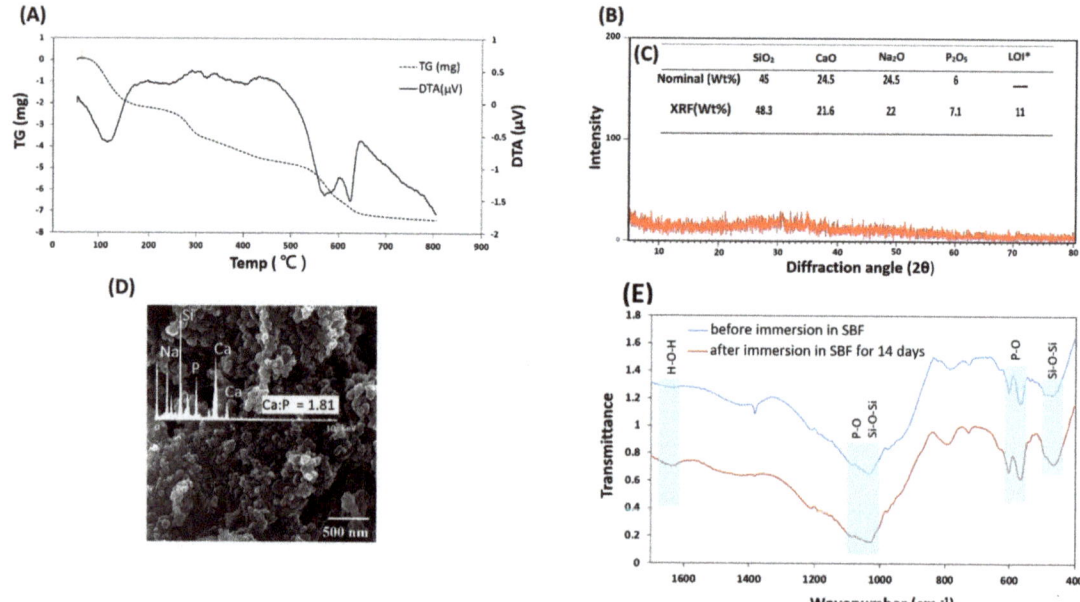

**Figure 2.** (**A**) DTA-TGA thermogram of 45S5 powder; showing a three-stage mass loss in TG and two exothermic and endothermic peaks in DTA curve. (**B**) XRD pattern of heat-treated BG; demonstrating amorphous structure. (**C**) Elemental analysis of 45S5 powder revealing similarity of the weight percentages to standard weight percentages of 45S5. * LOI: loss on ignition. (**D**) FESEM micrograph and EDS spectrum of synthesized 45S5 Bioglass powder after immersion in SBF for 14 days. (**E**) FTIR spectra of 45S5 Bioglass powder before and after immersion in SBF for 14 days. Abbreviations used in this Figure are as follows: Ca: calcium, P: phosphor, SBF: simulated body fluid.

The results of XRD analysis on heat treated BG powders are demonstrated in Figure 2B. The presence of a broad hump at around 30° has been known as a hallmark of amorphous materials [44]. Consequently, this feature demonstrates the amorphous nature of the synthesized Bioglass and confirms the synthesis of Bioglass powder.

The mass oxide concentrations obtained by XRF and nominal amounts are shown in Figure 2C. The weight percentage of the element oxides is consistent with standard weight percentages of 45S5 Bioglass [45,46]. The results verify 45S5 Bioglass was produced with desired weight percentages.

In order for the scaffold to bond with native tissue, there needs to be a hydroxyapatite layer at the interface. The formation of such a layer is one of the main results of using 45S5 Bioglass in the scaffolds. This happens due to a process of glass dissolution when in contact with SBF, during which the remainder of the dissolution process leads to a change in chemical composition and environment pH, causing nucleation of hydroxyapatite. As shown in Figure 2D, formation of an apatite layer was observed on the samples' surfaces after 14 days of immersion in SBF. Presence of an apatite phase was also validated by the EDS spectrum, as shown in Figure 2D. The ratio of Ca to P ions was approximately 1.81, which is known to be a characteristic of non-stoichiometric hydroxyapatite phase [47].

FTIR spectroscopic imaging was performed on the samples to detect the hydroxyapatite signal on the surface of SBF-treated BG powder. Figure 2E shows the spectra of BG powder before and after immersion in SBF for 14 days. The P–O bending vibration peaks at 560 and 604 cm$^{-1}$ and the P–O asymmetric stretching vibration bands between 1000 and 1150 cm$^{-1}$ represented the hydroxyapatite layer [48,49]. The most widely used peaks to differentiate hydroxyapatite and bioactive material are the ones corresponding to bending vibration. This is because of the superimposition of P–O stretching band and the Si–O stretching band of Bioglass, while the peak corresponding to Si-O bending was observed at 400–500 cm$^{-1}$ [50]. The magnitude of hydroxyapatite peaks at 560 and 604 cm$^{-1}$ increased after immersion in SBF which indicates the formation of an HA layer on the surface of the BG powder immersed in SBF solution for 14 days. The spectra exhibited bands at 1030 and 470 cm$^{-1}$ corresponding to the Si–O–Si stretch and Si–O–Si bend, respectively [50–52]. The appearance of the shoulder at around 1630 cm$^{-1}$ resulted from the H-O-H bond bending vibrations attributed to the absorbed water by the hydroxyapatite layer [53].

### 3.2. Physicocheimical Characterization of Scaffolds

#### 3.2.1. Morphology Observations

FESEM images of PCL/BG, PCL/HyA, and bilayer scaffolds are shown in Figure 3. As observed by FESEM, the 3D printing strategy leads to the precise production of pre-designed scaffolds (Figure 3A,D,G). The 0°/90° design was chosen for both PCL/HyA and PCL/BG scaffolds, because this pattern can be mechanically the strongest 3D-printed architecture [54]. The obtained results indicated that the scaffolds have the strut diameter of approximately 400 ± 5 μm and pore sizes of 200 ± 5 μm and 300 ± 5 μm for PCL/HyA and PCL/BG, respectively. Regarding the bilayer scaffolds, a clear transition from the PCL/HyA phase to the PCL/BG phase was observed (Figure 3H,I). FESEM observation confirmed the presence of well-distributed BG particles on both the surface and the inner part of the struts (Figure 3E,F). Furthermore, the upper and lower parts of Figure 3G–I, represent the PCL/BG phase (aimed at dentin regeneration) and the PCL/HyA phase (aimed at pulp regeneration), respectively. However, after showing the recuperation of the biological behavior of each phase, further investigation of the bilayer PCL/BG-PCL/HyA should be done in future studies.

#### 3.2.2. Atomic Force Microscopy (AFM)

To analyze and quantify the surface roughness of hyaluronic acid-grafted samples, AFM analysis was utilized (Figure 4A). A statistically significant increase in the surface roughness of the PCL/HyA scaffolds was demonstrated by AFM resulting from HyA grafting compared with PCL scaffolds ($p < 0.01$, $n = 3$). The untreated pure PCL surface showed an average roughness ($R_a$) of 42.8 nm. However, after the plasma treatment and hyaluronic acid coating, the PCL/HyA surface became rougher with an $R_a$ of 140 nm. These values confirmed the observations made through FESEM. It is generally recognized that an increase in roughness may drastically increase the biological response due to the higher surface/volume ratio [55,56].

#### 3.2.3. Static Water Contact Angle

PCL, PCL/BG, PCL/HyA scaffolds were subjected to contact angle measurements to evaluate the effect of the composition and surface treatment. As shown in Figure 4B, the contact angles for pure PCL, PCL/BG, and PCL/HyA scaffolds were 86 ± 2°, 80 ± 1°, and 63 ± 1°, respectively. For all samples, the contact angle values were below 90°, showing a hydrophilic tendency [57]. However, a slight decrease in the contact angle was observed after the addition of BG to the PCL, which is agreement with the findings in other studies [58,59]. The reason for this subtle change in contact angle could be that BG causes a local increase of pH when dissolved, and hydroxide ions can accelerate the cleavage of ester linkages [60]. In addition, the surface treatment of PCL scaffolds with plasma and HyA immobilization resulted a significant decrease in contact angle, that showed the plasma treatment can increase the surface wettability of the PCL-based scaffolds. Therefore, the results demonstrated that oxygen plasma treatment and HyA immobilization can affect hydrophilicity more than adding 45S5 Bioglass which probably increases cell adhesion. Similar results were obtained by Bruyas et al., who found that the addition of calcium phosphate-based materials did not significantly affect the contact angle of the PCL scaffold [58].

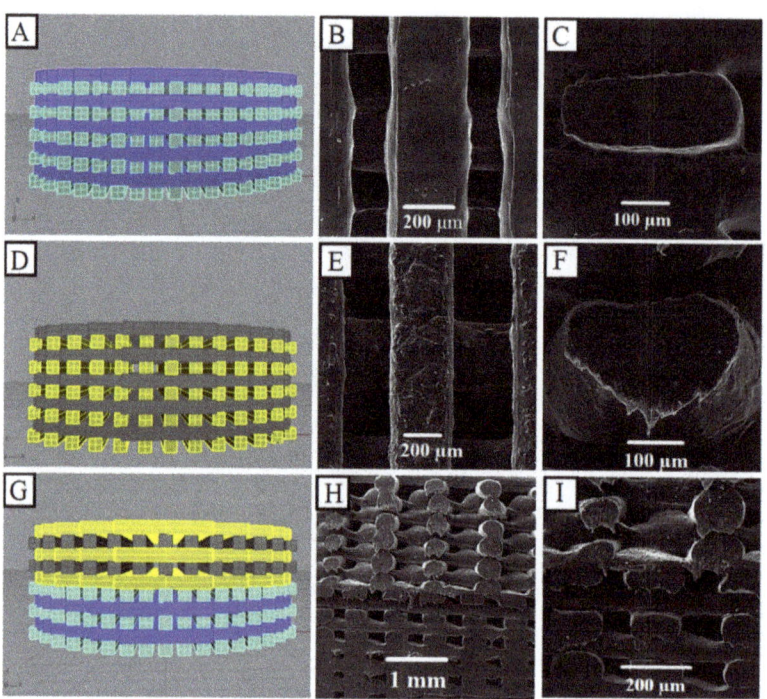

**Figure 3.** Morphological characterization of the 3D-printed PCL/BG, PCL/HyA, and bilayer scaffolds (**A,D,G**) CAD models of the scaffolds; (**B,E**) top view and (**C,F,H,I**) cross sectional FESEM images.

### 3.2.4. Mechanical Properties of 3D-Printed Scaffolds

In order to determine the impact of 45S5 Bioglass on the structural integrity of the scaffolds, the mechanical properties of porous PCL and PCL/BG scaffolds were characterized using compression strength tests. Figure 4C demonstrates the representative compressive stress versus strain responses of PCL and PCL/BG. The sample behaves as an elastomeric or elastic-plastic solid as shown by the three regimes: (i) a linear elastic regime, (ii) a plateau of stress resulting from macropores collapsing progressively, and finally (iii) an area of densification after the pores have totally collapsed throughout the material. The Young's modulus increased from 51.6 ± 0.62 MPa to 67.4 ± 0.54 MPa by the addition of 45S5 Bioglass to the composition (Figure 4D), which is in the range of the average value of the Young's modulus of PCL-based 3D-printed scaffolds characterized by other researchers [61,62]. Roohani et al. also obtained the range of 19.3–49.4 MPa for Young's modulus by adding BG to PCL [63]. The average yield stress value was 6.1 MPa and 9.16 MPa for the PCL and PCL/BG scaffolds, respectively. It can be concluded that by adding BG to a PCL-based scaffold's composition, the mechanical strength is increased [64].

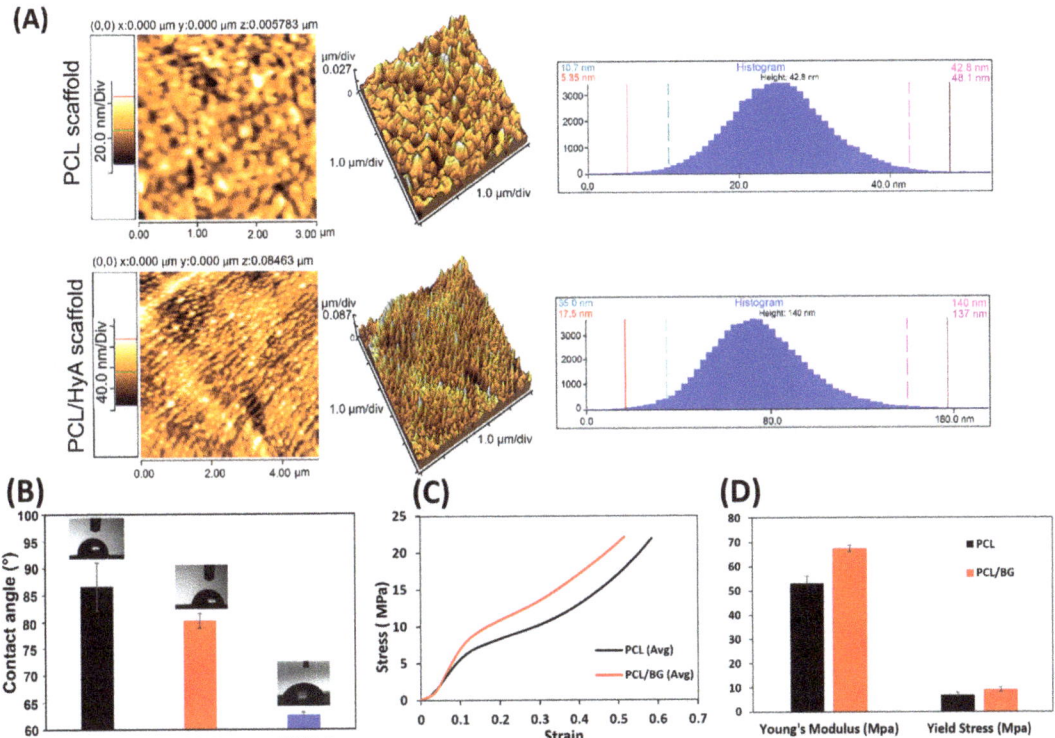

**Figure 4.** (**A**) Surface topography of 3D-printed PCL scaffolds with differential morphologies PCL and PCL/HyA obtained with AFM analysis; (**B**) contact angle measurement of 3D-printed PCL, PCL/BG, and PCL/HyA scaffolds; (**C**) compressive stress versus strain responses of PCL and PCL/BG 3D-printed scaffolds; (**D**) Young's modulus and yield strength values of PCL and PCL/BG 3D-printed scaffolds.

### 3.3. Cytotoxicity Assay

The cytotoxicity and cell viability of PCL/BG and PCL/HyA scaffolds were analyzed by MTT assay. The basis of this assay is the reduction reaction initiated by living cells' enzymes turning a yellow MTT to purple MTT-formazan crystal [65,66]. As displayed in Figure 5A, the relative cell viability of all the samples was higher than 90%, which confirms that they are cytocompatible and suitable for use in tissue engineering applications. Hyaluronic acid-coated PCL scaffolds (PCL/HyA) presented higher cell viability than pure PCL and PCL/BG scaffolds. The lowest cell viability was found for pure PCL samples, indicating that coating PCL with a hydrophilic material like HyA or compositing with 45S5 Bioglass can improve cell adhesion by increasing the hydrophilicity of the PCL scaffolds, which is in confirmation with the result obtained by Jensen et al. [67] and Kandelousi et al. [68].

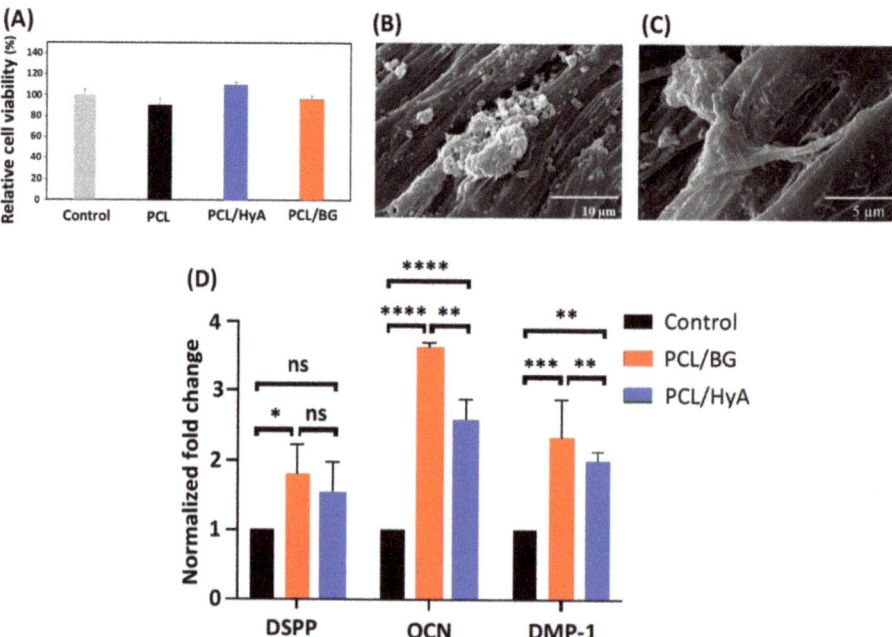

**Figure 5.** Viability of HGF cells, and morphology and differentiation ability of hDPSCs on 3D-printed scaffolds. (**A**) MTT cell viability analysis; (**B**) hDPSCs attachment on 3D-printed PCL/BG scaffold; (**C**) hDPSCs attachment on 3D-printed PCL/HyA scaffold; (**D**) Gene expression levels of DSPP, OCN, and DMP-1 in hDPSCs cultured in control conditions. ns (not significant), $p$-value > 0.05; * $p$-value ≤ 0.05; ** means $p$-value ≤ 0.01; *** means $p$-value ≤ 0.001; **** $p$-value ≤ 0.0001. Abbreviations used in this Figure are as follows: PCL: polycaprolactone, BG: 45S5 bioactive glass, HyA: hyaluronic acid, DSPP: dentin sialophosphoprotein, OCN: osteocalcin, and DMP-1: dentin matrix protein 1.

### 3.4. Cell Adhesion and Morphology Assay

Adhesion and morphology of the cells on 3D-printed PCL/BG and PCL/HyA scaffolds were analyzed using FESEM. As shown in Figure 5B,C, after two days' culture, the hDPSCs adhered to the surface of PCL/BG and PCL/HyA scaffolds, demonstrating a uniform dispersion. The presence of long cytoplasmic prolongations on both PCL/BG and PCL/HyA scaffolds revealed an appropriate cytocompatibility of the material and positive interaction between stem cells and 3D-printed scaffolds. However, both scaffolds promoted cellular adhesion, and hDPSC felt comfortable on the surface of both PCL/BG and PCL/HyA scaffolds; it seems that PCL/Hya scaffold provides the most favorable environment for hDPSC, which is due to the effective surface modification of PCL/HyA scaffold with plasma and HyA. This finding was in agreement with the obtained results of static water contact angle. In addition, Kudryavtseva et al. [69] also proved that the surface modification with plasma and HyA enhances cell attachment. Overall, the results of this section verified the role of these modified bioactive and hydrophilic scaffolds in supporting cellular adhesion of hDPSCs and indicated the impressive potential of these scaffolds for pulp-dentin regeneration applications.

### 3.5. Gene Expression Analysis

The odontogenic differentiation ability of hDPSCs seeded on PCL/BG and PCL/HyA scaffolds as well as on a petri dish as a 2D culture was investigated after 21 days of culture in differentiation medium. The expression level of the culture dish was set to baseline (=1.0). The gene expression of dentin-associated genes including dentin sialophosphoprotein (DSPP), dentin matrix protein-1 (DMP-1), and osteocalcin (OCN) were analyzed through

PCR. The results of gene expression are presented in Figure 5D with statistical differences among the groups. A very significant upregulation was observed in cells cultured on 3D-printed PCL/BG and PCL/HyA scaffolds compared to the 2D culture dish. Furthermore, the PCL/BG resulted in a significantly higher OCN and DMP-1 expression compared to PCL/HyA scaffolds ($p < 0.01$). The reason for the higher expression of DSPP, DMP-1, and OCN genes in PCL/BG scaffold compared to PCL/HyA scaffold is the presence of bioactive glass in PCL/BG, which makes the structure more mineralized and these markers are related to dentin which is a mineralized hard tissue. This difference in the expression of factors between the two scaffold types, along with the upregulation of odontoblast related genes DSPP, DMP-1 and OCN demonstrates the potential for PCL/HyA and PCL/BG scaffolds to be used in pulp and dentin regeneration, respectively, in conjunction with hDPSC.

## 4. Conclusions

With the failure of existing approaches in addressing many dental tissue complications, regenerative medicine seems like a promising approach to improve the regeneration of the dentin-pulp complex. However, traditional strategies producing scaffolds for tissue engineering in dental tissue have been ineffective. For this reason, this study presented a new strategy to fabricate 3D-printed tissue-engineering scaffolds. Polycaprolactone supplemented with 45S5 Bioglass and HyA was used to produce the scaffolds. It was shown that the coating of scaffolds with HyA had a significant impact on increasing the hydrophilicity of the scaffolds, resulting in a more favorable environment for the cells. At the same time, the addition of 45S5 Bioglass also resulted in a slightly more hydrophilic surface. The 45S5 Bioglass was found to increase the mechanical strength of the material. Furthermore, it was shown that both HyA-coated and 45S5 Bioglass-supplemented scaffolds present high cell viability. Moreover, the cellular attachment observed through FESEM and the significant upregulation of differentiation of the odontoblast-related markers DSPP, DMP-1, and OCN in both scaffold groups represent an environment assisting cellular activities. Overall, under the conditions of the present study, it might be concluded that PCL/HyA and PCL/BG scaffolds can induce an organized matrix formation similar to that of pulp and dentin tissues, respectively. The findings in this study show that the properties obtained through combining PCL with either 45S5 Bioglass or hyaluronic acid lead to the opportunity of producing a bilayer scaffold capable of assisting the regeneration of the dentin-pulp complex.

**Author Contributions:** Conceptualization, A.Z.; investigation, Z.M.N.; writing—original draft preparation, Z.M.N.; writing—review and editing, H.R.V., M.S. and M.S.A.; supervision, M.S. All authors have read and agreed to the published version of the manuscript.

**Funding:** This research received no external funding.

**Institutional Review Board Statement:** Not applicable.

**Data Availability Statement:** Not applicable.

**Acknowledgments:** Thanks to the Department of Nanotechnology and Advance Materials, Materials and Energy Research Center for supporting this research.

**Conflicts of Interest:** The authors declare no conflict of interest.

## References

1. Jung, C.; Kim, S.; Sun, T.; Cho, Y.-B.; Song, M. Pulp-dentin regeneration: Current approaches and challenges. *J. Tissue Eng.* **2019**, *10*, 2041731418819263. [CrossRef]
2. EzEldeen, M.; Loos, J.; Mousavi Nejad, Z.; Cristaldi, M.; Murgia, D.; Braem, A.; Jacobs, R. 3D-printing-assisted fabrication of chitosan scaffolds from different sources and cross-linkers for dental tissue engineering. *Eur. Cell Mater.* **2021**, *41*, 485–501. [CrossRef] [PubMed]
3. Alshehadat, S.A.; Thu, H.A.; Hamid, S.S.A.; Nurul, A.A.; Rani, S.A.; Ahmad, A. Scaffolds for dental pulp tissue regeneration: A review. *Int. Dent. Med J. Adv. Res.* **2016**, *2*, 1–12.

4. Shahrezaie, M.; Moshiri, A.; Shekarchi, B.; Oryan, A.; Maffulli, N.; Parvizi, J. Effectiveness of tissue engineered three-dimensional bioactive graft on bone healing and regeneration: An in vivo study with significant clinical value. *J. Tissue Eng. Regen. Med.* **2018**, *12*, 936–960. [CrossRef]
5. Moshiri, A.; Oryan, A.; Shahrezaee, M. An overview on bone tissue engineering and regenerative medicine: Current challenges, future directions and strategies. *J. Sports Med. Doping Stud.* **2015**, *5*, e144.
6. Sohrabian, M.; Vaseghi, M.; Khaleghi, H.; Dehrooyeh, S.; Kohan, M.S.A. Structural Investigation of Delicate-Geometry Fused Deposition Modeling Additive Manufacturing Scaffolds: Experiment and Analytics. *J. Mater. Eng. Perform.* **2021**, *30*, 6529–6541. [CrossRef]
7. Ghorbani, F.; Sahranavard, M.; Mousavi Nejad, Z.; Li, D.; Zamanian, A.; Yu, B. Surface functionalization of three dimensional-printed polycaprolactone-bioactive glass scaffolds by grafting GelMA under UV irradiation. *Front. Mater.* **2020**, *7*, 348. [CrossRef]
8. Mousavi, S.-M.; Nejad, Z.M.; Hashemi, S.A.; Salari, M.; Gholami, A.; Ramakrishna, S.; Chiang, W.-H.; Lai, C.W. Bioactive Agent-Loaded Electrospun Nanofiber Membranes for Accelerating Healing Process: A Review. *Membranes* **2021**, *11*, 702. [CrossRef]
9. Wang, F.; Xie, C.; Ren, N.; Bai, S.; Zhao, Y. Human Freeze-dried Dentin Matrix as a Biologically Active Scaffold for Tooth Tissue Engineering. *J. Endod.* **2019**, *45*, 1321–1331. [CrossRef]
10. Kanimozhi, K.; Basha, S.K.; Kaviyarasu, K.; SuganthaKumari, V. Salt leaching synthesis, characterization and in vitro cytocompatibility of chitosan/poly (vinyl alcohol)/methylcellulose–ZnO nanocomposites scaffolds using L929 fibroblast cells. *J. Nanosci. Nanotechnol.* **2019**, *19*, 4447–4457. [CrossRef] [PubMed]
11. Sola, A.; Bertacchini, J.; D'Avella, D.; Anselmi, L.; Maraldi, T.; Marmiroli, S.; Messori, M. Development of solvent-casting particulate leaching (SCPL) polymer scaffolds as improved three-dimensional supports to mimic the bone marrow niche. *Mater. Sci. Eng. C* **2019**, *96*, 153–165. [CrossRef] [PubMed]
12. Rao, F.; Yuan, Z.; Li, M.; Yu, F.; Fang, X.; Jiang, B.; Wen, Y.; Zhang, P. Expanded 3D nanofibre sponge scaffolds by gas-foaming technique enhance peripheral nerve regeneration. *Artif. Cells Nanomed. Biotechnol.* **2019**, *47*, 491–500. [CrossRef] [PubMed]
13. Pourhaghgouy, M.; Zamanian, A.; Shahrezaee, M.; Masouleh, M.P. Physicochemical properties and bioactivity of freeze-cast chitosan nanocomposite scaffolds reinforced with bioactive glass. *Mater. Sci. Eng. C* **2016**, *58*, 180–186. [CrossRef]
14. Namdarian, P.; Zamanian, A.; Asefnejad, A.; Saeidifar, M. Evaluation of Freeze-Dry Chitosan-Gelatin Scaffolds with Olibanum Microspheres Containing Dexamethasone for Bone Tissue Engineering. *Polym. Korea* **2018**, *42*, 982–993. [CrossRef]
15. Rashtchian, M.; Hivechi, A.; Bahrami, S.H.; Milan, P.B.; Simorgh, S. Fabricating alginate/poly (caprolactone) nanofibers with enhanced bio-mechanical properties via cellulose nanocrystal incorporation. *Carbohydr. Polym.* **2020**, *233*, 115873. [CrossRef] [PubMed]
16. Mirzaei, Z.; Kordestani, S.; Kuth, S.; Schubert, D.W.; Detsch, R.; Roether, J.A.; Blunk, T.; Boccaccini, A.R. Preparation and Characterization of Electrospun Blend Fibrous Polyethylene Oxide: Polycaprolactone Scaffolds to Promote Cartilage Regeneration. *Adv. Eng. Mater.* **2020**, *22*, 2000131. [CrossRef]
17. Sahranavard, M.; Zamanian, A.; Ghorbani, F.; Shahrezaee, M.H. A critical review on three dimensional-printed chitosan hydrogels for development of tissue engineering. *Bioprinting* **2019**, *17*, e00063. [CrossRef]
18. McGivern, S.; Boutouil, H.; Al-Kharusi, G.; Little, S.; Dunne, N.J.; Levingstone, T.J. Translational application of 3D bioprinting for cartilage tissue engineering. *Bioengineering* **2021**, *8*, 144. [CrossRef]
19. El Magri, A.; Vanaei, S.; Shirinbayan, M.; Vaudreuil, S.; Tcharkhtchi, A. An Investigation to Study the Effect of Process Parameters on the Strength and Fatigue Behavior of 3D-Printed PLA-Graphene. *Polymers* **2021**, *13*, 3218. [CrossRef]
20. Vanaei, H.R.; Shirinbayan, M.; Deligant, M.; Khelladi, S.; Tcharkhtchi, A. In-Process Monitoring of Temperature Evolution during Fused Filament Fabrication: A Journey from Numerical to Experimental Approaches. *Thermo* **2021**, *1*, 332–360. [CrossRef]
21. Vanaei, H.R.; Shirinbayan, M.; Vanaei, S.; Fitoussi, J.; Khelladi, S.; Tcharkhtchi, A. Multi-scale damage analysis and fatigue behavior of PLA manufactured by fused deposition modeling (FDM). *Rapid Prototyp. J.* **2021**, *27*, 371–378. [CrossRef]
22. Vanaei, S.; Parizi, M.S.; Vanaei, S.; Salemizadehparizi, F.; Vanaei, H.R. An Overview on Materials and Techniques in 3D Bioprinting Toward Biomedical Application. *Eng. Regen.* **2021**, *2*, 1–18. [CrossRef]
23. Hilkens, P.; Bronckaers, A.; Ratajczak, J.; Gervois, P.; Wolfs, E.; Lambrichts, I. The angiogenic potential of DPSCs and SCAPs in an in vivo model of dental pulp regeneration. *Stem Cells Int.* **2017**, *2017*, 2582080. [CrossRef] [PubMed]
24. Wu, Y.; Azmi, D.F.; Rosa, V.; Fawzy, A.S.; Fuh, J.Y.; Wong, Y.S.; Lu, W.F. Fabrication of dentin-like scaffolds through combined 3D printing and bio-mineralisation. *Cogent Eng.* **2016**, *3*, 1222777. [CrossRef]
25. Athirasala, A.; Tahayeri, A.; Thrivikraman, G.; França, C.M.; Monteiro, N.; Tran, V.; Ferracane, J.; Bertassoni, L.E. A dentin-derived hydrogel bioink for 3D bioprinting of cell laden scaffolds for regenerative dentistry. *Biofabrication* **2018**, *10*, 024101. [CrossRef] [PubMed]
26. Nyberg, E.; Rindone, A.; Dorafshar, A.; Grayson, W.L. Comparison of 3D-printed poly-ε-caprolactone scaffolds functionalized with tricalcium phosphate, hydroxyapatite, bio-oss, or decellularized bone matrix. *Tissue Eng. Part A* **2017**, *23*, 503–514. [CrossRef] [PubMed]
27. Ghorbani, F.; Zamanian, A.; Sahranavard, M. Mussel-inspired polydopamine-mediated surface modification of freeze-cast poly (ε-caprolactone) scaffolds for bone tissue engineering applications. *Biomed. Eng. /Biomed. Tech.* **2020**, *65*, 273–287. [CrossRef] [PubMed]

28. Nejad, Z.M.; Torabinejad, B.; Davachi, S.M.; Zamanian, A.; Garakani, S.S.; Najafi, F.; Nezafati, N. Synthesis, physicochemical, rheological and in-vitro characterization of double-crosslinked hyaluronic acid hydrogels containing dexamethasone and PLGA/dexamethasone nanoparticles as hybrid systems for specific medical applications. *Int. J. Biol. Macromol.* **2019**, *126*, 193–208. [CrossRef]
29. Ahmadian, E.; Eftekhari, A.; Dizaj, S.M.; Sharifi, S.; Mokhtarpour, M.; Nasibova, A.N.; Khalilov, R.; Samiei, M. The effect of hyaluronic acid hydrogels on dental pulp stem cells behavior. *Int. J. Biol. Macromol.* **2019**, *140*, 245–254. [CrossRef]
30. Hench, L.L. The story of Bioglass®. *J. Mater. Sci. Mater. Med.* **2006**, *17*, 967–978. [CrossRef]
31. Kraxner, J.; Michalek, M.; Romero, A.R.; Elsayed, H.; Bernardo, E.; Boccaccini, A.R.; Galusek, D. Porous bioactive glass microspheres prepared by flame synthesis process. *Mater. Lett.* **2019**, *256*, 126625. [CrossRef]
32. Bertuola, M.; Aráoz, B.; Gilabert, U.; Gonzalez-Wusener, A.; Pérez-Recalde, M.; Arregui, C.O.; Hermida, É.B. Gelatin–alginate–hyaluronic acid inks for 3D printing: Effects of bioglass addition on printability, rheology and scaffold tensile modulus. *J. Mater. Sci.* **2021**, *56*, 15327–15343. [CrossRef]
33. Singh, B.N.; Veeresh, V.; Mallick, S.P.; Jain, Y.; Sinha, S.; Rastogi, A.; Srivastava, P. Design and evaluation of chitosan/chondroitin sulfate/nano-bioglass based composite scaffold for bone tissue engineering. *Int. J. Biol. Macromol.* **2019**, *133*, 817–830. [CrossRef] [PubMed]
34. Badr-Mohammadi, M.-R.; Hesaraki, S.; Zamanian, A. Mechanical properties and in vitro cellular behavior of zinc-containing nano-bioactive glass doped biphasic calcium phosphate bone substitutes. *J. Mater. Sci. Mater. Med.* **2014**, *25*, 185–197. [CrossRef]
35. Feng, S.; Liu, J.; Ramalingam, M. 3D printing of stem cell responsive ionically-crosslinked polyethylene glycol diacrylate/alginate composite hydrogels loaded with basic fibroblast growth factor for dental pulp tissue engineering: A preclinical evaluation in animal model. *J. Biomater. Tissue Eng.* **2019**, *9*, 1635–1643. [CrossRef]
36. Monteiro, N.; Smith, E.E.; Angstadt, S.; Zhang, W.; Khademhosseini, A.; Yelick, P.C. Dental cell sheet biomimetic tooth bud model. *Biomaterials* **2016**, *106*, 167–179. [CrossRef]
37. Kokubo, T.; Hata, K.; Nakamura, T.; Yamamuro, T. Apatite formation on ceramics, metals and polymers induced by a CaO SiO2 based glass in a simulated body fluid. In *Bioceramics*; Elsevier: Amsterdam, The Netherlands, 1991; pp. 113–120.
38. Kılıç, S.; Okullu, S.Ö.; Kurt, Ö.; Sevinç, H.; Dündar, C.; Altınordu, F.; Türkoğlu, M. Efficacy of two plant extracts against acne vulgaris: Initial results of microbiological tests and cell culture studies. *J. Cosmet. Dermatol.* **2019**, *18*, 1061–1065. [CrossRef] [PubMed]
39. Salehi, G.; Behnamghader, A.; Pazouki, M.; Houshmand, B.; Mozafari, M. Synergistic reinforcement of glass-ionomer dental cements with silanized glass fibres. *Mater. Technol.* **2020**, *35*, 433–445. [CrossRef]
40. Chatzistavrou, X.; Zorba, T.; Kontonasaki, E.; Chrissafis, K.; Koidis, P.; Paraskevopoulos, K. Following bioactive glass behavior beyond melting temperature by thermal and optical methods. *Phys. Status Solidi* **2004**, *201*, 944–951. [CrossRef]
41. ElBatal, H.; Azooz, M.; Khalil, E.; Monem, A.S.; Hamdy, Y. Characterization of some bioglass–ceramics. *Mater. Chem. Phys.* **2003**, *80*, 599–609. [CrossRef]
42. El-Ghannam, A.; Hamazawy, E.; Yehia, A. Effect of thermal treatment on bioactive glass microstructure, corrosion behavior, ζ potential, and protein adsorption. *J. Biomed. Mater. Res. Off. J. Soc. Biomater. Jpn. Soc. Biomater. Aust. Soc. Biomater. Korean Soc. Biomater.* **2001**, *55*, 387–395.
43. Lefebvre, L.; Chevalier, J.; Gremillard, L.; Zenati, R.; Thollet, G.; Bernache-Assolant, D.; Govin, A. Structural transformations of bioactive glass 45S5 with thermal treatments. *Acta Mater.* **2007**, *55*, 3305–3313. [CrossRef]
44. Kumar, P.; Dehiya, B.S.; Sindhu, A.; Kumar, V. Synthesis and Characterization of Nano Bioglass for the Application of Bone Tissue Engineering. *J. Nanosci. Technol.* **2018**, *4*, 471–474. [CrossRef]
45. Baino, F.; Hamzehlou, S.; Kargozar, S. Bioactive glasses: Where are we and where are we going? *J. Funct. Biomater.* **2018**, *9*, 25. [CrossRef] [PubMed]
46. Schnettler, R.; Alt, V.; Dingeldein, E.; Pfefferle, H.-J.; Kilian, O.; Meyer, C.; Heiss, C.; Wenisch, S. Bone ingrowth in bFGF-coated hydroxyapatite ceramic implants. *Biomaterials* **2003**, *24*, 4603–4608. [CrossRef]
47. Chatzistavrou, X.; Velamakanni, S.; DiRenzo, K.; Lefkelidou, A.; Fenno, J.C.; Kasuga, T.; Boccaccini, A.R.; Papagerakis, P. Designing dental composites with bioactive and bactericidal properties. *Mater. Sci. Eng. C* **2015**, *52*, 267–272. [CrossRef] [PubMed]
48. Ohtsuki, C.; Kushitani, H.; Kokubo, T.; Kotani, S.; Yamamuro, T. Apatite formation on the surface of Ceravital-type glass-ceramic in the body. *J. Biomed. Mater. Res.* **1991**, *25*, 1363–1370. [CrossRef] [PubMed]
49. Pereira, M.d.M.; Clark, A.; Hench, L. Calcium phosphate formation on sol-gel-derived bioactive glasses in vitro. *J. Biomed. Mater. Res.* **1994**, *28*, 693–698. [CrossRef] [PubMed]
50. Notingher, I.; Jones, J.; Verrier, S.; Bisson, I.; Embanga, P.; Edwards, P.; Polak, J.; Hench, L. Application of FTIR and Raman spectroscopy to characterisation of bioactive materials and living cells. *J. Spectrosc.* **2003**, *17*, 275–288. [CrossRef]
51. Theodorou, G.; Goudouri, O.; Kontonasaki, E.; Chatzistavrou, X.; Papadopoulou, L.; Kantiranis, N.; Paraskevopoulos, K. Comparative bioactivity study of 45S5 and 58S bioglasses in organic and inorganic environment. *Bioceram. Dev. Appl.* **2011**, *1*, 1–4. [CrossRef]
52. Brauer, D.S.; Karpukhina, N.; O'Donnell, M.D.; Law, R.V.; Hill, R.G. Fluoride-containing bioactive glasses: Effect of glass design and structure on degradation, pH and apatite formation in simulated body fluid. *Acta Biomater.* **2010**, *6*, 3275–3282. [CrossRef]
53. Felisberto, M.D.; Laranjeira, M. Preparation and characterization of hydroxyapatite-coated iron oxide particles by spray-drying technique. *An. Da Acad. Bras. De Ciências* **2009**, *81*, 179–186.

54. Fernandes, J.; Deus, A.M.; Reis, L.; Vaz, M.F.; Leite, M. Study of the influence of 3D printing parameters on the mechanical properties of PLA. In Proceedings of the 3rd International Conference on Progress in Additive Manufacturing (Pro-AM 2018), Singapore, 14–17 May 2018; pp. 547–552.
55. Guarino, V.; Gloria, A.; Raucci, M.G.; De Santis, R.; Ambrosio*, L. Bio-inspired composite and cell instructive platforms for bone regeneration. *Int. Mater. Rev.* **2012**, *57*, 256–275. [CrossRef]
56. Khiabani, A.B.; Ghanbari, A.; Yarmand, B.; Zamanian, A.; Mozafari, M. Improving corrosion behavior and in vitro bioactivity of plasma electrolytic oxidized AZ91 magnesium alloy using calcium fluoride containing electrolyte. *Mater. Lett.* **2018**, *212*, 98–102. [CrossRef]
57. Syakur, A.; Sutanto, H. Determination of Hydrophobic Contact Angle of Epoxy Resin Compound Silicon Rubber and Silica. In Proceedings of the IOP Conference Series: Materials Science and Engineering, Semarang, Indonesia, 23–25 November 2016; p. 012025.
58. Bruyas, A.; Lou, F.; Stahl, A.M.; Gardner, M.; Maloney, W.; Goodman, S.; Yang, Y.P. Systematic characterization of 3D-printed PCL/β-TCP scaffolds for biomedical devices and bone tissue engineering: Influence of composition and porosity. *J. Mater. Res.* **2018**, *33*, 1948–1959. [CrossRef] [PubMed]
59. Keivani, F.; Shokrollahi, P.; Zandi, M.; Irani, S.; Shokrolahi, F.; Khorasani, S. Engineered electrospun poly (caprolactone)/polycaprolactor g-hydroxyapatite nano-fibrous scaffold promotes human fibroblasts adhesion and proliferation. *Mater. Sci. Eng. C* **2016**, *68*, 78–88. [CrossRef] [PubMed]
60. Bossard, C.; Granel, H.; Wittrant, Y.; Jallot, É.; Lao, J.; Vial, C.; Tiainen, H. Polycaprolactone/bioactive glass hybrid scaffolds for bone regeneration. *Biomed. Glasses* **2018**, *4*, 108–122. [CrossRef]
61. Yu, H.S.; Park, J.; Lee, H.-S.; Park, S.A.; Lee, D.-W.; Park, K. Feasibility of polycaprolactone scaffolds fabricated by three-Dimensional printing for tissue engineering of tunica albuginea. *World J. Men's Health* **2018**, *36*, 66–72. [CrossRef] [PubMed]
62. Seyedsalehi, A.; Daneshmandi, L.; Barajaa, M.; Riordan, J.; Laurencin, C.T. Fabrication and characterization of mechanically competent 3D printed polycaprolactone-reduced graphene oxide scaffolds. *Sci. Rep.* **2020**, *10*, 22210. [CrossRef]
63. Roohani-Esfahani, S.; Nouri-Khorasani, S.; Lu, Z.; Appleyard, R.; Zreiqat, H. Effects of bioactive glass nanoparticles on the mechanical and biological behavior of composite coated scaffolds. *Acta Biomater.* **2011**, *7*, 1307–1318. [CrossRef] [PubMed]
64. Tamjid, E. Three-dimensional polycaprolactone-bioactive glass composite scaffolds: Effect of particle size and volume fraction on mechanical properties and in vitro cellular behavior. *Int. J. Polym. Mater. Polym. Biomater.* **2018**, *67*, 1005–1015. [CrossRef]
65. Anderson, A.S. MTT Proliferation Assay. *Proc. West Va. Acad. Sci.* **2020**, *92*, 12498–12508.
66. Kamiloglu, S.; Sari, G.; Ozdal, T.; Capanoglu, E. Guidelines for cell viability assays. *Food Front.* **2020**, *1*, 332–349. [CrossRef]
67. Jensen, J.; Kraft, D.C.E.; Lysdahl, H.; Foldager, C.B.; Chen, M.; Kristiansen, A.A.; Rölfing, J.H.D.; Bünger, C.E. Functionalization of polycaprolactone scaffolds with hyaluronic acid and β-TCP facilitates migration and osteogenic differentiation of human dental pulp stem cells in vitro. *Tissue Eng. Part A* **2015**, *21*, 729–739. [CrossRef] [PubMed]
68. Kandelousi, P.S.; Rabiee, S.M.; Jahanshahi, M.; Nasiri, F. The effect of bioactive glass nanoparticles on polycaprolactone/chitosan scaffold: Melting enthalpy and cell viability. *J. Bioact. Compat. Polym.* **2019**, *34*, 97–111. [CrossRef]
69. Kudryavtseva, V.; Stankevich, K.; Kozelskaya, A.; Kibler, E.; Zhukov, Y.; Malashicheva, A.; Golovkin, A.; Mishanin, A.; Filimonov, V.; Bolbasov, E.; et al. Magnetron plasma mediated immobilization of hyaluronic acid for the development of functional double-sided biodegradable vascular graft. *Appl. Surf. Sci.* **2020**, *529*, 147196. [CrossRef]

*Review*

# 3D Printed and Conventional Membranes—A Review

Baye Gueye Thiam [1], Anouar El Magri [1,*], Hamid Reza Vanaei [2,*] and Sébastien Vaudreuil [1]

[1] Euromed Research Center, Euromed Polytechnic School, Euromed University of Fes, Route de Meknès (Rond point Bensouda), Fez 30000, Morocco; b.thiam@ueuromed.org (B.G.T.); s.vaudreuil@ueuromed.org (S.V.)

[2] Arts et Métiers Institute of Technology, CNAM, LIFSE, HESAM University, 75013 Paris, France

\* Correspondence: a.elmagri@ueuromed.org (A.E.M.); hamidreza.vanaei@ensam.eu (H.R.V.)

**Abstract:** Polymer membranes are central to the proper operation of several processes used in a wide range of applications. The production of these membranes relies on processes such as phase inversion, stretching, track etching, sintering, or electrospinning. A novel and competitive strategy in membrane production is the use of additive manufacturing that enables the easier manufacture of tailored membranes. To achieve the future development of better membranes, it is necessary to compare this novel production process to that of more conventional techniques, and clarify the advantages and disadvantages. This review article compares a conventional method of manufacturing polymer membranes to additive manufacturing. A review of 3D printed membranes is also done to give researchers a reference guide. Membranes from these two approaches were compared in terms of cost, materials, structures, properties, performance. and environmental impact. Results show that very few membrane materials are used as 3D-printed membranes. Such membranes showed acceptable performance, better structures, and less environmental impact compared with those of conventional membranes.

**Keywords:** 3D-printed membranes; additive manufacturing; membrane process

## 1. Introduction

Membrane technology, particularly polymer membranes, has multiple applications, including water treatment, electrodialysis, in batteries, and in the food and pharmaceutical industries [1–3]. A polymer membrane is a physical barrier separating two environments, endowed with selective permeability to certain species. In all applications, it is desirable that membranes possess high selectivity and stability, and low cost. Membrane choice depends on application type. Membranes can be of the following types: microporous, asymmetric composite thin-film, dense, or ion-exchange [1,4,5]. A microporous membrane is very similar in function to a conventional filter, where it rejects large particles (greater than 10 µm) while allowing for the smallest particles to pass [4]. For a dense membrane, permeants are transported by diffusion under the driving force of pressure, concentration, or electric potential gradient. A thin-film composite asymmetric membrane (TFC) is a microporous membrane featuring a dense thin selective layer. Ion exchange membranes can be either dense or microporous, and carry positively or negatively charged fixed ions in their polymer matrix. Their operating principle is based on the exclusion of ions of the same charge as the fixed ions of the membrane structure and the passage of ions of opposite charge.

Polymer membranes are produced using one of several approaches. Common approaches include phase inversion, stretching, track etching, sintering, electrospinning, and surface coatings of a support [1,4]. Manufacturing methods play an important role in membrane technology and its applications. Not only can membrane performance be significantly affected, but also their cost. Commercial activities and urgent needs have led to a rapid increase in membrane R&D to optimize performance, cost, and durability.

Although conventional methods offer efficient membranes, the precise control of preparation parameters remains problematic. To overcome these challenges, some researchers have been adopting the additive manufacturing (AM) of membranes. AM, also called 3D printing (3DP), is considered to be a possible approach to produce custom membranes with more manufacturing control than any other method of membrane manufacturing available today [5]. Membrane 3DP has thus attracted much interest, with many research and development studies on 3D-printed membranes. Review articles attempted to provide specific discussions in this regard [5–7], focusing only on the discussion of 3D printing technologies [5] and their water-related applications [6,7]. However, the difference between 3D-printed membranes and conventional membranes has not been studied. Some questions remain to be clarified. Do 3D-printable materials include common materials used for membranes? Do 3D-printed membranes offer the required properties to compete with conventional membranes? Do these 3DP membranes have lower cost and environmental impact than those of conventional membranes? All these questions can lead to many thoughts about conventional and additive membranes. These are some of the topics that this article attempts to elucidate while highlighting differences between conventionally and 3D-fabricated membranes. This comparison is inevitable to evaluate the potential of 3D membranes compared to membranes produced with methods that had undergone decades of optimization. Recent developments in AM membrane production is also summarized to highlight the current research areas. This paper briefly overviews conventional and 3DP membrane fabrication methods, followed by a critical review of 3DP membranes compared to conventionally produced membranes. Prospects for developing high-performance polymer membranes highlight the potential of such manufacturing techniques.

## 2. Membrane Manufacturing Techniques

When developing high-performance membranes, researchers focus much more on materials, while paying little attention to the used manufacturing processes. These processes, however, significantly affect membrane characteristics. This section presents conventional membrane production methods and 3D printing methods.

*2.1. Conventional Methods*

Conventional manufacturing methods are based on phase-inversion techniques, stretching, track etching, sintering, electrospinning, and layer by layer (Figure 1). Phase inversion, being a simple and fast method, is the most widely used for manufacturing membranes in which different kinds of polymers can be used for different applications. In such an approach, a polymer is first dissolved in a solvent to form a more or less viscous solution. This solution is then spread onto a glass plate and solidified [8,9]. This solidification can occur either through thermally or nonsolvent-induced phase separation.

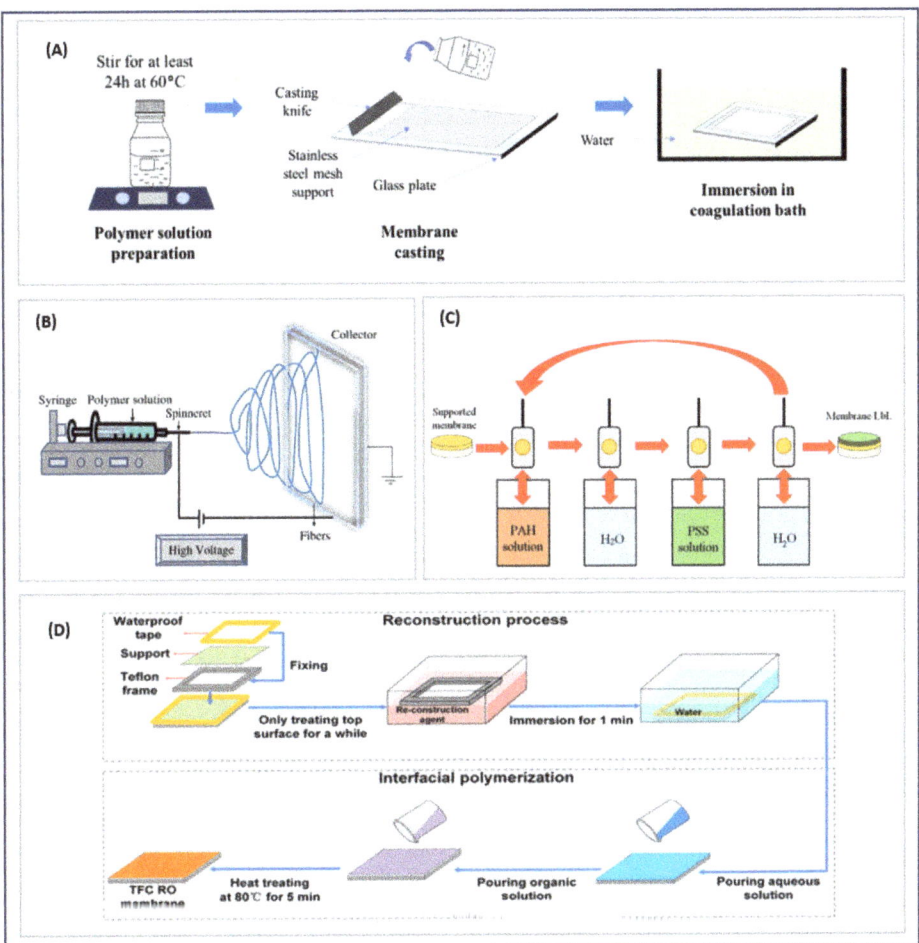

**Figure 1.** Schematic illustrations of membrane techniques. (**A**) Phase inversion [9] Reproduced from Doyan, A.; Leong, C.L.; Bilad, M.R.; Kurnia, K.A.; Susilawati, S.; Prayogi, S.; Narkkun, T.; Faungnawakij, K. Cigarette Butt Waste as Material for Phase Inverted Membrane Fabrication Used for Oil/Water Emulsion Separation. Polymers; published by MDPI, 2021. (**B**) Electrospinning [8] Reproduced from Tan, X. and Rodrigue, D., A Review on Porous Polymeric Membrane Preparation. Part I: Production Techniques with Polysulfone and Poly (Vinylidene Fluoride) Polymers; published by MDPI, 2019. (**C**) Layer by layer [10] Reproduced from Dmitrenko, M.; Kuzminova, A.; Zolotarev, A.; Ermakov, S.; Roizard, D.; Penkova, A. Enhanced Pervaporation Properties of PVA-Based Membranes Modified with Polyelectrolytes. Application to IPA Dehydration, Polymers; published by MDPI, 2021. (**D**) TFC manufacturing [11] Reproduced with permission from Shi, M.; Wang, Z.; Zhao, S.; Wang, J.; Wang, S. A Support Surface Pore Structure Re-Construction Method to Enhance the Flux of TFC RO Membrane; published by Journal of Membrane Science: published by Elsevier, 2017.

Another approach to produce porous membranes is by stretching dense extruded films [4,12]. Stretching a dense film perpendicularly to its extrusion direction creates small breaks that result in pore formation. The stretching technique is generally used to prepare microfiltration (MF), ultrafiltration (UF), and membrane-distillation (MD) membranes, and is preferred for highly crystalline polymers [13].

Track etching is also a technique to fabricate porous membranes for various applications including filtration and cell culture [14]. Track etching instead relies on the irradiation of the dense film perpendicularly to the surface [4,14]. The radiation-damaged material is then removed by postprocessing to create straight cylindrical pores. It is an expensive technique due to the use of high-energy radiation [15]. The most commonly used materials for track etched membranes are polyethylene naphthalate (PET), polypropylene (PP), and polycarbonate (PC) [13].

Membranes can also be produced by sintering powders of polymeric materials. Compressing and heating particles slightly below their melting temperature induce bonding [8,15], with spaces between the sintered particles becoming pores. Sintering is mainly used for the preparation of microfiltration membranes. The used polymers must have excellent resistance to chemicals and high temperatures [15].

Membranes are also produced from polymer nanofibers obtained through electrospinning. Polymers such as polyvinylidene fluoride (PVDF), polyacrylonitrile (PAN), or polystyrene (PS) are electrospinable. In the process, a viscoelastic polymer solution is loaded into a syringe placed at an optimal distance from a target (or collector). A strong electrical voltage is applied between syringe and manifold to stretch droplets from the syringe tip. It generates jets of nanofibers which then settle on the collector to form an electrospun membrane [1,8], which can be used for filtration and MD processes [13].

Support coatings are conventional methods for the surface treatment of membranes. For example, the fabrication of a TFC membrane relies on interfacial polymerization. In the process, an aqueous polyamine solution is first deposited on a microporous support; then, this amine-loaded support is immersed in a diacid chloride solution. The amine and acid chloride react at the interface between the two solutions to form an extremely thin and densely cross-linked membrane layer [11,16]. Membrane surfaces can also be modified by a layer-by-layer (LBL) process where electrostatic interaction between charged surfaces are exploited through a simple immersion process. LBL can also be used to fabricate multilayer thin films [10].

*2.2. Additive Manufacturing Method*

Additive manufacturing is a layer-by-layer manufacturing process capable of easily building complex, real custom objects. Various 3D printing techniques are available such as stereolithography, digital light processing (DLP), fused deposition modeling (FDM), multi-jet printing (MJP), and selective laser sintering (SLS) [17,18]. All these processes work on the same basic concept to produce the final object. The whole process begins with a computer-aided design (CAD) model, which is then converted into the stereolithography format (STL). The obtained 3D file is then preprocessed by specific software, where process parameters such as 3D part orientation into the build volume and slicing parameters are defined. The information is then sent to the 3D printer that carries out layer-by-layer manufacturing.

The FDM 3D printing process (or fused filament fabrication (FFF)) consists of filament extrusion that is deposited layer by layer through a printing nozzle (Figure 2A) [19]. This deposit is produced according to the X, Y and Z coordinates of the 3D model to be printed.

Stereolithography (SLA) consists of solidifying a photosensitive liquid resin layer by layer using an ultraviolet (UV) laser beam [18,20]. As shown in Figure 2B, the build platform is initially positioned in the tank with the photopolymer resin, one layer height away from the build window. The laser beam follows a predetermined path based on the cross-section of the 3D model. After one layer is hardened, the build platform is then raised to expose a new layer of liquid polymer. The laser again traces the cross section of the object, which instantly sticks to the hardened part. A digital light-processing (DLP) projector can replace the UV laser to achieve resin hardening, enabling a cost reduction system and faster processing. However, this results in reduced XY resolution.

SLS relies on a powerful laser beam to fuse powder at very precise points of the 3D file [17,20] (Figure 2C). A new layer of fine powder is then spread before fusing the laser onto the previous layer.

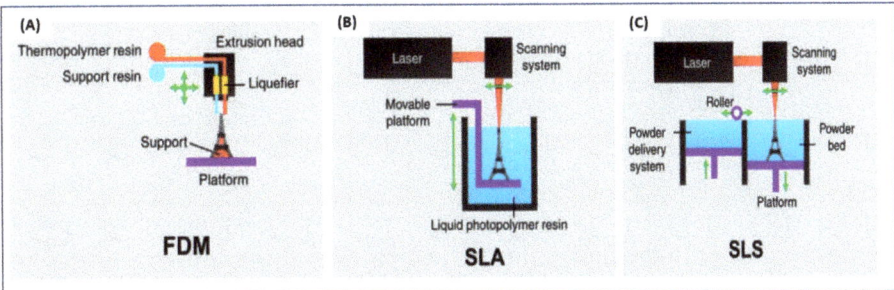

**Figure 2.** Schematic illustrations of 3D printing technologies. (**A**) FDM printing. (**B**) SLA printing. (**C**) SLS printing. [5]. Adapted from Low, Z.-X.; Chua, Y.T.; Ray, B.M.; Mattia, D.; Metcalfe, I.S.; Patterson, D.A. Perspective on 3D Printing of Separation Membranes and Comparison to Related Unconventional Fabrication Techniques, Journal of Membrane Science; Published by Elsevier, 2017.

## 3. Comparison of Conventional and 3DP Membranes

A comparison of 3DP polymer membranes with conventional membranes relies on available information from the literature, using a common basis. It includes material, structure, properties, performance, and cost. For example, the cost of a 3DP membrane is compared with the cost of a conventional membrane in the same application. Only some values of 3DP membrane properties were compared with those of conventional membranes due to the lack of available data for some 3DP membranes.

*3.1. Membrane Materials*

For material comparison, only the base polymer of the membrane is considered, as production of 3D membranes is usually carried out in the form of a composite membrane, i.e., 3D printing is used to manufacture the support, while other techniques are used to produce a selective layer. Materials are listed on the basis of reviews of conventional membranes [1,21–25], 3DP polymers [17,18,20,26], and reviewed articles on 3DP membranes.

Some 3D membrane materials are directly produced using common 3D printing technologies. FDM facilitates directly obtaining membranes from poro-lay [27,28], polylactic acid (PLA) [29], polylactide-co-glycolide (PLGA) [30], and polyethylene terephthalate (PET) [31]. The SLS technique is used to print polyamide 12 [32,33] and polysulfone [34] membranes, while SLA is used for diurethane dimethacrylate-co-polyethylene glycol diacrylate (DUDA-co-PEGDA) [35] and tangoplus [36] membranes. MJP can produce acrylonitrile butadiene styrene membranes (ABS-Like) [37,38]. Approaches based on solution casting printing allow for the direct production of PDMS [39], poly (vinyl alcohol) (PVA), polybenzimidazole (PBI) [40] and polyvinylidene fluoride (PVDF) [41] membranes. TFC membranes are also fabricated using 3D technology [42].

Figure 3 illustrates the materials used in conventional and 3DP membranes. Acronyms for those materials are listed and explained in the Supplementary Materials. Conventional membranes can be produced from a wide range of either natural or synthetic polymers, including vinyls, polyesters, fluorinated or chlorinated halogens, and acrylates. Very few materials are available for 3DP membranes, representing only 12% of those used in conventional membranes. The wide choice of polymers in conventional manufacturing is due to the expertise and increased development of new materials. As most polymers are soluble in solvents required to prepare cast or electrospinable solutions, this facilitates their use in phase inversion or electrospinning processes. Polymers can also be processed even without a suitable solvent, relying on sintering, track etching, or drawing processes to transform the extruded state into membranes. On the other hand, 3D printing systems are limited regarding membrane materials, as they are not compatible with all types of polymers. While printable polymers for membranes are gaining ground, the number of printed membranes remains very small. Solution casting printing can, however, allow for the printing of a wide range of currently not printable polymers [40]. If these polymers cannot be dissolved in an appropriate solvent, 3DP system development with extended printing materials is necessary.

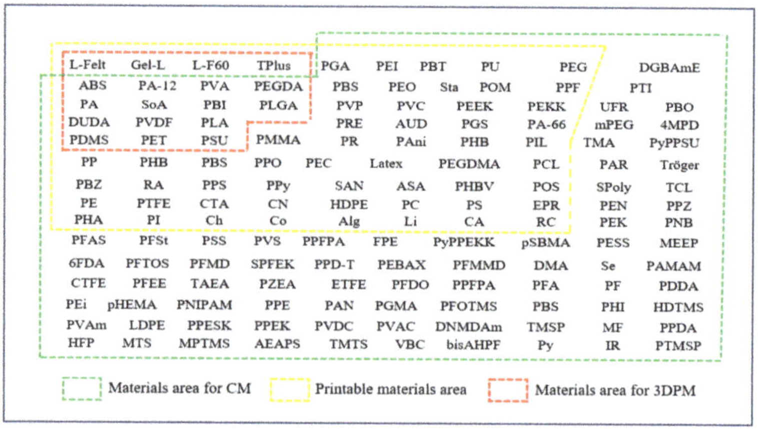

| L-Felt | Gel-L | L-F60 | TPlus | PGA | PEI | PBT | PU | PEG | | DGBAmE |
|---|---|---|---|---|---|---|---|---|---|---|
| ABS | PA-12 | PVA | PEGDA | PBS | PEO | Sta | POM | PPF | | PTI |
| PA | SoA | PBI | PLGA | PVP | PVC | PEEK | PEKK | UFR | | PBO |
| DUDA | PVDF | PLA | | PRE | AUD | PGS | PA-66 | mPEG | | 4MPD |
| PDMS | PET | PSU | PMMA | PR | PAni | PHB | PIL | TMA | | PyPPSU |
| PP | PHB | PBS | PPO | PEC | Latex | PEGDMA | PCL | PAR | | Tröger |
| PBZ | RA | PPS | PPy | SAN | ASA | PHBV | POS | SPoly | | TCL |
| PE | PTFE | CTA | CN | HDPE | PC | PS | EPR | PEN | | PPZ |
| PHA | PI | Ch | Co | Alg | Li | CA | RC | PEK | | PNB |
| PFAS | PFSt | PSS | PVS | PPFPA | FPE | PyPPEKK | pSBMA | PESS | | MEEP |
| 6FDA | PFTOS | PFMD | SPFEK | PPD-T | PEBAX | PFMMD | DMA | Se | | PAMAM |
| CTFE | PFEE | TAEA | PZEA | ETFE | PFDO | PPFPA | PFA | PF | | PDDA |
| PEi | pHEMA | PNIPAM | PPE | PAN | PGMA | PFOTMS | PBS | PHI | | HDTMS |
| PVAm | LDPE | PPESK | PPEK | PVDC | PVAC | DNMDAm | TMSP | MF | | PPDA |
| HFP | MTS | MPTMS | AEAPS | TMTS | VBC | bisAHPF | Py | IR | | PTMSP |

Materials area for CM   Printable materials area   Materials area for 3DPM

**Figure 3.** Polymers used for the manufacture of membranes: conventional membrane (CM) vs. 3D-printed membrane (3DPM) materials.

*3.2. Membrane Structures*

According to the nature of FFF process, each deposition has its own strong influence on different aspects of the constructed parts. This issue clearly means that the final parts' thermal, mechanical, and rheological characteristics are affected by different deposition mechanisms. There are various mechanisms of deposition based on the filling of layers, namely, counter fill, raster fill, counter, and raster fill.

The structure of a membrane influences its properties, hence the need for proper control during preparation. Structures of 3D-printed and conventional membranes are shown in Figure 4A,B, respectively. Conventional membranes generally have smooth surface morphologies (i.e., low roughness), as shown in Figure 4Bb1. Pore structure in conventional membranes, including porosity, interconnectivity, distribution, and size, is often asymmetric or unordered. For example, membranes formed by phase inversion exhibit structures characterized by their fingerlike pores under a thin layer of dense skin (Figure 4Bb2). For membranes obtained through electrospinning, a scaffold structure with disorganized but interconnected pores and low tortuosity is observed (Figure 4Bb3). This lack of uniformity in the pore structure of conventional membranes can be attributed to difficulties in controlling the preparation parameters. Although pore size can be controlled in the stretching technique, this pore formation mechanism only applies to high crystallinity polymer membranes [12]. On the other hand, 3DP membranes result from a CAD object (Figure 4Aa1), enabling the control of all parameters to achieve the desired structure. Figure 4Aa2,a3 show images of such 3DP patterned membranes. The 3DP membranes with embossed or grooved structures can easily be produced, resulting in larger surfaces than those of flat membranes. Patterned membranes are of great interest to researchers, as such membranes can exhibit improved transport performance and reduced concentration polarization while alleviating fouling [35]. The technique of 3D printing offers great manufacturing flexibility while enabling easier fabrication of complex structures than conventional methods can. The resolution limits of 3D printing are, however, limiting in membrane production. While available 3D printing methods are capable of high resolution in the z dimension, the same precision cannot be obtained for the x and y axes [7].

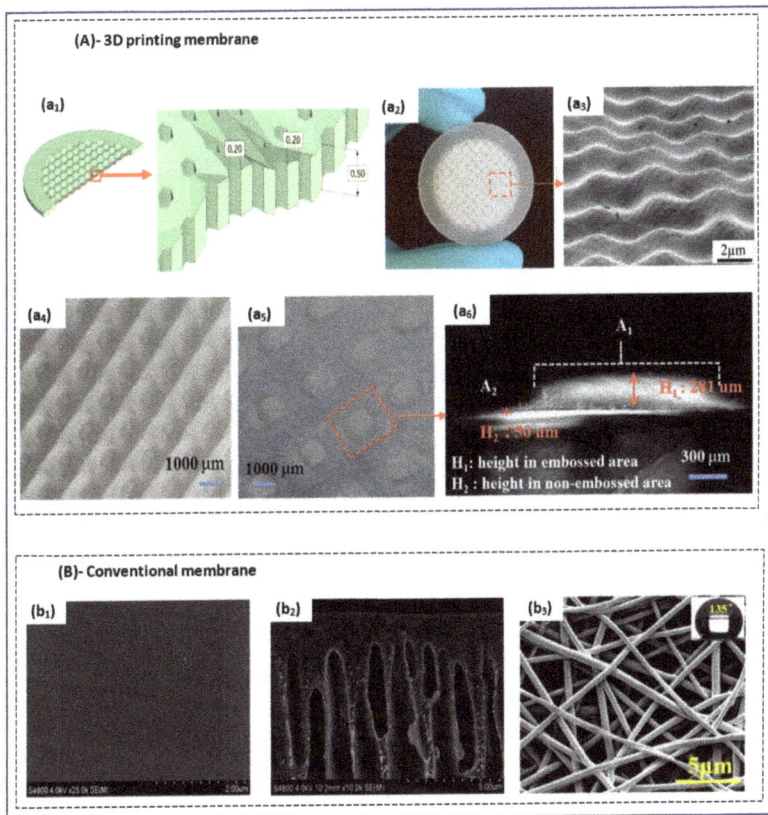

Figure 4. Structures of (A) 3D-printed and (B) conventional membranes. (a1,a2) 3D membrane support and its CAD, respectively; (a3–a6) 3D-printed membranes surface structures [31,37]; Reproduced with permission from Koh, E.; Lee, Y.T. Development of an Embossed Nanofiber Hemodialysis Membrane for Improving Capacity and Efficiency via 3D Printing and Electrospinning Technology, Separation and Purification Technology; published by Elsevier, 2020. Reproduced with permission from Al-Shimmery, A.; Mazinani, S.; Ji, J.; Chew, Y.M.J.; Mattia, D., 3D Printed Composite Membranes with Enhanced Anti-Fouling Behaviour, Journal of Membrane Science; published by Elsevier, 2019. (b1) surfaces of a conventional membrane, (b2) SEM micrographs of cross-sections of conventional membranes (phase inversion) [43] Reproduced with permission from Zhu, L.-J.; Liu, F.; Yu, X.-M.; Gao, A.-L.; Xue, L.-X. Surface Zwitterionization of Hemocompatible Poly(Lactic Acid) Membranes for Hemodiafiltration. Journal of Membrane Science; Elsevier 2015. (b3) SEM images of the surface of electrospinning membrane [44] Reproduced with permission from Zhang, Z.-M.; Gan, Z.-Q.; Bao, R.-Y.; Ke, K.; Liu, Z.-Y.; Yang, M.-B.; Yang, W. Green and Robust Superhydrophilic Electrospun Stereocomplex Polylactide Membranes: Multifunctional Oil/Water Separation and Self-Cleaning, Journal of Membrane Science; Elsevier, 2020.

If we look at TFC membranes used in desalination, the formation of the polyamide layer by interfacial polymerization is more successful for 3D printing than with the traditional method. Although conventional TFC membranes exhibit excellent permeability selectivity, their fabrication procedure is inherently limiting [42,45]. The intrinsic roughness of polyamide films has long been associated with a high fouling propensity in reverse-osmosis (RO) processes. Moreover, one cannot precisely control membrane thickness during fabrication, as the process simply self-terminates during film formation, yielding thickness of 100–200 nm [46]. The 3D printing can instead be used to deposit monomers as nanoscale droplets that forms polyamide onto a substrate. A thickness of 37 nm was achieved for 3D TFC membranes [42], meaning that the 3D membrane offers controllable roughness and independence during the in situ formation of an active polyamide film on a support. Figure 5 illustrates examples of conventional and 3DP polyamide layers.

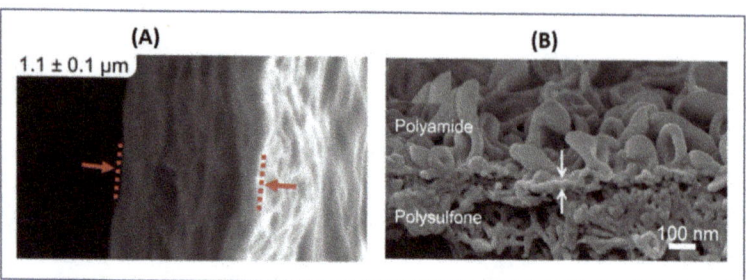

**Figure 5.** SEM images of polyamide TFC membranes with a polyamide layer: (**A**) printed [7]. Reproduced from Yanar, N.; Kallem, P.; Son, M.; Park, H.; Kang, S.; Choi, H. A New Era of Water Treatment Technologies: 3D Printing for Membranes, Journal of Industrial and Engineering Chemistry; published by Elsevier, 2020 and (**B**) conventional [47]. Reproduced with permission from Perera, D.H.N.; Song, Q.; Qiblawey, H.; Sivaniah, E. Regulating the Aqueous Phase Monomer Balance for Flux Improvement in Polyamide Thin Film Composite Membranes, Journal of Membrane Science; published by Elsevier, 2015.

Most 3DP technologies do not produce membranes with the flexibility of traditional methods. However, the configuration of membranes using traditional methods is limited to simple structures (e.g., flat). This limitation can benefit the increased use of 3DP techniques where almost any complex geometric shape can be designed and produced. Examples of complex-shaped membranes are shown in Figure 6. The technology of 3D printing can create a one-print system that incorporates both the membrane and other components (Figure 6A).

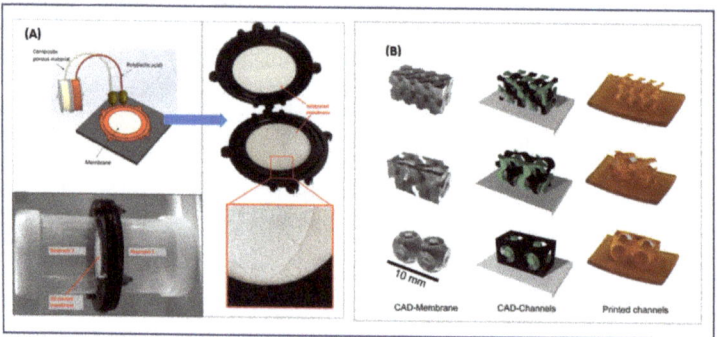

**Figure 6.** Structures of complex shapes of 3D-printed membranes. (**A**) Design of an integrated membrane device [27]. Reproduced with permission from Kalsoom, U.; Hasan, C.K.; Tedone, L.; Desire, C.; Li, F.; Breadmore, M.C.; Nesterenko, P.N.; Paull, B., Low-Cost Passive Sampling Device with Integrated Porous Membrane Produced Using Multimaterial 3D Printing; Anal. Chem., American Chemical Society, 2018. (**B**) Sheetlike triply periodic minimal-surface architecture (TPMS)-like 3D membrane [48]. Reproduced with permission from Femmer, T.; Kuehne, A.J.C.; Wessling, M. Print Your Own Membrane: Direct Rapid Prototyping of Polydimethylsiloxane, Lab Chip; published by Royal Society of Chemistry, 2014.

### 3.3. Properties and Performance

Some of the major properties of 3D-printed membranes are given in Table 1. Thickness is a key factor in determining membrane performance. A thicker membrane generally exhibits lower permeability but higher surface resistance, thus affecting performance. The thickness of conventional membranes can reach values of 150–250 µm (for separation: e.g., water–oil), and 150 µm (for RO) [49]. The thickness of 3DP membranes is more significant, with values of 800 µm (for water–oil separation) [39] or 500 µm (water–oil separation) [37]. The thickness of a cation exchange membrane fabricated by FDM for use in microbial fuel cells reached 2000 µm [28]. It is much thicker than conventional membranes for such application, where an average thickness of 142.75 µm is found [50]. It results from the layer-by-layer operation of additive manufacturing, where the lower single layer height cannot go below 25 µm (example of SLA and DLP). The need for multiple layers to achieve structural integrity results in thicker membranes.

**Table 1.** Properties of 3D-printed membranes compared with conventional membranes.

| Application | Membrane | Preparation Method | Thickness (µm) | Pore Size (µm) | Roughness (µm) | WCA (°) | Tensile Strength (MPa) | Reference |
|---|---|---|---|---|---|---|---|---|
| hemodialysis | PET(PMMA-g-PDMS) | FDM combined with Electrospinning | 150 | 0.14 | 0.500 | 50 | 12 | [31] |
| hemodialysis | PLA/PDA-g-PSBMA | Nonsolvent induced phase separation (NIPS) | 35 | - | - | 55.1 | | [43] |
| oil–water separation | (PDMS)/SiO$_2$ | FDM using ink | 800 | 370 | - | 160 | - | [39] |
| oil–water separation | ABS–PES | MultiJet 3D Printing | 500 | 200 | 73 | 83 ± 2 | - | [37] |
| oil–water separation | PLA/polystyrène (PS) | FDM | - | 250 | - | 151.7 | - | [29] |
| oil–water separation | polysulfone (PSU) | SLS | 355 | 51.8 | 0.135 | 161 | 17.3 | [34] |
| ultrafiltration | PSU/Fe$_3$O$_4$ | Electrospinning | 234–241 | 0.07362 | - | 21.78 | 1.75 | [51] |
| wastewater treatment | PA6 | Electrospinning | | 0.753 | - | 123 | 0.047 | [52] |
| filtration | PVDF | 3D printing near-field electrospinning (NFES) | - | 250 | - | 130 | ~50 | [53] |
| filtration | PVDF | Melt spinning and stretching | - | 0.550 | 3.617 | 92.6 | 27.9 | [12] |

Pores of 3DP membranes are generally larger than those of conventional membranes for a given application. For example, 3DP membranes applied to water–oil separation have pores diameters of 370 µm [39], 200 µm [37], 250 µm [29] and 51.8 µm [34], while those in conventional membranes are generally less than 1 µm [51,53,54]. Pore size in 3DP membranes varies according to the desired structure and depends on the resolution of the used printing technology. The actual product resolution is usually lower than the nominal 3D printer resolution [5]. While most available 3D printers are not yet able to print below submicron resolution [6], two-photon polymerization (TPP) technology has achieved a resolution currently capped at ~100 nm [5]. Technologies with finer resolution are required to achieve smaller pore size without post modification.

Thickness and pores are not the only factors influencing membrane performance. Surface roughness also has positive or negative influence during application. For 3DP membranes, surface roughness depends on the 3D production technology. Conventional membranes can exhibit a rougher or smoother surface than 3DP membranes, depending on the process used. Chowdhury et al. [42] confirmed that their 3DP TFC membrane had a lower controlled roughness (~4.3 nm) than conventional TFC membranes. Reduction in roughness helps reduce the risk of membrane fouling.

Hydrophobicity or hydrophilicity are properties that could be advantageous or disadvantageous to membranes depending on the application. This depends on the used materials and/or the surface structure of the membrane. The use of a hydrophobic polymer, for example, likely results in a hydrophobic membrane. This membrane hydrophobicity is characterized by its water contact angle (WCA). Conventional membranes for separation have contact angles of 92.6° (PVDF) [12] or 21.87° (PSU) [51]. The 3DP membranes, using the same base materials and the same applications, exhibit higher contact angles at 130° (PVDF) [53] and 161° (PSU) [34]. The surface structure can also affect membrane behavior against water. 3D printing can produce superhydrophobic membranes inspired by the leaves of plants [29,54], with a structure behaving like a leaf to achieve high hydrophobicity at the surface.

Mechanical properties are also important in membrane applications. While the tensile strength of a conventional membrane used for water–oil separation can reach 1.75 MPa (PSU) [51] or 27.9 MPa (PVDF) [12], values of 17.3 (PSU) [34] and 50 MPa (PVDF) [53] were achieved for 3DP membranes. The improved mechanical properties of 3DP membranes against traditional membranes can be explained by their higher thickness. The 3DP membranes can nevertheless experience mechanical anisotropy that depends on the printing technology used and the raster orientation (layer) [6].

All membrane properties influence application performance. A PLA 3DP membrane decorated with polystyrene (PS) nanospheres [29], denoted 3DP-M1, was compared with conventional membranes used for water/oil separation. The performance of this membrane was compared with that of the conventional membranes of similar materials. The chosen systems include a nanofiber membrane based on PLA modified with $SiO_2$ (P-2) [55], an electrospun stereocomplex PLA membrane (sc-PLA) [44], a membrane in fibrous Janus in PLA containing carbon nanotubes (PLA/CNT) [56], and another containing $SiO_2$ (PLA/$SiO_2$) [56]. Results of water/hexane separation efficiency and the flux of the membranes are shown in Figure 7. The separation efficiency of the membranes, including 3DP membranes, were all equal to or greater than 99%. The 3DP-M1 membrane exhibited a higher flux (60,000 LMH) than that of conventional membranes (Table 2). This flux was almost stable after 10 cycles, similar to conventional membranes. The water contact angle value for these membranes is also given in Table 2. Surface wettability has crucial influence on the oil/water separation performance of materials. A 151.7° WCA value was observed for the 3DP-M1 membrane, revealing hydrophobic behavior, while conventional membranes P-2, sc-PLA, PLA/CNT and PLA/$SiO_2$ exhibited WCA of 135°, 141°, 142°, and 0°, respectively. Modifying pure PLA is thus a way to achieve a superhydrophobic surface in PLA membrane. Manufacturing membranes with lotus leaf structures can also increase hydrophobicity, feasible through a 3D printing approach.

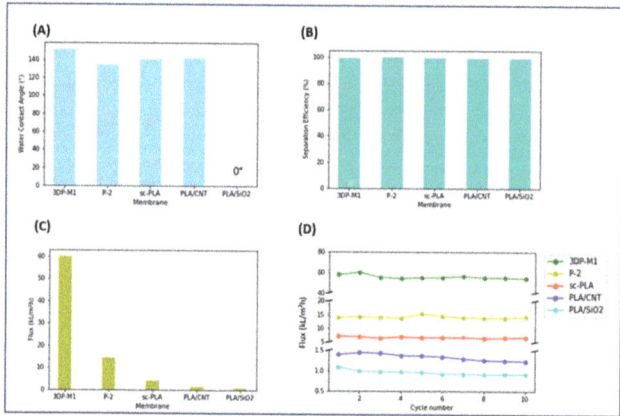

**Figure 7.** Separation performance of 3DP-M1 and conventional membranes. (**A**) Water contact angle. (**B**) Separation efficiency (n-hexane/water). (**C**) Permeation flux. (**D**) Permeation flux of n-hexane/water mixture for 10 separation cycles.

Table 2. Comparison of 3DP-M1 performance with conventional membranes.

| Membrane | WCA (°) | Flux (LMH) | Separation Efficiency (%) | Reference |
|---|---|---|---|---|
| 3DP-M1 | 151.7 | 60,000 | 99.4% | [29] |
| P-2 | 135 | 14,379 | 100 | [55] |
| sc-PLA | 141 | 4200 | 99.6 | [44] |
| PLA/CNT | 142 | 1435 | 99 | [56] |
| PLA/SiO2 | 0 | 1025 | 99 | [56] |

The water flux and salt rejection efficiency of a TFC membrane with a 3D printing deposited polyamide layer [42] were compared with those of conventional membranes using information collected from [46,47,57–70]. Figure 8 shows the performance of these membranes. The 3DP TFC membrane exhibited a >96% rejection of salt and high permeance (>3 LMH. Bar$^{-1}$) at the same time (Figure 8, colored area). Surface roughness of ~100 nm was observed for conventional membranes [46], which is much higher than the 4.3 nm obtained for 3DP membranes. The technology of 3D printing, unlike the conventional method, can achieve a controlled polyamide layer formation, explaining the good performance achieved by 3DP TFC membrane.

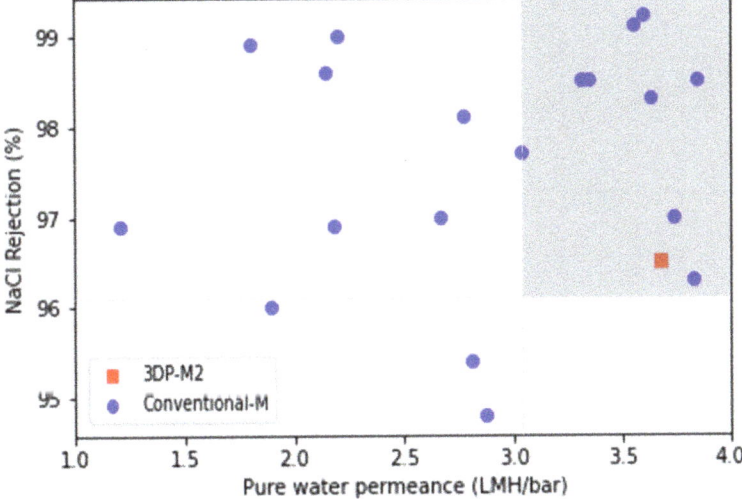

Figure 8. Desalination performance of printed polyamide vs. conventional membranes [46,47,57–70]. NaCl salt rejection and pure water permeance.

A 3DP membrane offers acceptable performance in desalination and water–oil separation applications. This membrane type has also been tested in other applications with promising results. The 3DP PDMS membranes applied for gas–liquid contact showed higher $CO_2$ transport in water than that of common hollow fiber membranes [48]. Philamore et al. [36] compared a conventional cation exchange membrane (CEM) of a microbial fuel cell to a 3DP membrane. The conventional CEM produced the highest power at 11.39 mW, against the 0.92 mW achieved by the Tangoplus 3DP membrane. A hemodialysis membrane fabricated via 3D printing and electrospinning technology showed a blood water removal capacity of 27% [31], while the removal of urea and NaCl during 4 h of hemodialysis reached ~17% (from 1.45 to 1.21 mg/L) and ~14% (from 0.9% to 0.8%), respectively. Isozyme clearance approached 68%.

## 3.4. Cost and Environmental Impact

Manufacturing methods affect not only membrane performance, but also the production costs. In the current circumstances, it is challenging to compare production costs of conventional and 3DP membranes, mainly due to the lack of price information in the case of 3DP membranes. Some studies nevertheless confirmed that their 3DP membrane is less expensive than conventional membranes. In the case of 3DP membranes, manufacturing costs include 3D printer purchase costs (investment) and used printing materials (consumables, operating). According to Low et al. [5,6], 3D printers are more expensive than most conventional manufacturing techniques such as solution casting, LBL, and phase inversion. This is reversed in the case of the material used during 3D printing, as Philamore et al. [36] reported significantly lower raw material costs to produce a 20 $cm^2$ Tangoplus 3DP membrane compared to the equivalent area of conventional material. The Tangoplus resin used to produce a membrane costs USD 0.16, while an equivalent area of conventional membrane costs between USD 0.22 and 0.40. You et al. [28] mentioned that, while their materials are cheaper than conventional membrane material, membrane production costs (in Lay-Fomm, Gel-Lay, and Lay-Felt) were EUR 0.58–0.60 (USD 0.65–0.67). This is higher than the EUR 0.30–0.56 (USD 0.33–0.62) costs of conventional membranes [28]. Their study used 30 $cm^2$ membranes, resulting in production costs of around 0.022 USD. $cm^{-2}$. This is low compared to the 0.25 USD. $cm^{-2}$ cost of the commonly used Nafion membrane [71]. These costs are shown in Table 3. Membrane cost would also depend on production volume. Although 3D printers are more expensive, a large production volume with inexpensive raw material results in inexpensive 3DP membranes.

Table 3. Cost comparison of 3D and conventional membranes for microbial fuel cell.

|  | Cost | Reference |
| --- | --- | --- |
| Material cost to produce a 3D membrane (USD/membrane) | 0.16 | [36] |
| Material cost to produce a conventional membrane (USD/membrane) | 0.22–0.40 | [36] |
| Production cost of a 3D membrane (USD $cm^{-2}$) | 0.022 | [28] |
| Production cost of a conventional Nafion membrane (USD $cm^{-2}$) | 0.25 | [71] |

One of the hazards of 3D printing processes is emissions from used materials, such as ultrafine particles (UFP) and volatile organic compound (VOC) [6]. Additive manufacturing could nevertheless be greatly significant for green environments, as waste is reduced or recycled. Large-scale conventional membrane production, on the other hand, can have potential environmental impacts because most preparations require toxic products such as N-methyl-2-pyrrolidone (NMP), N,N-dimethylformamide (DMF), or N,N-dimethylacetamide (DMAc). Although there are regulations (e.g., Registration, Evaluation, Authorization and Restriction of Chemicals (REACH)) [72] currently aimed at reducing solvent emissions and the harmful use of toxic solvents, stricter regulations require either more environmentally friendlier solvents or alternative solutions. Nonuniform manufacturing associated with these conventional methods also results in high amounts of waste [7]. As polymers used to manufacture conventional or 3D membranes are mostly derived from fossil sources, environmentally friendlier products are also needed to reduce the environmental impact.

## 4. Conclusions

In this study, conventionally prepared polymer membranes and 3D printed membranes were compared, accounting for recent developments in membrane production by additive manufacturing. Results showed differences between 3DP and conventional membranes in terms of materials, properties, performance cost, and environmental impact. This study showed that common materials for membranes are not well-adapted to additive manufacturing. This can be observed by the low number of suitable printing materials in comparison with conventional membrane materials, explainable by the inability of 3D printing technologies to use a wide range of materials. The 3DP membranes exhibited, however, a much better structure than that of conventional membranes. It can be attributed to the possibility of printing complex shapes in a controlled manner, whereas parameters are not measured or precisely controlled in conventional processes. The 3D membranes were shown to exhibit properties and performance approaching conventional membranes, and 3D printing has shown its ability of creating nature-resembling structures to improve performance. Another benefit of 3D printing is the ease in customizing membrane design to satisfy customer needs within a short turnaround time. The advantages associated with additive manufacturing could thus revolutionize the manufacture of low-cost high-performance membranes.

To achieve this, key areas must be further developed, including improvement in XY resolution and development of printers able to process a wide range of materials. The introduction of hybrid materials could be advantageous for the properties of 3DP membranes. An issue to address is the long-term stability and performance of 3D membranes, something not fully known due to the limited number of research groups working on 3D membranes. Further investigations are thus needed to demonstrate their suitability in membrane applications. Another essential area of 3DP research is the creation of a unique printing system incorporating both membrane and other components. Such development would greatly benefit membrane production. Another research direction with exciting possibilities is 4D printing, where the element of time is added to 3D printing. This enables changes in properties, function, or shape to a 3D-printed part with time [6]. Such 4D approaches could enable the production of more efficient membranes. All these eventualities, combined with larger 3D printers having very high printing speeds, can increase the potential for industrial use. Environmental considerations, including fees associated to proper waste disposal, can encourage traditional membrane manufacturing to switch to 3D printing to reduce the amount of produced waste.

**Supplementary Materials:** The following are available online at https://www.mdpi.com/article/10.3390/polym14051023/s1, List of acronyms of materials.

**Author Contributions:** Conceptualization, S.V. and A.E.M.; methodology, B.G.T., A.E.M., H.R.V. and S.V.; formal analysis, B.G.T., A.E.M., H.R.V. and S.V; investigation, B.G.T., A.E.M. and S.V.; resources, B.G.T., A.E.M., H.R.V. and S.V.; writing—original draft preparation, B.G.T., A.E.M. and S.V.; writing—review and editing, B.G.T., A.E.M., H.R.V. and S.V; visualization, B.G.T., A.E.M., H.R.V. and S.V.; supervision, S.V.; project administration, S.V.; funding acquisition, S.V. All authors have read and agreed to the published version of the manuscript.

**Funding:** This research received no external funding.

**Institutional Review Board Statement:** Not applicable.

**Informed Consent Statement:** Not applicable.

**Data Availability Statement:** Not applicable.

**Conflicts of Interest:** The authors declare no conflict of interest.

# References

1. Thiam, B.G.; Vaudreuil, S. Review—Recent Membranes for Vanadium Redox Flow Batteries. *J. Electrochem. Soc.* **2021**, *168*, 070553. [CrossRef]
2. Dong, X.; Lu, D.; Harris, T.A.L.; Escobar, I.C. Polymers and Solvents Used in Membrane Fabrication: A Review Focusing on Sustainable Membrane Development. *Membranes* **2021**, *11*, 309. [CrossRef]
3. Pandele, A.M.; Oprea, M.; Dutu, A.A.; Miculescu, F.; Voicu, S.I. A Novel Generation of Polysulfone/Crown Ether-Functionalized Reduced Graphene Oxide Membranes with Potential Applications in Hemodialysis. *Polymers* **2021**, *14*, 148. [CrossRef]
4. Baker, R.W. Membrane Technology. In *Encyclopedia of Polymer Science and Technology*; John Wiley & Sons, Ltd.: Hoboken, NJ, USA, 2001; ISBN 978-0-471-44026-0.
5. Low, Z.-X.; Chua, Y.T.; Ray, B.M.; Mattia, D.; Metcalfe, I.S.; Patterson, D.A. Perspective on 3D Printing of Separation Membranes and Comparison to Related Unconventional Fabrication Techniques. *J. Membr. Sci.* **2017**, *523*, 596–613. [CrossRef]
6. Tijing, L.D.; Dizon, J.R.C.; Ibrahim, I.; Nisay, A.R.N.; Shon, H.K.; Advincula, R.C. 3D Printing for Membrane Separation, Desalination and Water Treatment. *Appl. Mater. Today* **2020**, *18*, 100486. [CrossRef]

7. Yanar, N.; Kallem, P.; Son, M.; Park, H.; Kang, S.; Choi, H. A New Era of Water Treatment Technologies: 3D Printing for Membranes. *J. Ind. Eng. Chem.* **2020**, *91*, 1–14. [CrossRef]
8. Tan, X.; Rodrigue, D. A Review on Porous Polymeric Membrane Preparation. Part I: Production Techniques with Polysulfone and Poly (Vinylidene Fluoride). *Polymers* **2019**, *11*, 1160. [CrossRef] [PubMed]
9. Doyan, A.; Leong, C.L.; Bilad, M.R.; Kurnia, K.A.; Susilawati, S.; Prayogi, S.; Narkkun, T.; Faungnawakij, K. Cigarette Butt Waste as Material for Phase Inverted Membrane Fabrication Used for Oil/Water Emulsion Separation. *Polymers* **2021**, *13*, 1907. [CrossRef]
10. Dmitrenko, M.; Kuzminova, A.; Zolotarev, A.; Ermakov, S.; Roizard, D.; Penkova, A. Enhanced Pervaporation Properties of PVA-Based Membranes Modified with Polyelectrolytes. Application to IPA Dehydration. *Polymers* **2019**, *12*, 14. [CrossRef]
11. Shi, M.; Wang, Z.; Zhao, S.; Wang, J.; Wang, S. A Support Surface Pore Structure Re-Construction Method to Enhance the Flux of TFC RO Membrane. *J. Membr. Sci.* **2017**, *541*, 39–52. [CrossRef]
12. Ji, D.; Xiao, C.; Chen, K.; Zhou, F.; Gao, Y.; Zhang, T.; Ling, H. Solvent-Free Green Fabrication of PVDF Hollow Fiber MF Membranes with Controlled Pore Structure via Melt-Spinning and Stretching. *J. Membr. Sci.* **2021**, *621*, 118953. [CrossRef]
13. Lalia, B.S.; Kochkodan, V.; Hashaikeh, R.; Hilal, N. A Review on Membrane Fabrication: Structure, Properties and Performance Relationship. *Desalination* **2013**, *326*, 77–95. [CrossRef]
14. Apel, P. Track Etching Technique in Membrane Technology. *Radiat. Meas.* **2001**, *34*, 559–566. [CrossRef]
15. Remanan, S.; Sharma, M.; Bose, S.; Das, N.C. Recent Advances in Preparation of Porous Polymeric Membranes by Unique Techniques and Mitigation of Fouling through Surface Modification. *ChemistrySelect* **2018**, *3*, 609–633. [CrossRef]
16. Yang, Z.; Zhou, Y.; Feng, Z.; Rui, X.; Zhang, T.; Zhang, Z. A Review on Reverse Osmosis and Nanofiltration Membranes for Water Purification. *Polymers* **2019**, *11*, 1252. [CrossRef] [PubMed]
17. Ligon, S.C.; Liska, R.; Stampfl, J.; Gurr, M.; Mülhaupt, R. Polymers for 3D Printing and Customized Additive Manufacturing. *Chem. Rev.* **2017**, *117*, 10212–10290. [CrossRef]
18. Herzberger, J.; Sirrine, J.M.; Williams, C.B.; Long, T.E. Polymer Design for 3D Printing Elastomers: Recent Advances in Structure, Properties, and Printing. *Progress Polymer Sci.* **2019**, *97*, 101144. [CrossRef]
19. Fico, D.; Rizzo, D.; Casciaro, R.; Esposito Corcione, C. A Review of Polymer-Based Materials for Fused Filament Fabrication (FFF): Focus on Sustainability and Recycled Materials. *Polymers* **2022**, *14*, 465. [CrossRef]
20. Zhou, L.; Fu, J.; He, Y. A Review of 3D Printing Technologies for Soft Polymer Materials. *Adv. Funct. Mater.* **2020**, *30*, 2000187. [CrossRef]
21. Yang, J.; An, X.; Liu, L.; Tang, S.; Cao, H.; Xu, Q.; Liu, H. Cellulose, Hemicellulose, Lignin, and Their Derivatives as Multi-Components of Bio-Based Feedstocks for 3D Printing. *Carbohydr. Polym.* **2020**, *250*, 116881. [CrossRef]
22. Sidhikku Kandath Valappil, R.; Ghasem, N.; Al-Marzouqi, M. Current and Future Trends in Polymer Membrane-Based Gas Separation Technology: A Comprehensive Review. *J. Ind. Eng. Chem.* **2021**, *98*, 103–129. [CrossRef]
23. Bandehali, S.; Sanaeepur, H.; Ebadi Amooghin, A.; Shirazian, S.; Ramakrishna, S. Biodegradable Polymers for Membrane Separation. *Separat. Purif. Technol.* **2021**, *269*, 118731. [CrossRef]
24. Han, Y.; Ho, W.S.W. Polymeric Membranes for $CO_2$ Separation and Capture. *J. Membr. Sci.* **2021**, *628*, 119244. [CrossRef]
25. Thiam, B.G.; El Magri, A.; Vaudreuil, S. An Overview on the Progress and Development of Modified Sulfonated Polyether Ether Ketone Membranes for Vanadium Redox Flow Battery Applications. *High Perform. Polym.* **2022**, *34*, 131–148. [CrossRef]
26. Chakraborty, S.; Biswas, M.C. 3D Printing Technology of Polymer-Fiber Composites in Textile and Fashion Industry: A Potential Roadmap of Concept to Consumer. *Compos. Struct.* **2020**, *248*, 112562. [CrossRef]
27. Kalsoom, U.; Hasan, C.K.; Tedone, L.; Desire, C.; Li, F.; Breadmore, M.C.; Nesterenko, P.N.; Paull, B. Low-Cost Passive Sampling Device with Integrated Porous Membrane Produced Using Multimaterial 3D Printing. *Anal. Chem.* **2018**, *90*, 12081–12089. [CrossRef]
28. You, J.; Preen, R.J.; Bull, L.; Greenman, J.; Ieropoulos, I. 3D Printed Components of Microbial Fuel Cells: Towards Monolithic Microbial Fuel Cell Fabrication Using Additive Layer Manufacturing. *Sustain. Energy Technol. Assess.* **2017**, *19*, 94–101. [CrossRef]
29. Xing, R.; Huang, R.; Qi, W.; Su, R.; He, Z. Three-Dimensionally Printed Bioinspired Superhydrophobic PLA Membrane for Oil-Water Separation. *AIChE J* **2018**, *64*, 3700–3708. [CrossRef]
30. Gao, D.; Wang, Z.; Wu, Z.; Guo, M.; Wang, Y.; Gao, Z.; Zhang, P.; Ito, Y. 3D-Printing of Solvent Exchange Deposition Modeling (SEDM) for a Bilayered Flexible Skin Substitute of Poly (Lactide-Co-Glycolide) with Bioorthogonally Engineered EGF. *Mater. Sci. Eng. C* **2020**, *112*, 110942. [CrossRef] [PubMed]
31. Koh, E.; Lee, Y.T. Development of an Embossed Nanofiber Hemodialysis Membrane for Improving Capacity and Efficiency via 3D Printing and Electrospinning Technology. *Separat. Purif. Technol.* **2020**, *241*, 116657. [CrossRef]
32. Yuan, S.; Zhu, J.; Li, Y.; Zhao, Y.; Li, J.; Puyvelde, P.V.; Van der Bruggen, B. Structure Architecture of Micro/Nanoscale ZIF-L on a 3D Printed Membrane for a Superhydrophobic and Underwater Superoleophobic Surface. *J. Mater. Chem. A* **2019**, *7*, 2723–2729. [CrossRef]
33. Yuan, S.; Strobbe, D.; Kruth, J.-P.; Van Puyvelde, P.; Van der Bruggen, B. Production of Polyamide-12 Membranes for Microfiltration through Selective Laser Sintering. *J. Membr. Sci.* **2017**, *525*, 157–162. [CrossRef]
34. Yuan, S.; Strobbe, D.; Kruth, J.-P.; Puyvelde, P.V.; Van der Bruggen, B. Super-Hydrophobic 3D Printed Polysulfone Membranes with a Switchable Wettability by Self-Assembled Candle Soot for Efficient Gravity-Driven Oil/Water Separation. *J. Mater. Chem. A* **2017**, *5*, 25401–25409. [CrossRef]

35. Seo, J.; Kushner, D.I.; Hickner, M.A. 3D Printing of Micropatterned Anion Exchange Membranes. *ACS Appl. Mater. Interfaces* **2016**, *8*, 16656–16663. [CrossRef]
36. Philamore, H.; Rossiter, J.; Walters, P.; Winfield, J.; Ieropoulos, I. Cast and 3D Printed Ion Exchange Membranes for Monolithic Microbial Fuel Cell Fabrication. *J. Power Sources* **2015**, *289*, 91–99. [CrossRef]
37. Al-Shimmery, A.; Mazinani, S.; Ji, J.; Chew, Y.M.J.; Mattia, D. 3D Printed Composite Membranes with Enhanced Anti-Fouling Behaviour. *J. Membr. Sci.* **2019**, *574*, 76–85. [CrossRef]
38. Mazinani, S.; Al-Shimmery, A.; Chew, Y.M.J.; Mattia, D. 3D Printed Fouling-Resistant Composite Membranes. *ACS Appl. Mater. Interfaces* **2019**, *11*, 26373–26383. [CrossRef] [PubMed]
39. Lv, J.; Gong, Z.; He, Z.; Yang, J.; Chen, Y.; Tang, C.; Liu, Y.; Fan, M.; Lau, W.-M. 3D Printing of a Mechanically Durable Superhydrophobic Porous Membrane for Oil–Water Separation. *J. Mater. Chem. A* **2017**, *5*, 12435–12444. [CrossRef]
40. Singh, M.; Haring, A.P.; Tong, Y.; Cesewski, E.; Ball, E.; Jasper, R.; Davis, E.M.; Johnson, B.N. Additive Manufacturing of Mechanically Isotropic Thin Films and Membranes via Microextrusion 3D Printing of Polymer Solutions. *ACS Appl. Mater. Interfaces* **2019**, *11*, 6652–6661. [CrossRef]
41. Liang, Y.; Zhao, J.; Huang, Q.; Hu, P.; Xiao, C. PVDF Fiber Membrane with Ordered Porous Structure via 3D Printing near Field Electrospinning. *J. Membr. Sci.* **2021**, *618*, 118709. [CrossRef]
42. Chowdhury, M.R.; Steffes, J.; Huey, B.D.; McCutcheon, J.R. 3D Printed Polyamide Membranes for Desalination. *Science* **2018**, *361*, 682–686. [CrossRef]
43. Zhu, L.-J.; Liu, F.; Yu, X.-M.; Gao, A.-L.; Xue, L.-X. Surface Zwitterionization of Hemocompatible Poly(Lactic Acid) Membranes for Hemodiafiltration. *J. Membr. Sci.* **2015**, *475*, 469–479. [CrossRef]
44. Zhang, Z.-M.; Gan, Z.-Q.; Bao, R.-Y.; Ke, K.; Liu, Z.-Y.; Yang, M.-B.; Yang, W. Green and Robust Superhydrophilic Electrospun Stereocomplex Polylactide Membranes: Multifunctional Oil/Water Separation and Self-Cleaning. *J. Membr. Sci.* **2020**, *593*, 117420. [CrossRef]
45. Badalov, S.; Arnusch, C.J. Ink-Jet Printing Assisted Fabrication of Thin Film Composite Membranes. *J. Membr. Sci.* **2016**, *515*, 79–85. [CrossRef]
46. Zhang, Z.; Qin, Y.; Kang, G.; Yu, H.; Jin, Y.; Cao, Y. Tailoring the Internal Void Structure of Polyamide Films to Achieve Highly Permeable Reverse Osmosis Membranes for Water Desalination. *J. Membr. Sci.* **2020**, *595*, 117518. [CrossRef]
47. Perera, D.H.N.; Song, Q.; Qiblawey, H.; Sivaniah, E. Regulating the Aqueous Phase Monomer Balance for Flux Improvement in Polyamide Thin Film Composite Membranes. *J. Membr. Sci.* **2015**, *487*, 74–82. [CrossRef]
48. Femmer, T.; Kuehne, A.J.C.; Wessling, M. Print Your Own Membrane: Direct Rapid Prototyping of Polydimethylsiloxane. *Lab Chip* **2014**, *14*, 2610. [CrossRef]
49. Chen, J.P.; Mou, H.; Wang, L.K.; Matsuura, T. Membrane Filtration. In *Advanced Physicochemical Treatment Processes*; Wang, L.K., Hung, Y.-T., Shammas, N.K., Eds.; Handbook of Environmental Engineering; Humana Press: Totowa, NJ, USA, 2006; ISBN 978-1-59745-029-4.
50. Palanisamy, G.; Jung, H.-Y.; Sadhasivam, T.; Kurkuri, M.D.; Kim, S.C.; Roh, S.-H. A Comprehensive Review on Microbial Fuel Cell Technologies: Processes, Utilization, and Advanced Developments in Electrodes and Membranes. *J. Clean. Prod.* **2019**, *221*, 598–621. [CrossRef]
51. Al-Husaini, I.S.; Yusoff, A.R.M.; Lau, W.-J.; Ismail, A.F.; Al-Abri, M.Z.; Wirzal, M.D.H. Iron Oxide Nanoparticles Incorporated Polyethersulfone Electrospun Nanofibrous Membranes for Effective Oil Removal. *Chem. Eng. Res. Des.* **2019**, *148*, 142–154. [CrossRef]
52. Yalcinkaya, F.; Yalcinkaya, B.; Hruza, J. Electrospun Polyamide-6 Nanofiber Hybrid Membranes for Wastewater Treatment. *Fibers Polym.* **2019**, *20*, 93–99. [CrossRef]
53. Liang, J.-W.; Prasad, G.; Wang, S.-C.; Wu, J.-L.; Lu, S.-G. Enhancement of the Oil Absorption Capacity of Poly(Lactic Acid) Nano Porous Fibrous Membranes Derived via a Facile Electrospinning Method. *Appl. Sci.* **2019**, *9*, 1014. [CrossRef]
54. Yang, Y.; Li, X.; Zheng, X.; Chen, Z.; Zhou, Q.; Chen, Y. 3D-Printed Biomimetic Super-Hydrophobic Structure for Microdroplet Manipulation and Oil/Water Separation. *Adv. Mater.* **2018**, *30*, 1704912. [CrossRef] [PubMed]
55. Ye, B.; Jia, C.; Li, Z.; Li, L.; Zhao, Q.; Wang, J.; Wu, H. Solution-blow Spun PLA/$SiO_2$ Nanofiber Membranes toward High Efficiency Oil/Water Separation. *J. Appl. Polym. Sci.* **2020**, *137*, 49103. [CrossRef]
56. Qin, Y.; Shen, H.; Han, L.; Zhu, Z.; Pan, F.; Yang, S.; Yin, X. Mechanically Robust Janus Poly(Lactic Acid) Hybrid Fibrous Membranes toward Highly Efficient Switchable Separation of Surfactant-Stabilized Oil/Water Emulsions. *ACS Appl. Mater. Interfaces* **2020**, *12*, 50879–50888. [CrossRef] [PubMed]
57. Lee, T.H.; Roh, J.S.; Yoo, S.Y.; Roh, J.M.; Choi, T.H.; Park, H.B. High-Performance Polyamide Thin-Film Nanocomposite Membranes Containing ZIF-8/CNT Hybrid Nanofillers for Reverse Osmosis Desalination. *Ind. Eng. Chem. Res.* **2020**, *59*, 5324–5332. [CrossRef]
58. ElSherbiny, I.M.A.; Ghannam, R.; Khalil, A.S.G.; Ulbricht, M. Isotropic Macroporous Polyethersulfone Membranes as Competitive Supports for High Performance Polyamide Desalination Membranes. *J. Membr. Sci.* **2015**, *493*, 782–793. [CrossRef]
59. Lee, T.H.; Lee, M.Y.; Lee, H.D.; Roh, J.S.; Kim, H.W.; Park, H.B. Highly Porous Carbon Nanotube/Polysulfone Nanocomposite Supports for High-Flux Polyamide Reverse Osmosis Membranes. *J. Membr. Sci.* **2017**, *539*, 441–450. [CrossRef]
60. García, A.; Rodríguez, B.; Oztürk, D.; Rosales, M.; Diaz, D.I.; Mautner, A. Incorporation of CuO Nanoparticles into Thin-Film Composite Reverse Osmosis Membranes (TFC-RO) for Antibiofouling Properties. *Polym. Bull.* **2018**, *75*, 2053–2069. [CrossRef]

61. Yan, W.; Wang, Z.; Wu, J.; Zhao, S.; Wang, J.; Wang, S. Enhancing the Flux of Brackish Water TFC RO Membrane by Improving Support Surface Porosity via a Secondary Pore-Forming Method. *J. Membr. Sci.* **2016**, *498*, 227–241. [CrossRef]
62. Duan, J.; Pan, Y.; Pacheco, F.; Litwiller, E.; Lai, Z.; Pinnau, I. High-Performance Polyamide Thin-Film-Nanocomposite Reverse Osmosis Membranes Containing Hydrophobic Zeolitic Imidazolate Framework-8. *J. Membr. Sci.* **2015**, *476*, 303–310. [CrossRef]
63. Lee, H.D.; Kim, H.W.; Cho, Y.H.; Park, H.B. Experimental Evidence of Rapid Water Transport through Carbon Nanotubes Embedded in Polymeric Desalination Membranes. *Small* **2014**, *10*, 2653–2660. [CrossRef] [PubMed]
64. Shan, X.; Li, S.-L.; Fu, W.; Hu, Y.; Gong, G.; Hu, Y. Preparation of High Performance TFC RO Membranes by Surface Grafting of Small-Molecule Zwitterions. *J. Membr. Sci.* **2020**, *608*, 118209. [CrossRef]
65. Ma, R.; Ji, Y.-L.; Weng, X.-D.; An, Q.-F.; Gao, C.-J. High-Flux and Fouling-Resistant Reverse Osmosis Membrane Prepared with Incorporating Zwitterionic Amine Monomers via Interfacial Polymerization. *Desalination* **2016**, *381*, 100–110. [CrossRef]
66. Kim, H.J.; Choi, K.; Baek, Y.; Kim, D.-G.; Shim, J.; Yoon, J.; Lee, J.-C. High-Performance Reverse Osmosis CNT/Polyamide Nanocomposite Membrane by Controlled Interfacial Interactions. *ACS Appl. Mater. Interfaces* **2014**, *6*, 2819–2829. [CrossRef]
67. Kim, H.J.; Lim, M.-Y.; Jung, K.H.; Kim, D.-G.; Lee, J.-C. High-Performance Reverse Osmosis Nanocomposite Membranes Containing the Mixture of Carbon Nanotubes and Graphene Oxides. *J. Mater. Chem. A* **2015**, *3*, 6798–6809. [CrossRef]
68. Yin, J.; Zhu, G.; Deng, B. Graphene Oxide (GO) Enhanced Polyamide (PA) Thin-Film Nanocomposite (TFN) Membrane for Water Purification. *Desalination* **2016**, *379*, 93–101. [CrossRef]
69. Xu, Y.; Gao, X.; Wang, X.; Wang, Q.; Ji, Z.; Wang, X.; Wu, T.; Gao, C. Highly and Stably Water Permeable Thin Film Nanocomposite Membranes Doped with MIL-101 (Cr) Nanoparticles for Reverse Osmosis Application. *Materials* **2016**, *9*, 870. [CrossRef]
70. Di Vincenzo, M.; Barboiu, M.; Tiraferri, A.; Legrand, Y.M. Polyol-Functionalized Thin-Film Composite Membranes with Improved Transport Properties and Boron Removal in Reverse Osmosis. *J. Membr. Sci.* **2017**, *540*, 71–77. [CrossRef]
71. Wang, Y.-P.; Liu, X.-W.; Li, W.-W.; Li, F.; Wang, Y.-K.; Sheng, G.-P.; Zeng, R.J.; Yu, H.-Q. A Microbial Fuel Cell–Membrane Bioreactor Integrated System for Cost-Effective Wastewater Treatment. *Appl. Energy* **2012**, *98*, 230–235. [CrossRef]
72. Bergkamp, L.; Herbatschek, N. Regulating Chemical Substances under REACH: The Choice between Authorization and Restriction and the Case of Dipolar Aprotic Solvents: Regulating Chemical Substances Under Reach. *Rev. Eur. Comp. Int. Environ. Law* **2014**, *23*, 221–245. [CrossRef]

MDPI
St. Alban-Anlage 66
4052 Basel
Switzerland
Tel. +41 61 683 77 34
Fax +41 61 302 89 18
www.mdpi.com

*Polymers* Editorial Office
E-mail: polymers@mdpi.com
www.mdpi.com/journal/polymers

www.ingramcontent.com/pod-product-compliance
Lightning Source LLC
LaVergne TN
LVHW070653100526
838202LV00013B/950